ELECTROANALYTICAL CHEMISTRY

ELECTROANALYTICAL CHEMISTRY

A SERIES OF ADVANCES

edited by

Allen J. Bard

Department of Chemistry
University of Texas
Austin, Texas

Israel Rubinstein

Department of Materials
and Interfaces
Weizmann Institute of Science
Rehovot, Israel

VOLUME 21

MARCEL DEKKER, INC. NEW YORK • BASEL

The Library of Congress Cataloged the First Issue of This Title as Follows:

Electroanalytic chemistry: a series of advances, v. 1
New York, M. Dekker, 1966–
v. 23 cm.
Editors: 1966–1995 A. J. Bard
1996– A. J. Bard and I. Rubinstein
1. Electromechanical analysis—Addresses, essays, lectures
1. Bard, Allen J., ed.
QD115E499 545.3 66-11287
Library of Congress
0-8247-7399-3 (v. 21)

This book is printed on acid-free paper.

Headquarters
Marcel Dekker, Inc.
270 Madison Avenue, New York, NY 10016
tel: 212-696-9000; fax: 212-685-4540

Eastern Hemisphere Distribution
Marcel Dekker AG
Hutgasse 4, Postfach 812, CH-4001 Basel, Switzerland
tel: 41-61-261-8482; fax: 41-61-261-8896

World Wide Web
http://www.dekker.com

The publisher offers discounts on this book when ordered in bulk quantities. For more information, write to Special Sales/Professional Marketing at the headquarters address above.

Current printing (last digit):
10 9 8 7 6 5 4 3 2 1

PRINTED IN THE UNITED STATES OF AMERICA

INTRODUCTION TO THE SERIES

This series is designed to provide authoritative reviews in the field of modern electroanalytical chemistry defined in its broadest sense. Coverage is comprehensive and critical. Enough space is devoted to each chapter of each volume so that derivations of fundamental equations, detailed descriptions of apparatus and techniques, and complete discussions of important articles can be provided, so that the chapters may be useful without repeated reference to the periodical literature. Chapters vary in length and subject area. Some are reviews of recent developments and applications of well-established techniques, whereas others contain discussion of the background and problems in areas still being investigated extensively and in which many statements may still be tentative. Finally, chapters on techniques generally outside the scope of electroanalytical chemistry, but which can be applied fruitfully to electrochemical problems, are included.

Electroanalytical chemists and others are concerned not only with the application of new and classical techniques to analytical problems, but also with the fundamental theoretical principles upon which these techniques are based. Electroanalytical techniques are proving useful in such diverse fields as electro-organic synthesis, fuel cell studies, and radical ion formation, as well as with such problems as the kinetics and mechanisms of electrode reactions, and the effects of electrode surface phenomena, adsorption, and the electrical double layer on electrode reactions.

It is hoped that the series is proving useful to the specialist and non-specialist alike—that it provides a background and a starting point for graduate students undertaking research in the areas mentioned, and that it also proves valuable to practicing analytical chemists interested in learning about and applying electroanalytical techniques. Furthermore, electrochemists and industrial chemists with problems of electrosynthesis, electroplating, corrosion, and fuel cells, as well as other chemists wishing to apply electrochemical techniques to chemical problems, may find useful material in these volumes.

A. J. B.
I. R.

CONTRIBUTORS TO VOLUME 21

CHARLES R. MARTIN Colorado State University, Ft. Collins, Colorado

DAVID T. MITCHELL Colorado State University, Ft. Collins, Colorado

T. P. MOFFAT National Institute of Standards and Technology, Gaithersburg, Maryland

JOHN L. STICKNEY University of Georgia, Athens, Georgia

CONTENTS OF VOLUME 21

CONTENTS OF OTHER VOLUMES

VOLUME 4

VOLUME 5

VOLUME 6

VOLUME 7

VOLUME 8

VOLUME 13

VOLUME 14

VOLUME 15

VOLUME 16

VOLUME 17

VOLUME 18

VOLUME 19

VOLUME 20

TEMPLATE-SYNTHESIZED NANOMATERIALS IN ELECTROCHEMISTRY

Charles R. Martin and David T. Mitchell

Colorado State University
Ft. Collins, Colorado

I. INTRODUCTION

Nanomaterials constitute an emerging subdiscipline of the chemical and materials sciences that deals with the development of methods for synthesizing nanoscopic particles of a desired material and with scientific investigations of the nanomaterial obtained [1–5]. Nanomaterials have numerous possible commercial and technological applications, including uses in analytical chemistry [6–9], electronic, optical, and mechanical devices [4,5,10], drug delivery [11], and bioencapsulation [12]. In addition, this field poses an important fundamental question: How do the properties (e.g., electronic, optical, magnetic) of a nanoscopic particle of a material differ from the analogous properties for a macroscopic sample of the same material?

There are now numerous chemical methods for preparing nanomaterials [1,4,10]. Our research group and others have been exploring a method we call "template synthesis" [1–3]. This method entails synthesizing the desired material within the pores of a porous membrane or other solid (the template material). The template method has a number of interesting and useful features. First, it is a very general approach, amenable to nearly all material synthetic methodologies. Second, template materials with highly monodisperse pores (Fig. 1) are available, and thus highly monodisperse nanostructures can be obtained. Furthermore, these nanostructures can be extraordinarily small, with diameters routinely in the nm range and often as small as Å. Finally, the template material can be removed to leave either a dispersion [13] or an array of the synthesized nanostructures (see Fig. 2).

Nanomaterials and electrochemistry have a long shared history (e.g.,

FIG. 1. Electron micrographs of polycarbonate (A and B) and alumina (C and D) template membranes. For each type of membrane, an image of a larger pore membrane is presented (A and C) so that the characteristics of the pores can be clearly seen. An image of a membrane with nanopores is also presented (B and D). (A) Scanning electron micrograph of the surface of a polycarbonate membrane with 1 μm-diameter pores. (B) Transmission electron micrograph (TEM) of a graphite replica of the surface of a polycarbonate membrane with 30-nm-diameter pores. The pores appear "ragged." This is an artifact of the graphite replica. (C, D) TEMs of microtomed sections of alumina membranes with 70-nm-diameter (C) and ~10-nm-diameter (D) pores.

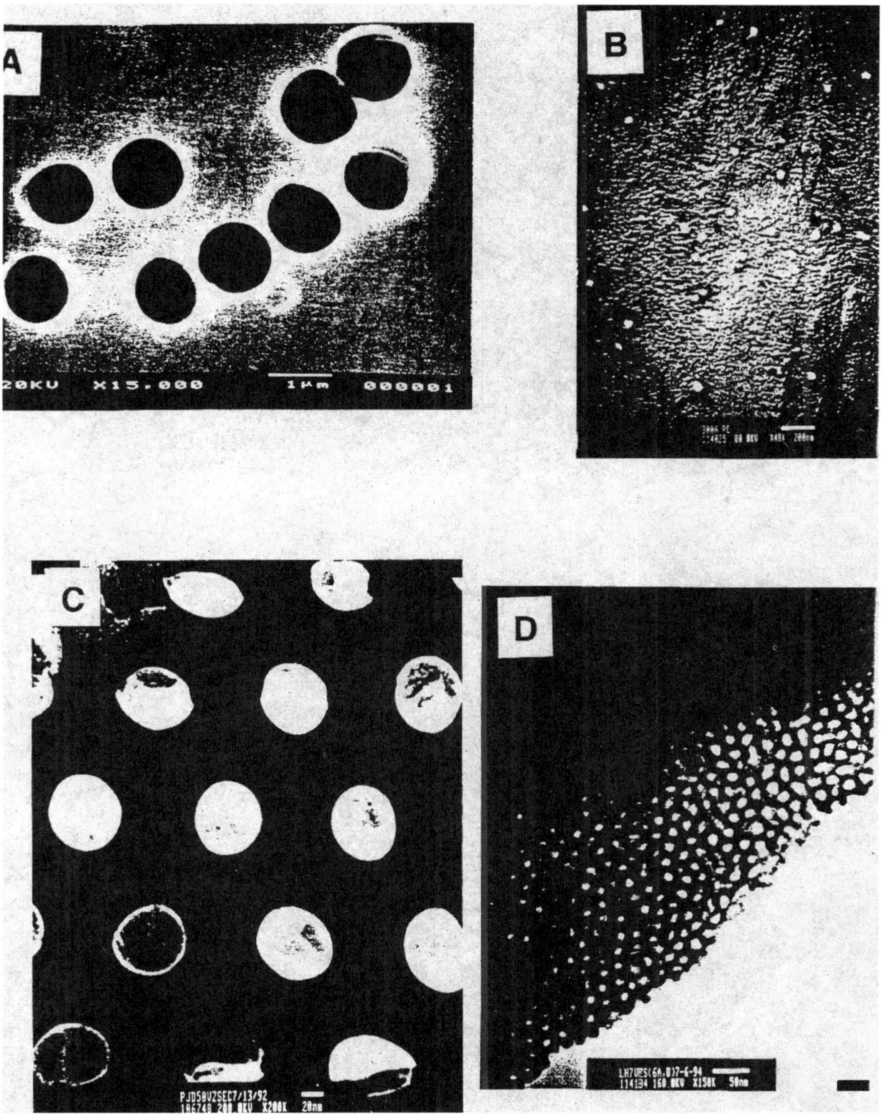
A
X15,000
1μm
B
C
D

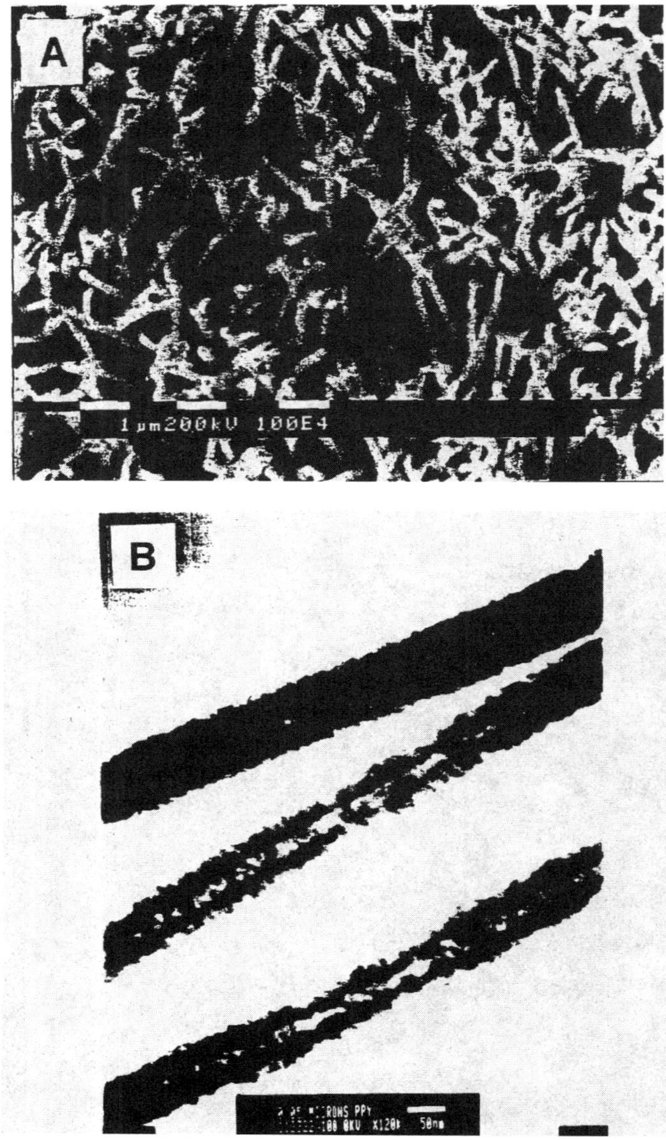

FIG. 2. (A) Scanning electron micrograph of an array of gold nanotubules protruding from a substrate surface. (B) Transmission electron micrograph of three polypyrrole nanotubules.

the use of finely dispersed Pt particles as catalysts in fuel cell electrodes). This chapter, however, deals specifically with applications of template-synthesized nanomaterials in electrochemistry. We begin with an overview of template materials. Three possible electrochemical applications of such materials are then discussed. The first entails use of the template method for preparing ensembles of nanoscopic electrodes. The second application concerns the development of a new type of ion-permselective membrane—the metal nanotubule membrane. These membranes can be viewed as universal ion exchangers because they can be electrochemically switched between cation-permselective, anion-permselective and nonpermselective states. The transport properties of metal nanotubule membranes can also be made selective on the basis of either the size or chemistry of the molecule to be transported; possible applications of these membranes in chemical separations are discussed. The final application reviewed here entails the use of the template method to prepare monodisperse nanoparticles of Li^+ intercalation materials for possible use as electrodes in Li ion batteries.

II. TEMPLATE SYNTHESIS

The template method involves using the pores in a microporous solid as nanoscopic beakers for the synthesis of nanoparticles of the desired material [1,3,10]. A wide variety of materials are available for use as template materials [1,10,14–19]. Pore diameter sizes range from Angstroms to many μm. Several of the more common materials used as templates are reviewed below.

Commercially available track-etched plastic filtration membranes have proven especially useful as template materials. In the early 1960s, Price and Walker discovered that damage tracks produced in mica by high-energy particles could be preferentially etched to yield pores with diameters dependent on the etching time [20–22]. The procedure for etching damage tracks was subsequently perfected for other minerals and plastics [20]. The resulting pores in plastics are cylindrical, randomly distributed, and highly monodisperse (Fig. 1A and 1B). Pore size typically ranges from tens of μm to 10 nm. Porosities are from 15 to 0.01%, depending on pore size and density. Track-etched polymer (polycarbonate and polyester) membranes are commercially available in a range of pore sizes and pore densities from suppliers like Corning, Nucleopore, Whatman, and Poretics.

Possin, in 1970, was the first to use the pores in track-etched mica membranes as templates to make nanomaterials [23]. This was accomplished by electrodepositing metal into the pores of the membrane; metal nanowires with diameters as small as 40 nm were prepared. Williams and Giordano improved on Possin's technique and were able to fabricate nanowires with diameters as small as 8 nm [24].

There has been extensive recent use of track-etched membranes as templates. As will be discussed in detail below, these membranes are ideal for producing parallel arrays of metal nanowires or nanotubules. This is usually done via electroless metal deposition [25], but many metals have also been deposited electrochemically [26]. For example, several groups have used track-etched templates for deposition of nanowires and segmented nanowires, which they then examined for giant magnetoresistance [27–29]. Other materials templated in the pores of track etch membranes include conducting polymers [30] and polymer-metal composites [31].

Anodically grown aluminum oxide (Al_2O_3) has also been used extensively as a template [3,32–37]. When grown on high-purity aluminum, this material has a hexagonal pattern of cylindrical pores, which extend through the thickness of the alumina (Fig. 1C and 1D). These microporous alumina films can be removed from the substrate A1 metal and collected as a free-standing membrane [37,38].

Microporous alumina membranes have many features that make them especially valuable as templates. One is the ability to control the pore size by varying growth conditions [37]. Pore diameters from ~10 to ~400 nm can be prepared. Another potentially useful feature is that alumina can be heated to high temperatures (~1000°C) without degradation. Furthermore, these membranes have pores that are highly monodisperse and porosities that can be greater than 50%. The membranes can also be highly transparent in the visible region. This allows for investigations of the optical properties of the nanomaterial deposited within the pore [37]. The amphoteric nature of aluminum oxide allows it to be dissolved away in acidic or basic solution to expose the nanostructures deposited within the pores. Another benefit is that such membranes are commercially available and relatively inexpensive.

A very early use of anodic alumina as a template involved colorization of the alumina by depositing nanometals in the pores [39]. Somewhat later, Kawai and Ueda templated cobalt and nickel in alumina by electrodeposition [40]. Other metals were deposited by Andersson et al. [41] and Patel et al. [42]. The use of anodic alumina as a template increased after Furneaux et al. developed a convenient voltage-reduction method for detaching the porous anodized alumina from the underlying aluminum [38].

More recently, the pores of anodic Al_2O_3 have been filled with carbon [34], conducting polymers [35], and semiconductors [33], as well as lithium insertion materials for battery applications (discussed in detail in a later section). Schmid and co-workers found that dispersions of gold colloids and clusters could be aligned in the nanotubular pores to form quantum wires [36]. Feldheim et al. also used Al_2O_3 templates to align Au nanoparticles, which they ''shrink-wrapped'' together by polymerizing with pyrrole [43]. The diversity of these applications shows the importance of anodic alumina as a template material.

Recently, a novel type of nanoporous membrane template has been developed called nanochannel glass [44]. These membranes are formed by heating and drawing an array of glass capillaries until the desired pore size and density is achieved. Pores as small as 17 nm have been fabricated by this method. Pore densities can be as high as 3×10^9 pores per cm^2. Like anodic alumina, nanochannel glass is a highly ordered array with parallel monodisperse pores running through the thickness of the membrane. To date, copper, platinum, and nickel have been deposited by electrodeposition into these templates [45].

It is important to note that in addition to microporous solids, other chemical systems have been used to template the growth of nanomaterials. For example, emulsions have been used to pattern both the pores in titania [14] and the packing of latex particles [46]. Reversed micelles have also been used as patterning agents. Examples include the syntheses of superparamagnetic ferrite nanoparticles [15] and $BaCO_3$ nanowires [47]. Finally, carbon nanotubules have also been used as templates [16,48,49]. A variety of nanomaterials including metal oxides [16,48,49] and GaN have been synthesized inside such tubules [50].

Synthetic lipids and peptides have been found to self-assemble into tubules [51,52]. Several groups have used these tubules as templates [17,51,53–56]. Much of this work has been the electroless deposition of metals [51,54]. Electrolessly plated Ni tubules were found to be effective field emission cathode sources [55]. Other materials templated in or on self-assembled lipid tubules include conducting polymer [56] and inorganic oxides [53]. Nanotubules from cellular cytoskeletons have also been used for electroless deposition of metals [57].

In 1992, Mobil researchers reported the discovery of a family of mesoporous molecular sieves prepared with liquid crystal templates [18]. One member of this family, designated MCM-41, has hexagonally packed pores, which are highly monodisperse. Moreover, the size of these pores can be varied from ~15 to 100 Å. This ability to control pore size combined with high pore monodispersity make MCM-41 attractive as a template material. Wu and Bein have prepared graphitic carbon nanowires [58] and conducting filaments of polyanaline [59] in these templates.

Molds prepared by lithographic methods are novel and versatile template materials [19,60,61]. There are many ways in which the mold pattern can be transferred to a substrate, including stamping to produce patterned voids into which materials can be cast [60], coating with a monolayer of thiols and then stamping to leave the monolayer (microcontact printing) [61], or contacting with the surface to produce microchannel capillaries, which are filled by capillary action [19]. These methods are typified by the work of Whitesides' group, which developed several of the innovative templating procedures utilizing molds [19,61].

III. NANOSCOPIC ELECTRODES AND ENSEMBLES

Electrochemistry at electrodes with microscopic dimensions (e.g., a disk of 10 μm diameter) and nanoscopic dimensions (e.g., a disk of <100 nm diameter) constitutes one of the most important frontiers in modern electrochemical science [25]. Such micro- and nanoscopic electrodes allow for electrochemical experiments that are impossible at electrodes of macroscopic dimensions (e.g., disks of mm diameter; we call such electrodes ''macroelectrodes''). Examples of unique opportunities afforded by micro- and nanoscopic electrodes include the possibility of doing electrochemistry in highly resistive media and the possibility of investigating the kinetics of redox processes that are too fast to study at electrodes of conventional dimensions (both are discussed in detail below). In addition, microscopic electrodes have proven extremely useful for in vivo electrochemistry [62].

The small size of nanoelectrodes also makes possible the detection of discrete electron transfer events. Fan and Bard have recently shown coulombic staircase response using electrodes of nanometer dimensions [63]. Ingram and co-workers have also shown coulombic staircase response, in their case while studying colloids and collections of colloids [64]. Fan and Bard have also applied nanoelectrodes to achieve high-resolution electrochemical imaging and single-molecule detection [65].

Electrochemistry of proteins is another case where electrode size affects the electrochemical results. Direct adsorption of proteins, such as enzymes, onto bulk metal surfaces frequently results in denaturation of the protein and loss of bioactivity. In contrast, when proteins are adsorbed to metal nanoparticles, bioactivity is often retained [66]. For example, Crumbliss et al. found that they could adsorb redox enzymes to colloidal gold with no loss of enzymatic activity. The enzyme-covered nanoparticles were then electrodeposited onto platinum gauze or glassy carbon to make an enzyme electrode [66]. Additionally, Natan and co-workers found that cytochrome *c* retained reversible cyclic voltammetry when deposited onto 12-nm-diameter gold particles attached to a conductive substrate [67]. In contrast, if the cytochrome *c* was deposited on larger surface features (aggregates of the gold nanoparticles), the cyclic voltammetry became quasi-reversible or irreversible, indicating denaturation of the protein [67].

In order to explore the effects of small electrode size, we have used the template method to prepare ensembles of disk-shaped nanoelectrodes with diameters as small as 10 nm. We have shown that these nanoelectrode ensembles (NEEs) demonstrate dramatically lower electroanalytical detection limits compared to analogous macroelectrodes. The experimental methods used to prepare these ensembles and some recent results are reviewed below.

A. Experimental

1. Template Membranes

Nanoelectrode ensembles were prepared by electroless deposition of Au within the pores of polycarbonate membrane filters (Poretics). Filters with pore diameters of 10 and 30 nm were used [25]. The pore densities and average center-to-center distances between pores for these membranes are shown in Table 1. Multiplying the pore density (pores cm^{-2}) by the cross-

TABLE 1

Characteristics of Membranes

Pore diameter (nm)	Pore density (cm^{-2})	Distance between pores (μm)	Fractional pore area[a]	Fractional electrode area[b]
10	6×10^8	0.2	0.00047	0.00094
30	6×10^8	0.2	0.0042	0.0042

[a] Determined from electron micrographs of membrane.
[b] Determined electrochemically.

sectional area of a single pore (cm^2 per pore) provides a parameter called the fractional pore area [25] (Table 1). This is an important parameter because, assuming that each pore produces an active nanodisk electrode, the fractional pore area is equivalent to the fractional electrode area, which is the sum of the areas of the Au nanodisk elements in the NEE divided by the geometric area of the NEE. We have shown that the fractional electrode area can be determined experimentally from the double layer charging currents obtained at the NEE [25,68–70]. Ideally, this experimental fractional electrode area and the fractional pore area should be equivalent (Table 1).

2. *Electroless Au Deposition*

Electroless metal deposition involves the use of a chemical reducing agent to plate a metal from solution onto a surface. The key requirement is to arrange the chemistry such that the kinetics of homogeneous electron transfer from the reducing agent to the metal ion are slow. A catalyst that accelerates the rate of metal ion reduction is then applied to the surface to be coated. In this way, metal ion is reduced only at the surface, and the surface becomes coated with the desired metal. The thickness of the metal film deposited can be controlled by varying the plating time.

The electroless plating process has been described previously [25]. Briefly, a ''sensitizer'' (Sn^{2+}) is first applied to the surfaces (pore walls plus faces) of the template membrane. This is accomplished by simply immersing the membrane into a solution containing $SnCl_2$. The Sn^{2+}-sensitized membrane is then activated by immersion into an aqueous $AgNO_3$ solution. This causes a redox reaction in which the surface-bound Sn(II) gets oxidized to Sn(IV) and the Ag^+ is reduced to elemental Ag. As a result, the pore walls and membrane faces become coated with discrete, nanoscopic Ag particles. The Ag-coated membrane is then immersed into an Au plating bath. The Ag particles are galvanically displaced by Au, and the pore walls and membrane faces become coated with Au particles. These particles catalyze the reduction of Au(I) on the membrane surfaces using formaldehyde as the reducing agent.

To produce NEEs, the plating process is continued until solid Au nanowires are obtained in each pore. In addition, both faces of the membrane become coated with Au films. For the NEE application, one of these surface Au films is removed using a simple Scotch-tape method [25]. This exposes the ends of the Au nanowires that are embedded within the pores of the membrane. These Au disks constitute the electrode elements of the NEE. Electrical contact is made to the Au surface film that was left intact on the opposite face of the membrane. By applying a potential to this Au surface film (relative to a reference electrode immersed into the same solution), redox reactions can be driven (in parallel) at the ensemble of Au nanodisks. Details of the NEE fabrication procedure have been described previously [25].

It is important to point out that if plating is terminated before solid Au nanowires are obtained, Au nanotubules that span the complete thickness of the template membrane are deposited within the pores. We have shown that these nanotubule membranes have interesting ion [71] and molecular [72] transport properties. This will be subject of the following section.

B. Characterization of NEEs

1. Electron Microscopy and Optical Absorption Spectroscopy

The key feature of the electroless deposition process is that Au deposition begins at the pore wall. As a result, after brief deposition times, a hollow Au tubule is obtained within each pore [71]. These tubules can be imaged by taking transmission electron micrographs of microtomed sections of the tubule-containing membrane. An image of this type for a membrane that contained 50-nm-diameter pores is shown in Fig. 3. The Au tubules (black rings) appear elliptical (and ragged) due to distortion by the microtoming process. Scanning electron microscopy can also be used to image the face of the membrane in order to see the individual Au nanodisk electrode elements [25]. However, the 10-nm Au disk size can push the limits of resolution of a conventional scanning electron microscope.

Nanometals have interesting optical properties [37,73,74]. For example, suspensions of nanoscopic Au particles can be pink, purple, or blue depending on the diameter of the particles [74]. These colors arise from the plasmon resonance absorption of the nanometal particle, a phenomenon we have explored in some detail [37,73]. We have shown that membranes containing Au nanowires like those described here also show this plasmon resonance band, and as a result such membranes can show a wide variety of colors. This absorption in the visible region provides an interesting optical approach for characterizing the Au nanowire–containing membranes.

Figure 4 compares absorption spectra for membranes containing 10- and 30-nm-diameter Au nanowires. The wavelength of maximum absorption intensity for the membrane containing the 10-nm-diameter nanowires is blueshifted relative to that for the membrane containing the 30-nm-diameter nanowires. This blueshift for the smaller-diameter nanowires is

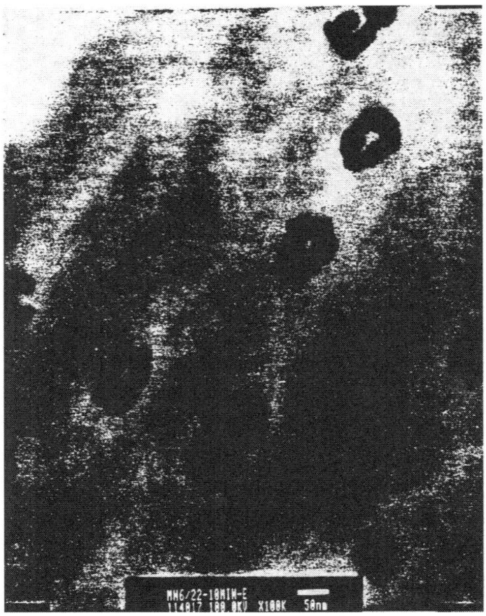

FIG. 3. Transmission electron micrograph of a microtomed section of a polycarbonate template membrane after deposition of Au tubules within the pores of the membrane. Pore diameter was 50 nm.

in qualitative agreement with the predictions of effective medium theory [37,73]. As would be expected from the spectra shown in Fig. 4, the membranes containing the 10-nm-diameter nanowires appear pink in color, whereas the membranes containing the 30-nm-diameter nanowires are purple. Because of these distinctive colors, it is easy to distinguish the 30-nm disk-diameter NEEs (30NEEs) from the 10-nm disk-diameter NEEs (10NEEs).

2. Double-Layer Charging Currents

A persistent problem with micro- and nanoelectrodes is the sealing of the conductive element to the insulating material that surrounds the element such that solution does not creep into this junction [25,68,75]. This solution creeping is undesirable because it causes the double layer charging currents

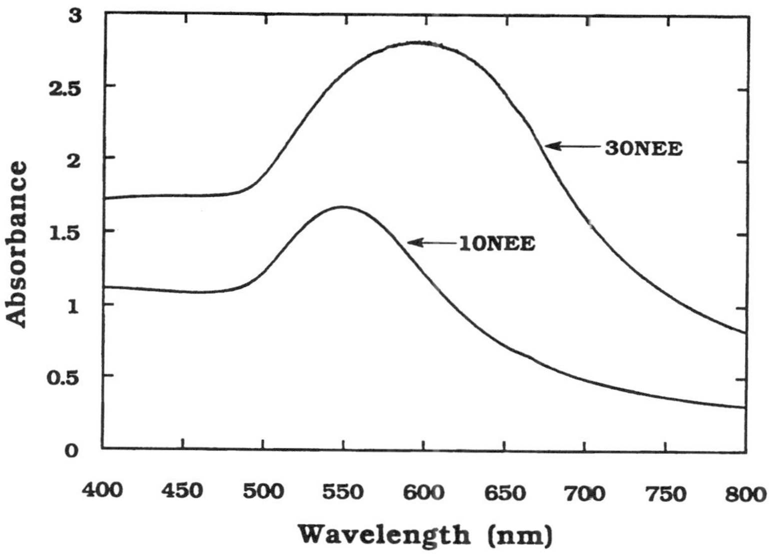

FIG. 4. UV-Visible spectra of a 30NEE and a 10NEE.

to be spuriously large. Previous methods for improving the seal have included silanization of the surrounding insulator [75] and impregnating the junction between the electrode and the insulating material with low molecular weight polyethylene [68]. However, neither of these methods has proven completely satisfactory. We have recently introduced a superior method for sealing the junction between the Au nanowires and the polycarbonate host membrane [25]. This method exploits the heat-shrinkability of this template membrane.

The polycarbonate membranes are stretch-oriented during fabrication in order to improve their mechanical properties. If the membrane is subsequently heated above its glass-transition temperature (~150°C), the polymer chains relax to their unstretched conformation and the membrane shrinks. This shrinking of the membrane around the Au nanowires in the pores causes the junction between the nanowire and the pore wall to be sealed. This is illustrated in Fig. 5, which shows voltammograms for trimethylaminomethylferrocene ($TMAFc^+$) before (Fig. 5A) and after (Fig.

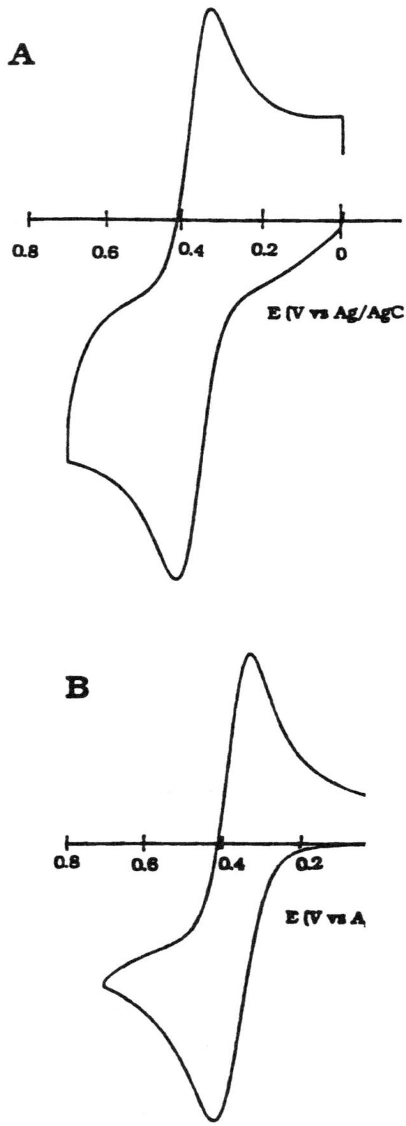

FIG. 5. Cyclic voltammograms at 100 mV s^{-1} at a 10NEE in 5 μM aqueous $TMAFc^+$, 1 mM $NaNO_3$: (A) before thermal treatment and (B) after thermal treatment of the NEE.

5B) the heat-treatment procedure. Before heat treatment, the double layer charging currents are pronounced. After heat treatment, the charging currents are not discernible at the current sensitivity setting used.

We have used voltammetric measurements in the absence of the electroactive species to quantitatively evaluate this heat-sealing procedure. The magnitude of the double layer charging current can be obtained from these voltammograms [25,68–70], which allows for a determination of the fractional electrode area (Table 1). This experimental fractional electrode area can then be compared to the fractional pore area calculated from the known pore diameter and density of the membrane (Table 1). In order to use this method, the double layer capacitance of the metal must be known. The double layer capacitance of Au was determined from measurements of charging currents at Au macro-disk electrodes of known area (Fig. 6, curve A). A value of 21 $\mu F\ cm^{-2}$ was obtained.

Figure 6 compares double layer charging currents obtained at a 10NEE, a 30NEE, and a Au macroelectrode with active area equal to the geometric areas of the NEEs. As would be expected, the charging currents at the NEEs are significantly lower than at the macroelectrode. The fractional electrode areas obtained from the double layer charging currents at the NEEs are shown in Table 1. The fractional electrode area for the 30NEE is, within experimental error, identical to the fractional pore area (Table 1). This indicates that each of the pores is filled with a Au nanowire and that the heat-shrinking procedure used to seal these 30-nm-diameter nanowires is quite effective.

The fractional electrode area at the 10NEE is within a factor of 2 of the fractional pore area. This larger-than-expected fractional electrode area may result from a small amount of solution leakage around the 10-nm-diameter Au nanowires. The alternative possibility is that the pore density and/or pore size determined by electron microscopy is incorrect. This is a likely possibility because, as indicated above, it is difficult to image 10-nm structures using a conventional scanning electron microscope. Given the general observation that the sealing problem for a micro- or nanoelectrode becomes worse as the diameter of the electrode decreases [68,75], the agreement between the fractional electrode areas and the fractional pore areas (Table 1) is satisfactory.

C. Faradaic Electrochemistry at the NEEs

The nature of the Faradaic currents observed at a NEE depend on the distance between the electrode elements and the time scale (e.g., scan rate)

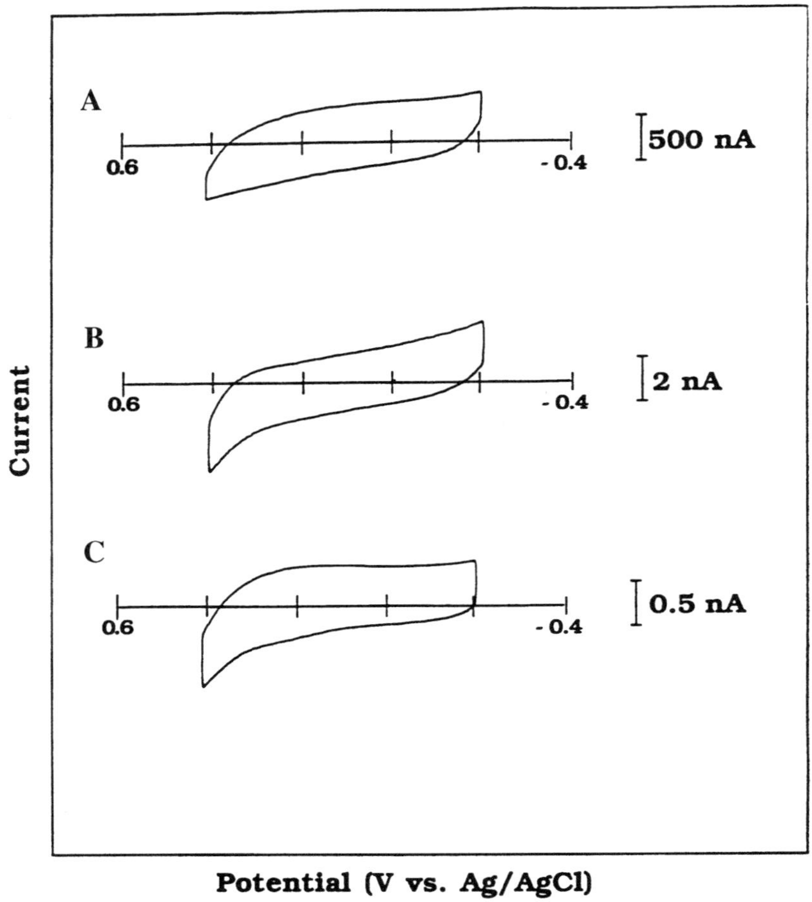

FIG. 6. Background cyclic voltammograms in 50 mM $NaNO_3$ at 100 mV s^{-1} for: (A) gold macro-disk electrode; (B) 30NEE; (C) 10NEE. The geometric area for all electrodes was 0.079 cm^2.

of the experiment [25,76]. These NEEs operate in the "total-overlap" response regime at the scan rates used here. In this total-overlap regime, the diffusion layers at the individual elements of the NEE have overlapped to produce a diffusion layer that is linear to the entire geometric area of the NEE [25,76]. As a result, conventional peaked-shaped voltammograms are obtained. Indeed, for the reversible case, the voltammogram at a NEE operating in this total-overlap regime should be identical to the voltammogram obtained at a macroelectrode with active area equivalent to the geometric area of the NEE.

Experimental and simulated cyclic voltammograms for a solution that was 5 μM in $TMAFc^+$ and 0.5 mM in supporting electrolyte (sodium nitrate) are shown in Fig. 7 [25]. The experimental data were obtained at a 10NEE. In agreement with the above discussion, the experimental voltammograms are peak shaped, and peak current increases with the square root

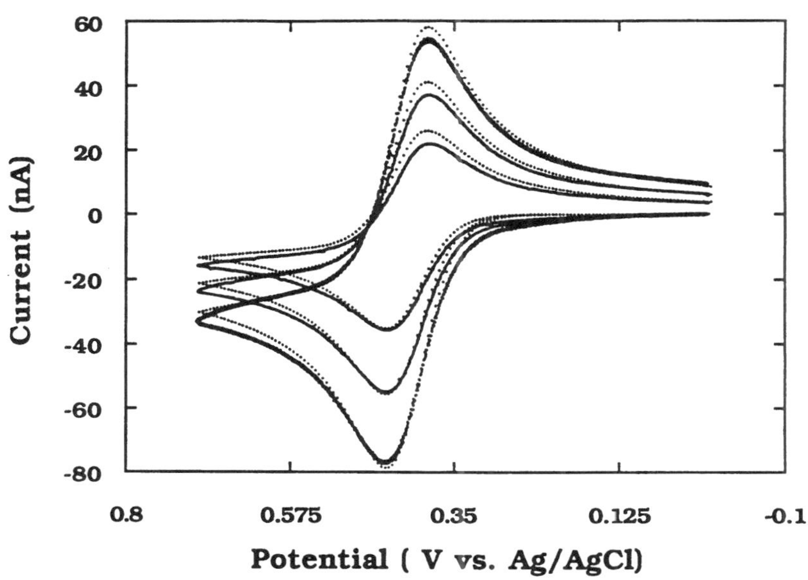

FIG. 7. Simulated (dotted curves) and experimental (solid curves) voltammograms at 100 mV s^{-1} at a 10NEE (0.079 cm^2 geometric area) in 5 μM $TMAFc^+$ and 0.5 mM sodium nitrate. Simulation assumes the total overlap limiting case (i.e., a macroelectrode with area = 0.079 cm^2).

of scan rate. This latter point is proven by the agreement between the experimental and simulated voltammograms. The simulated data were obtained by assuming reversible electrochemistry at a macroelectrode with active area equivalent to the geometric area of the NEE. Other than assuming an arbitrarily high value for the standard heterogeneous rate constant, there are no adjustable parameters in these simulations. The quantitative agreement between the experimental and simulated voltammograms indicates that the reversible, total-overlap–limiting case is, indeed, operative at this NEE.

Three other comments are worth making regarding the data shown in Fig. 7. First, the concentration of both the electroactive species (5 μM) and the supporting electrolyte (0.5 mM) are low. Low concentrations were used because we have discovered an interesting effect of supporting electrolyte concentration on the electrochemistry observed at the NEEs (see below). Second, it is possible that such low supporting electrolyte concentrations exacerbate problems of uncompensated solution resistance. At the concentration of electroactive species and the scan rates used here, the current at the NEEs is low (e.g., <100 nA in Fig. 7). As a result, uncompensated solution resistance distorts the voltammograms to a negligible extent, even at the mM supporting electrolyte concentrations used here. Indeed, voltammograms with and without application of 90% iR compensation are identical [25].

Third, the experimental voltammograms in Fig. 7 have not been corrected for background currents. Nevertheless, the agreement between the experimental and simulated voltammograms (where only Faradaic currents are simulated) is good. Background subtraction is not necessary at these scan rates, because the double layer charging currents at the NEEs are orders of magnitude lower than at a macroelectrode of equivalent geometric area (Fig. 6). Because the Faradaic currents at the NEE and the macroelectrode are, for the reversible case, identical (Fig. 7), this diminution in the background currents also means the signal-to-background ratio at the NEE is orders of magnitude larger than at the macroelectrode. This point will be explored further below. It is this enhancement in signal-to-background ratio that allows us to use such low concentrations of electroactive species.

So far we have discussed only the reversible case. The equivalence of the net Faradaic current at the NEE and at a macroelectrode of the same geometric area (Fig. 7) means that the flux at the individual elements of the NEE are many orders of magnitude larger than the flux at the macroelectrode. Indeed, the experimentally determined fractional electrode areas (Table 1) indicate that, for the reversible case, the flux at the elements of a

10NEE will be ~1000 times higher than at the macroelectrode; the flux at the elements of the 30NEE will be ~250 times higher. The higher fluxes at the NEE elements means that the NEEs will be more sensitive to the kinetics of electron transfer than a macroelectrode [25].

The simplest way to think about this situation is that for any redox couple, the quasireversible case can be observed at a NEE at much lower scan rates than at a macroelectrode. Indeed, because flux is related to the square root of scan rate, the 10^3-fold enhancement in flux at the 10NEE means that one would have to scan a macroelectrode at a scan rate 10^6 times higher in order to obtain the same kinetic information obtainable at the NEE. That is, if for a particular redox couple one observed quasireversible voltammetry at the 10NEE at scan rates above 1 V s^{-1}, one would have to scan at rates above 10^6 V s^{-1} to achieve the quasireversible case for this couple at a macroelectrode. This ability to obtain kinetic information at dramatically lower scan rates is an important advantage of a NEE.

The $Ru(NH_3)_6^{3+/2+}$ voltammograms (Fig. 8) illustrate this point. The standard heterogeneous rate constant for this couple has been measured by a number of groups; values of 0.26 [77], 1.8 [78], and 76 [79] cm s^{-1} have

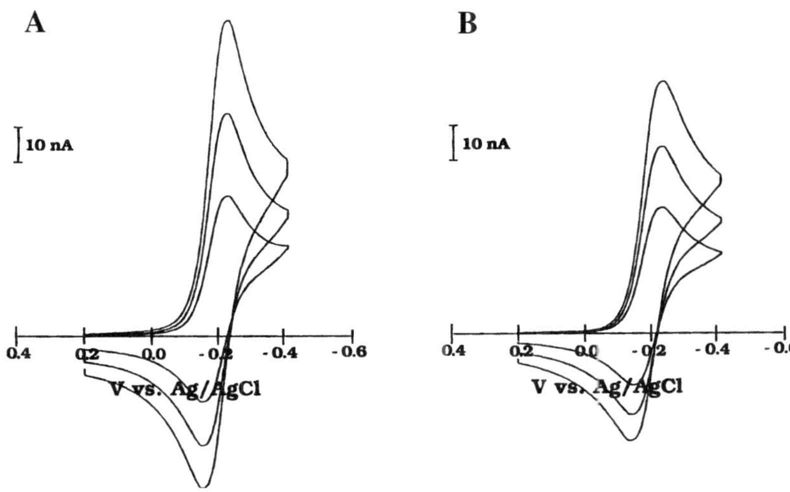

FIG. 8. Cyclic voltammograms for 5 μM $Ru(NH_3)_6Cl_3$ and 10 mM pH 7 phosphate buffer as supporting electrolyte: (A) 30NEE; (B) 10NEE. Scan rates are 20, 50, and 100 mV s^{-1}.

been reported. Assuming, for the sake of illustration, a value of 1 cm s^{-1}, Nicholson's theory shows that quasireversible $Ru(NH_3)_6^{3+/2+}$ voltammograms will be obtained, at a macroelectrode, at scan rates above ~5 V s^{-1} [80].

Figure 8A shows voltammograms at various scan rates for $Ru(NH_3)_6^{3+/2+}$ at a 30NEE. This couple shows reversible voltammetry (ΔE_{pk} ~59 mV) at the lowest scan rates shown, but the voltammograms become quasireversible at scan rates above 0.01 V s^{-1}. Therefore, as expected, the transition to quasireversible behavior is observed at dramatically lower scan rates at the 30NEE than would be observed at a macroelectrode. It is again important to emphasize that the increase in ΔE_{pk} observed is not due to uncompensated solution resistance [25].

Because the fractional electrode area at the 10NEE is lower than at the 30NEE (Table 1), the transition to quasireversible behavior would be expected to occur at even lower scan rates at the 10NEE. Voltammograms for $Ru(NH_3)_6^{3+/2+}$ at a 10NEE are shown in Fig. 8B. At the 10NEE it is impossible to obtain the reversible case, even at a scan rate as low as 5 mV s^{-1}. The effect of quasireversible electrochemistry is clearly seen in the larger ΔE_{pk} values and in the diminution of the voltammetric peak currents at the 10NEE (relative to the 30NEE; Fig. 8). This diminution in peak current is characteristic of the quasireversible case at an ensemble of nanoelectrodes [78,81]. These preliminary studies indicate that the response characteristics of the NEEs are in qualitative agreement with theoretical predictions [78,81].

D. Electroanalytical Detection Limits at the NEEs

For the reversible case, the voltammetric detection limit for a redox species at a NEE should be the detection limit for the same species at the corresponding macroelectrode multiplied by the fractional electrode area (Table 1) of the NEE [25]. Because the fractional electrode area for the 10NEE is approximately 10^{-3}, this suggests that the voltammetric detection limit at a 10NEE should be 3 orders of magnitude lower than the detection limit obtained at a macroelectrode. Figure 9A shows voltammograms at a macroelectrode at various low concentrations of $TMAFc^+$. As would be expected, the Faradaic signal ultimately vanishes into the background double layer–charging currents. Taking a total measured current that is twice the background charging current as the criterion for establishing the detection limit [25], these voltammograms show that the detection limit for $TMAFc^+$ at the macroelectrode is 1.6 μM.

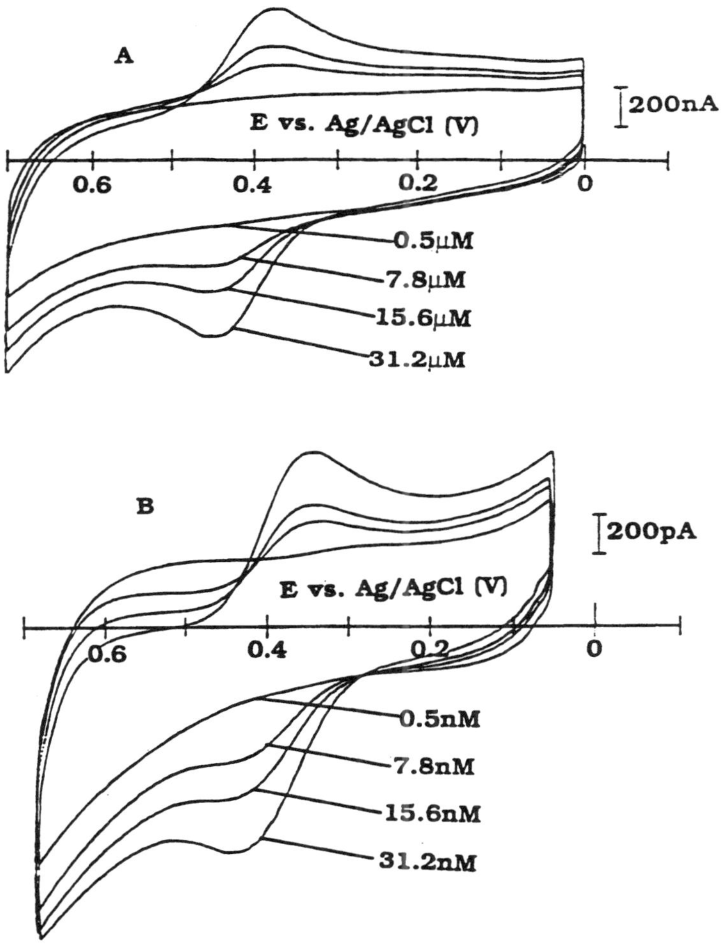

FIG. 9. Cyclic voltammograms at 100 mV s^{-1} in aqueous $TMAFc^+$ at: (A) a gold macro-disk electrode in 50 mM sodium nitrate; (B) a 10NEE in 1 mM sodium nitrate. $TMAFc^+$ concentrations are as indicated; electrode geometric area in both cases is 0.079 cm^2.

Voltammograms for various low concentrations of $TMAFc^+$ at a 10NEE are shown in Fig. 9B. While the voltammograms look nearly identical to those obtained at the macroelectrode, the concentrations are 3 orders of magnitude lower. Using the same criterion for the detection limit, we obtain a detection limit at the 10NEE that is 3 orders of magnitude lower (1.6 nM) than at the macroelectrode. This experimentally observed enhancement in detection limit at the NEE is exactly as would be predicted from the fractional electrode area data in Table 1.

Cyclic voltammetry is generally considered to be of limited use in ultratrace electrochemical analysis. This is because the high double layer–charging currents observed at a macroelectrode make the signal-to-background ratio low. The voltammograms in Fig. 9B clearly show that at the NEEs, cyclic voltammetry can be a very powerful electroanalytical technique. There is, however, a caveat. Because the NEEs are more sensitive to electron transfer kinetics, the enhancement in detection limit that is, in principle, possible could be lost for couples with low values of the heterogeneous rate constant. This is because one effect of slow electron transfer kinetics at the NEE is to lower the measured Faradaic currents (e.g., Fig. 8).

This sensitivity to slow electron transfer kinetics could, however, prove to be an advantage in sensor applications where a mediator, with fast electron transfer kinetics, is used to shuttle electrons to a redox enzyme [82]. Chemical species that are electroactive in the same potential region as the mediator can act as interferants at such sensors. If such an interfering electroactive species shows slow electron transfer kinetics, it might be possible to eliminate this interference at the NEE. This is because at the NEE, the redox wave for the kinetically slow interferant might be unobservable in the region where the kinetically fast mediator is electroactive. We are currently exploring this possibility.

E. The Effect of Supporting Electrolyte

As indicated above, all of the experimental data reported thus far were obtained at low concentrations of both supporting electrolyte (mM) and electroactive species (μM). This was done because we have observed an interesting effect of supporting electrolyte concentration on the shape of the voltammetric waves observed at the NEEs [25]. We have found that the reversibility of the voltammetric waves for all couples investigated to date ($TMAFc^{+/2+}$, $Ru(NH_3)_6^{3+/2+}$, $Mo(CN)_8^{4-/3-}$) improves as the concentration of supporting electrolyte decreases. This effect is illustrated for $TMAFc^{+/2+}$ in Fig. 10. Note that the peak currents decrease and the ΔE_{pk} values increase as the concentration of supporting electrolyte increases.

This effect has been observed in all of the supporting electrolytes we have investigated to date; these include $NaNO_3$, $NaClO_4$, Na_2SO_4, Et_4NBF_4, Et_4NClO_4, $Mg(NO_3)_2$, KNO_3, KPF_6, $ZnSO_4$, and pH = 7.0 phosphate buffer. Furthermore, this effect is reproducible and reversible. By reversible we mean that if the NEE is taken out of a solution with high supporting electrolyte concentration (e.g., 50 mM) and returned to a solution with low electrolyte concentration (e.g., 1 mM), the voltammetric wave immediately assumes the reversible appearance characteristic of the low electrolyte concentration solution (Fig. 10). Furthermore, we have observed this effect for NEEs prepared from both polycarbonate and polyester membranes, for NEEs prepared from membranes with and without the PVP that is used to improve membrane wettability, and for NEEs with low densities of elec-

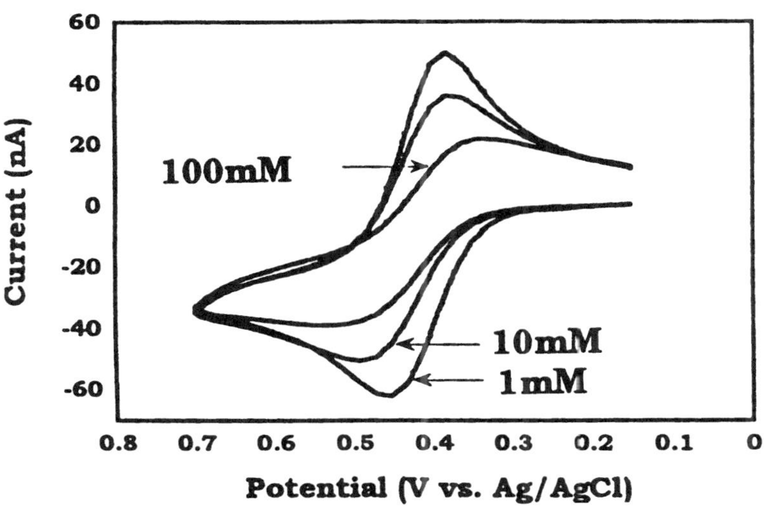

FIG. 10. Cyclic voltammograms illustrating the effect of supporting electrolyte concentration at a 10NEE. 5 μM $TMAFc^+$ in aqueous $NaNO_3$ at the indicated concentrations of $NaNO_3$. Scan rate = 100 mV s^{-1}.

trode elements [70] where total overlap does not occur [83]. We are currently exploring this interesting and unexpected effect of supporting electrolyte concentration.

F. Conclusions

We have demonstrated a new method for preparing electrodes with nanoscopic dimensions. We have used this method to prepare nanoelectrode ensembles with individual electrode element diameters as small as 10 nm. This method is simple, inexpensive, and highly reproducible. The reproducibility of this approach for preparing nanoelectrodes is illustrated by the fact that NEEs given to other groups yielded the same general electrochemical results as obtained in our laboratory [84]. These NEEs display cyclic voltammetric detection limits that are as much as 3 orders of magnitude lower than the detection limits achievable at a conventional macroelectrode.

IV. GOLD NANOTUBULE MEMBRANES WITH ELECTROCHEMICALLY SWITCHABLE ION-TRANSPORT SELECTIVITY

As discussed above, the electroless plating procedure used to prepare the NEEs initially produces Au tubules within the pores of the template membrane. We have shown that by carefully controlling the plating rate, the inside diameter of these tubules can be varied at will [71,72,85]. At long plating times Au tubules with inside diameters of molecular dimensions (<1 nm) are obtained [72]. We will show here that membranes containing such nanoscopic Au tubules can function as electrochemically switchable ion exchangers. In the following section we will show that these membranes can be used as "molecular filters" to cleanly separate small molecules on the basis of molecular size.

A. Preparation of the Au Nanotubule Membranes

The pores in a commercially available polycarbonate filtration membrane (Poretics) were used as templates to form the nanotubules (pore diameter $= 50$ nm; pore density $= 6 \times 10^8$ pores cm^{-2}; thickness $= 6$ μm). As before, the electrolessly plated Au deposits both on the pore walls and the membrane faces [71]. The gold surface layers on the membrane faces allow us to make electrical contact to the Au nanotubules within the pores. The thickness of the gold layers deposited on the pore walls can be controlled by varying the plating time. As a result, the inside diameter (i.d.) of the gold nanotubules can be varied at will (as determined from measurements of helium gas flux [86,87] across the tubule-containing membrane).

B. Results

We describe the results of three sets of experiments that demonstrate that these Au nanotubule membranes can show selective ion transport. All of these experiments involve a U-tube cell in which the membrane to be studied separates two aqueous solutions. The simplest experiment entails using a ''feed'' solution of a colored anionic or cationic species on one side of the membrane and a ''permeate'' solution that is initially devoid of the colored species on the other side of the membrane. A membrane that contained ~2.5-nm-radius [86,87] Au nanotubules was used for these experiments.

When the feed solution is 1 mM in KCl and 0.5 mM in the cationic dye methylene blue and the receiver solution is 1 mM KCl, the initially colorless receiver solution turns blue due to transport of the cationic dye across the membrane [71]. In contrast, when the feed solution is 1 mM in KCl and 5 mM in $KMnO_4$ (MnO_4^- is red) and the receiver solution is 1 mM KCl, the receiver solution remains colorless [71]. These experiments provide simple visual evidence that this membrane transports a large cation but does not transport a much smaller anion. We have used potentiometric measurements to explore the nature of this cation permselectivity.

The extent of ion permselectivity displayed by a membrane can be expressed quantitatively by the transference numbers [88] for cations (t_+) and anions (t_-) within the membrane. Transference numbers can be determined potentiometrically by using a concentration cell [88], in which the membrane to be evaluated separates two electrolyte solutions that contain the same salt but at different concentrations. For a 1 : 1 salt, the membrane potential (E_m) is given by

$$E_m = (2.303RT/nF)(t_+ - t_-) \log(a_h/a_l) \quad (1)$$

where a_h and a_l are the activities of the salt in the solution of high and low salt concentration, respectively [89]. Equation (1) indicates that for an ideal cation-permselective membrane ($t_+ = 1.0$), a plot of E_m versus $\log(a_h/a_l)$ would be linear, with an intercept of 0 and a slope of 59 mV (dashed line, Fig. 11).

A concentration cell was assembled in which a gold nanotubule mem-

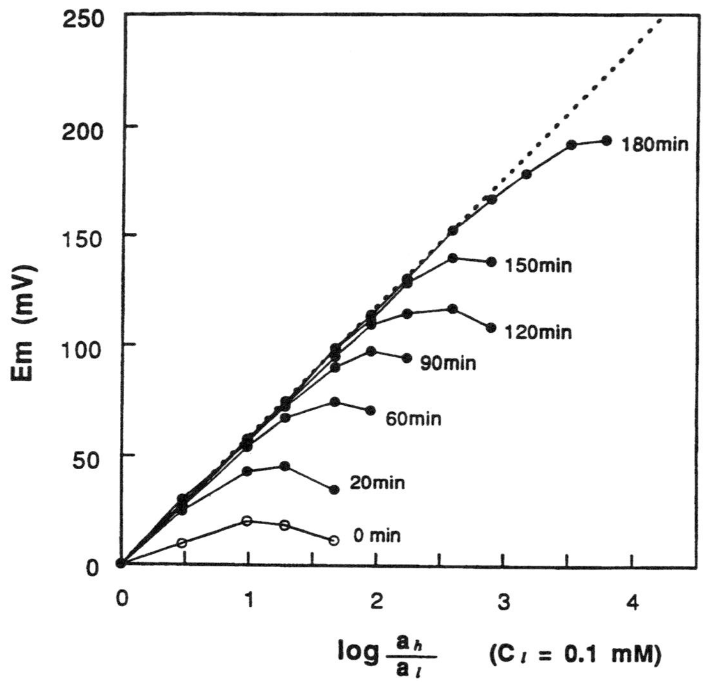

FIG. 11. E_m values obtained for membranes prepared for the indicated plating times. The solution on the *l* side of the membrane was 0.1 mM KCl. The solution on the *h* side of the membrane was varied from 0.1 mM KCl to 1 M KCl. The E_m values were measured by using two Ag/AgCl (KCl saturated) electrodes placed in each half-cell through agar salt-bridges. The dashed line is for ideal cation-permselective behavior.

brane separated two KCl solutions. The potential of the membrane was not controlled with a potentiostat. However, Cl^- chemisorbs to gold [90] and the gold films on the membrane faces and the inside walls of the gold tubules have excess negative (Cl^-) charge on their surfaces. This excess negative charge is balanced by a layer of excess positive charge (K^+) in the solution immediately adjacent to gold surfaces (the electrical double layer) [91].

Data obtained from this concentration cell (Fig. 11) show that these membranes can show ideal cation-permselective behavior and that the region over which ideal behavior is observed is extended to higher salt concentrations as plating time increases. These observations can be explained as follows: over the range of plating times used in Fig. 11, the average inside radii of the gold tubules varied from ~9.4 nm (60 min plating time) to ~0.8 nm (180 min plating time). Gouy-Chapman theory [91] predicts that over the salt concentration range used here, the thickness of the electrical double layer within the tubules (as approximated by the Debye length) varies from ~30 nm (lowest concentration) to ~0.3 nm (highest concentration). Figure 11 indicates that the gold nanotubule membranes show ideal cation permselectivity, provided the radius of the tubule is small relative to the thickness of the electrical double layer within the tubule.

To illustrate this point, consider the membrane plated for 60 minutes. The tubules in this membrane average ~9.4 nm in inside radius. At low concentrations of salt, the electrical double should be thicker than this tubule radius. Anions are excluded from the tubes, and ideal cation permselectivity is observed. At high salt concentrations, the electrical double layer is thin relative to the tubule radius. Anions can now enter the tubules, and ideal cation permselectivity is lost (Fig. 11). Finally, the membrane plated for 180 minute shows cation permselectivity almost identical to that of the ionomer Nafion® [92], which is a highly cation-permselective polymer used in industrial electrolytic processes [93].

We consider now the idea of controlling the permselectivity by potentiostatically injecting excess charge into the gold nanotubules. For these studies, it is essential to use an anion that does not chemisorb to gold because we do not want the excess charge to be determined by chemisorption. Because F^- does not chemisorb to Au [94], KF was chosen as the electrolyte. A concentration cell was assembled in which a gold nanotubule membrane separated solutions that were 1 mM and 10 mM in KF. This membrane was connected (through the Au surface layers) to the working electrode lead of a potentiostat, and the potential applied to the membrane was varied over the range from −0.5 to +0.5 V versus Ag/AgCl. The E_m values were measured at each applied potential (Fig. 12).

The dashed lines at the top and bottom of Fig. 12 are the E_m values that would be achieved if the nanotubule membrane showed ideal cation and ideal anion permselectivity, respectively [Eq. (1)]. At negative applied potentials, the nanotubule membrane shows ideal cation permselectivity, whereas at positive applied potentials the membrane shows ideal anion permselectivity. This selectivity occurs because at negative applied potentials, excess electrons are present on the walls of the tubes and excess posi-

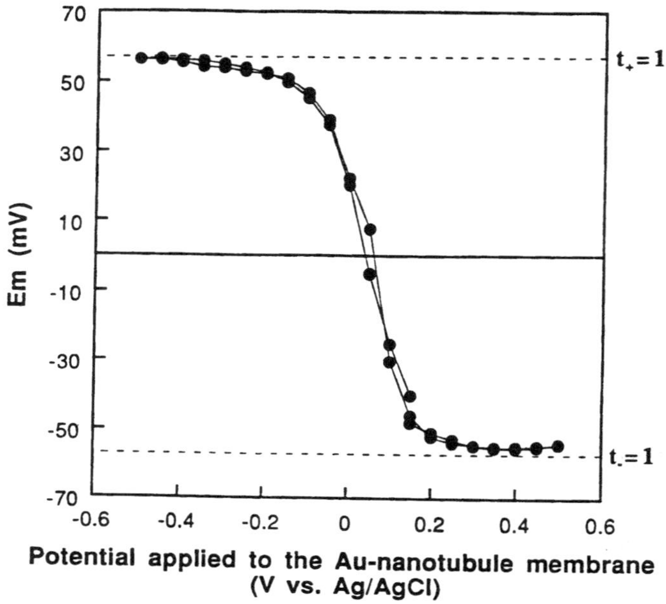

FIG. 12. Variation of E_m with potential applied to the membrane (1 mM KF on the low side and 10 mM KF on the high side of the membrane; tubule radius ~1.1 nm). The potential of the membrane was controlled with a potentiostat versus a Ag/AgCl reference electrode immersed in the high side solution.

tive charge (K^+) accumulates within the tubes. As a result, anions (F^-) are excluded and cations (K^+) are transported by the membrane. At positive applied potentials the opposite situation occurs—cations are excluded and anions are transported.

For any combination of metal and electrolyte, there is a potential called the potential of zero charge (pzc) where there is no excess charge on the metal. At this potential the nanotubule membranes should show neither cation nor anion permselectivity, and E_m should approach 0 mV.* E_m for the tubule-containing membrane does, indeed, go from the ideal cation-

* In fact, at the pzc, the membrane should show the liquid junction potential [88] based on the transference numbers of the ions in the bulk solution.

permselective value, through zero to the ideal anion-permselective value (Fig. 12). Furthermore, the potential at which E_m approaches zero is close to the reported pzc* (−4 mV for Au in 1 mM NaF [95]).

Figure 12 shows that the gold nanotubule membranes can function as electronically switchable ion exchange membranes. However, it would seem that this would only be possible if the electrolyte contained only non-adsorbing anions such as F^-. If a chemisorbing anion (such as Cl^- or Br^-) [90,96] were present, it would adsorb at positive applied potentials yielding a cation-permselective membrane. While the anion would not chemisorb at sufficiently negative applied potentials, the metal would have excess electron density at such potentials, and, again, cation permselectivity would be observed. Hence, in the presence of a chemisorbing anion, cation permselectivity will be observed at all applied potentials [71].

Such anion adsorption can be prevented by chemisorbing a monolayer of a strongly adherent thiol molecule to the Au surfaces [97,98]. 1-Propanethiol (PT) was used here because the gold nanotubules can still be wetted with water after chemisorption of the PT monolayer [97].† The E_m versus applied potential curves for an untreated and PT-treated gold nanotubule membrane, with KBr solutions present on either side of the membrane, are shown in Fig. 13. The untreated membrane shows only cation permselectivity, but the permselectivity of the PT-treated membrane can be switched, exactly as was the case with the nonadsorbing electrolyte (Fig. 12).

C. Conclusions

We have demonstrated that these Au nanotubule membranes can be cation permselective, anion permselective, or nonselective, depending on the potential applied to the membrane‡ [99]. These membranes can be as permse-

* This reported pzc is for polycrystalline Au. Electrochemists may debate the precise meaning of the pzc for a polycrystalline metal surface. However, for our purposes here it is the potential at which there is no net excess charge on the polycrystalline metal surface.

† This is undoubtedly because prior studies have shown that, because of its short alkyl chain, PT monolayers are disordered.

‡ Although polymeric membranes that show switchable transport properties have been described, the switchability was based on Faradaic electrolysis of the polymer.

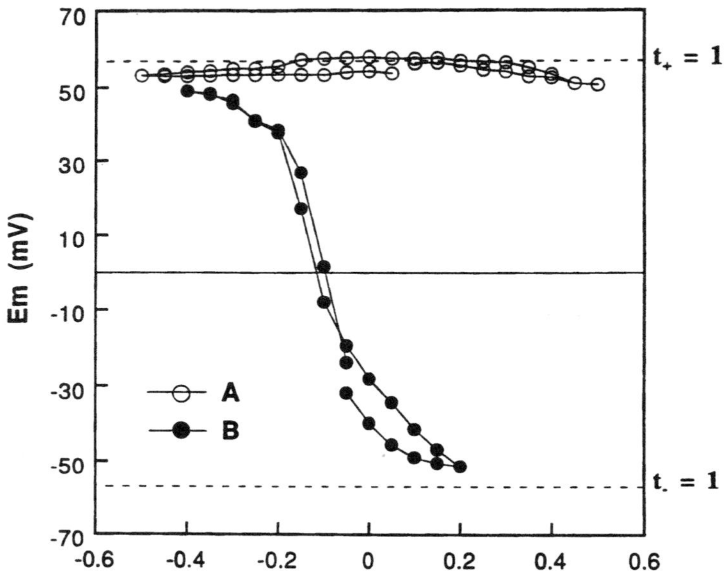

FIG. 13. Variation of E_m with potential applied to the membrane (1 mM KBr on *l* side and 10 mM KBr on *h* side; membrane as per Fig. 12) for an untreated Au nanotubule membrane (upper curve) and a PT-coated membrane (lower curve).

lective as the commercially important Nafion® polymer and may have applications in both fundamental and applied electrochemistry.

V. MOLECULAR FILTRATION AND CHEMICAL TRANSPORT SELECTIVITY IN THE AU NANOTUBULE MEMBRANES

The work discussed above shows that the Au nanotubule membranes can have one important type of transport selectivity—charge-based selectivity. It occurred to us that because the Au nanotubules can be of molecular dimensions, these membranes might show molecular size–based transport selectivity as well [72]. Finally, the thiol chemisorption chemistry introduced above provides a route for introducing chemically based transport selectivity [85]. Hence, the Au nanotubule membranes should, in principle, be able to show all three of the important transport selectivity paradigms—

charge, size, and chemical. If so, these membranes might be of interest in the area of membrane-based chemical separations, where a membrane is used to separate chemical species [100–104].

A. Size-Based Selectivity

The idea of using membranes to filter molecules on the basis of size is not without precedent. Dialysis is used routinely to separate low molecular weight species from macromolecules [105]. In addition, nanofiltration membranes are known for certain small molecule separations (such as water purification), but such membranes typically combine both size and chemical transport selectivity and are particularly designed for the separation involved. Hence, in spite of the importance of the concept, synthetic membranes that contain a collection of monodisperse, molecule-sized pores that can be used as molecular filters to separate small molecules on the basis of size are currently not available.

1. Experimental

The electroless plating method described above was used to prepare membranes that contain cylindrical gold nanotubules, which span the complete thickness of the membrane. As before, polycarbonate filtration membranes with cylindrical, monodisperse pores (Poretics, 6 μm thick, pore dia. = 30 nm) were used as the templates. The inside diameter (i.d.) of the nanotubules can be varied by varying the plating time (Fig. 14). At sufficiently long plating times, Au nanotubules with i.d.s of molecular dimensions (<1 nm) are obtained* [106].

We have discovered that the shape of the Au nanotubule can be changed by varying the rate of the plating reaction. When high plating rates are used,† Au is preferentially deposited on the faces of the membrane, and nanotubules with bottlenecks at both ends, but a larger i.d. in the middle (Fig. 15), are obtained. Such bottleneck tubules are a form of ultrathin film composite membrane [102,107] and should provide high permeate flux without sacrificing transport selectivity. This is because selectivity should

* We approximated the i.d.s of tubules using a gas permeation method (see Ref. 71).

† Our previous studies were done using a plating bath at pH 10. Higher plating rates can be achieved using the same plating bath at pH 12.

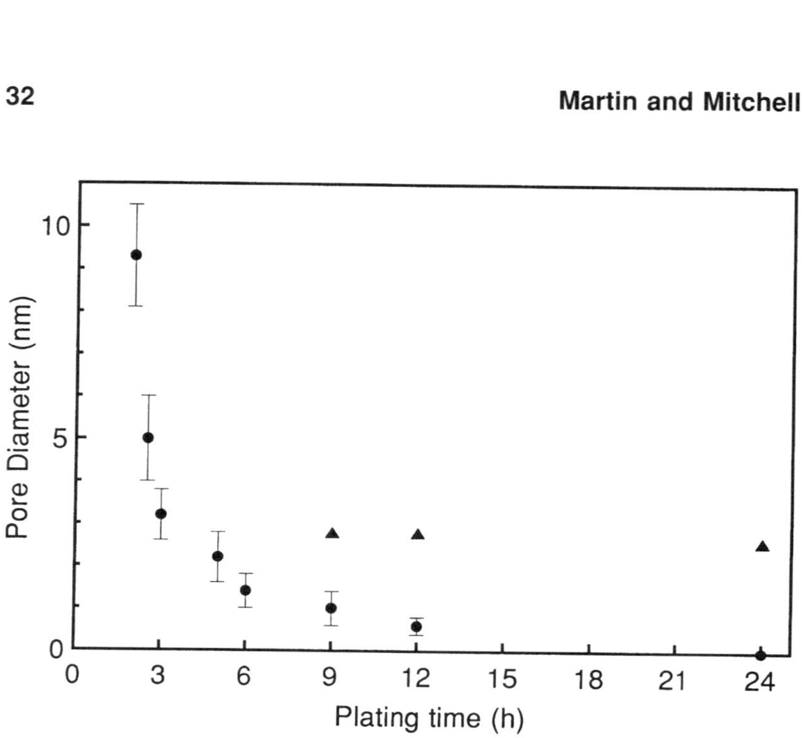

FIG. 14. Plot of nanotubule diameter versus plating time for the Au nanotubule membranes. The standard deviation reflects measurements on at least three different membranes prepared under identical conditions. The circles are for membranes that have the Au surface layers still present on the membrane faces. The triangles are after removal of both surface layers.

be determined by permeation through the bottleneck, but overall flux is determined by permeation in the larger-i.d. tubule that spans the membrane.

We show that membranes containing such bottleneck nanotubules can be used to cleanly separate small molecules on the basis of size. The nanotubule membrane was mounted in a U-tube permeation cell such that the membrane separated a feed solution from a permeate solution [108]. The feed solution was equimolar in two compounds of differing molecule size. We call these the "smaller" and the "larger" molecules; three smaller-molecule/larger-molecule pairs were investigated here (Fig. 16). The permeate solution, initially just pure water, was periodically assayed for the presence of both the smaller and larger molecules. In all three cases (Fig. 16), easily measurable quantities of the smaller molecule were obtained in the permeate solution, but the larger molecule was completely undetectable.

We began our studies, however, with simpler single-molecule perme-

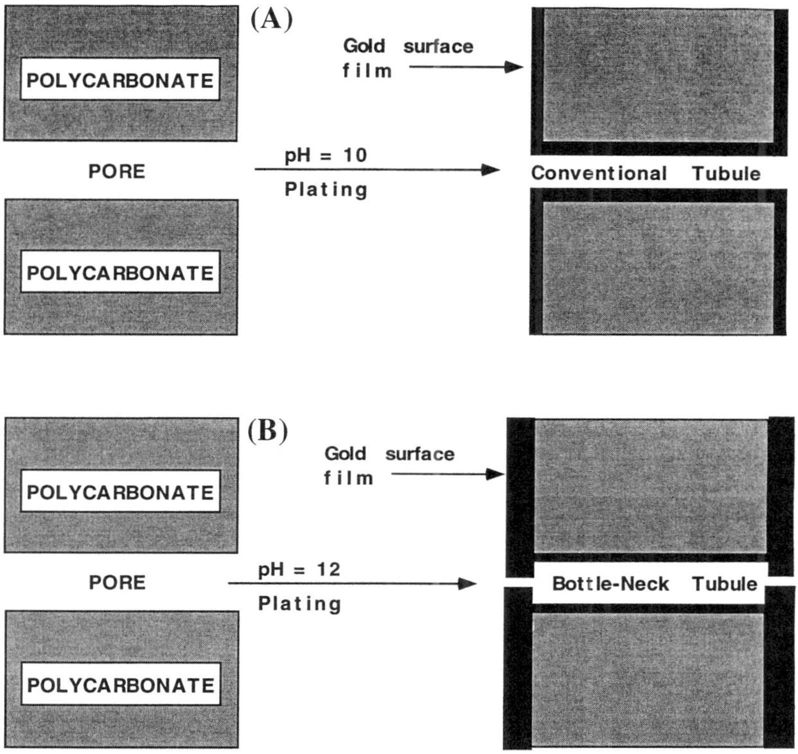

FIG. 15. Schematic illustrations of shapes of the Au nanotubule obtained by doing the electroless Au plating at (A) pH = 10 and (B) pH = 12. The higher pH causes bottleneck tubules. The tubules plated at the lower pH also have some of this bottleneck character. Hence, the depictions in both (A) and (B) are approximate and serve to illustrate the conceptual differences between the two types of nanotubules investigated here.

ation experiments* [109]. These experiments were done in the same U-tube permeation cell [108], but the feed solution contained only one of the molecules of the smaller-molecule/larger-molecule pair. The flux of this molecule across the nanotubule membrane was determined, and then in a separate experiment the flux of the other molecule of the pair was mea-

* Such single molecule permeation experiments are used routinely.

Anilinium chloride

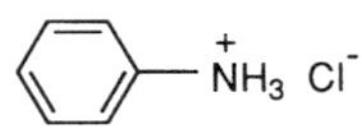

Rhodamine B

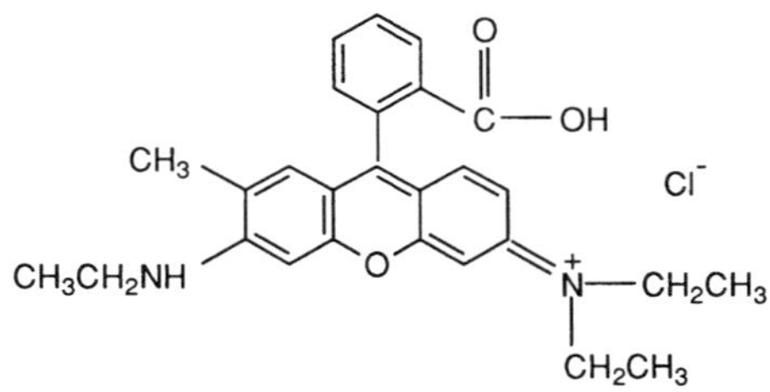

Methyl viologen

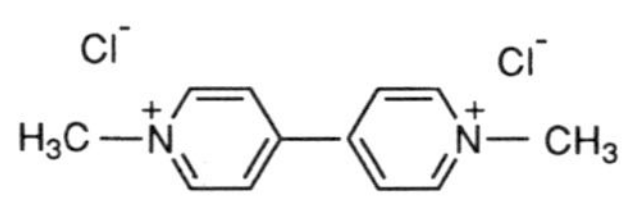

Ru(bpy)$_3$ Cl$_2$

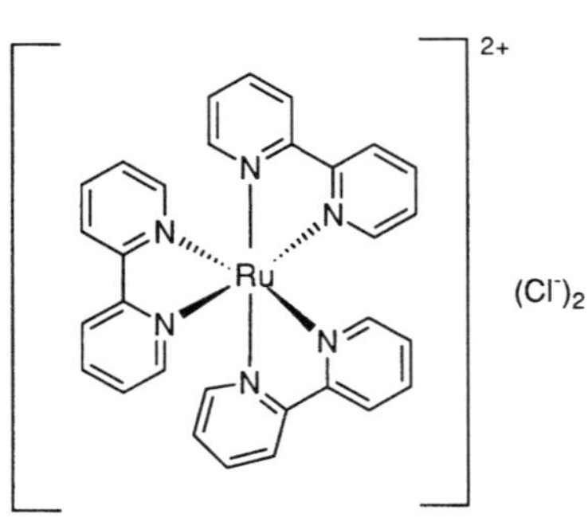

FIG. 16. Chemical structures and approximate relative sizes of the three pairs of molecules studied here. Note that the charges on the molecules of each pair are the same. Because the permeate solution was initially just pure water, the cationic molecules come across the membrane with their charge-balancing anions.

Pyridine

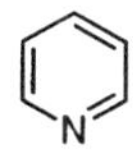

Quinine

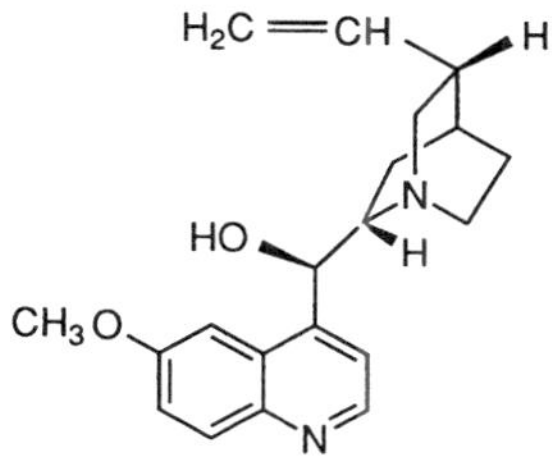

FIG. 16. Continued

sured.* The objective was to explore the effect of nanotubule i.d. on transport rate and selectivity. Nanotubules with the more conventional shape (Fig. 15A) were used for these preliminary studies.

2. *Results of Single-Molecule Permeation Experiments*

Results of such single-molecule permeation experiments, using the MV^{2+}/$Ru(bpy)_3^{2+}$ pair (Fig. 16), and membranes with four different nanotubule i.d.s, are shown in Fig. 17. The slopes of these permeation curves define the fluxes of MV^{2+} and $Ru(bpy)_3^{2+}$ across the membrane. A permeation selectivity coefficient (α_i)† can be obtained by dividing the MV^{2+} flux by the $Ru(bpy)_3^{2+}$ flux.

---

* The flux was determined by continuously monitoring the UV absorbance of the permeate solution by flowing the permeate through a UV detector.

† The subscript i is for "ideal" and signifies that these coefficients were obtained via single-molecule permeation experiments (see, e.g., Ref. 109).

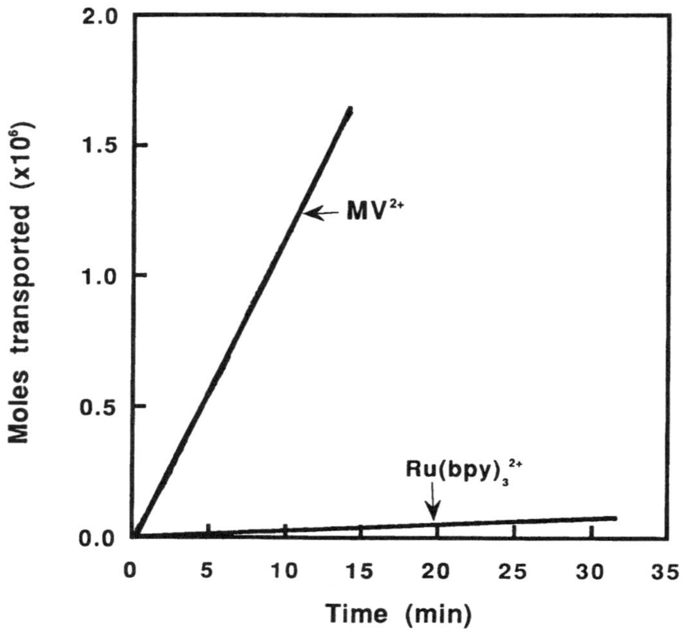

Moles transported ($x10^6$)
2.0
1.5
1.0
0.5
0.0
MV^{2+}
$Ru(bpy)_3^{2+}$
0
5
10
15
20
25
30
35
Time (min)

(A)

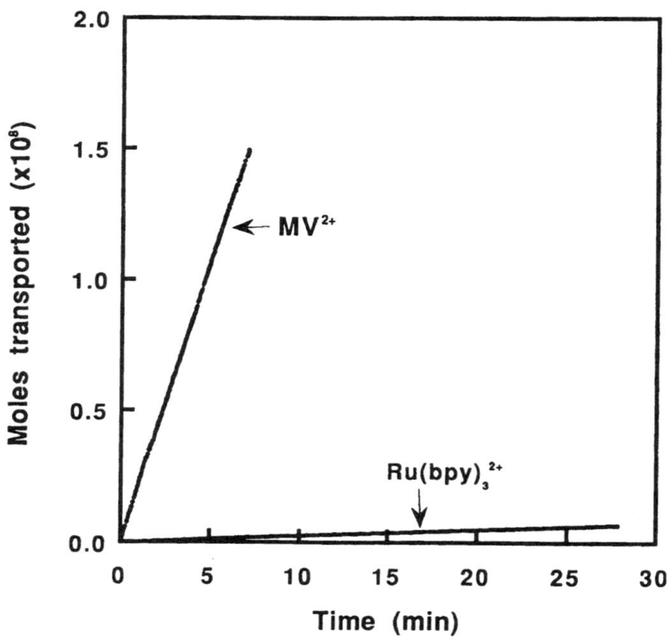

Moles transported ($x10^8$)
2.0
1.5
1.0
0.5
0.0
MV^{2+}
$Ru(bpy)_3^{2+}$
0
5
10
15
20
25
30
Time (min)

(B)

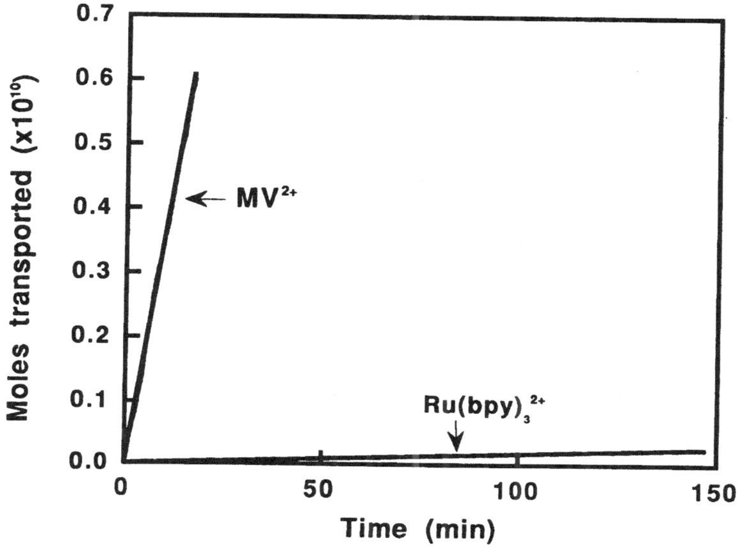

(C)

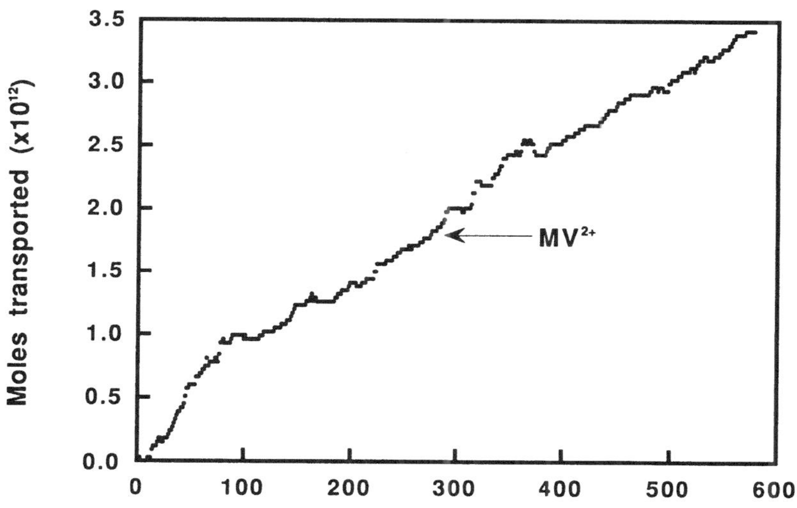

(D)

FIG. 17. Moles of MV^{2+} and $Ru(bpy)_3^{2+}$ transported versus time (see Fig. 16 for chemical structures). Membranes containing nanotubules as depicted in Fig. 15A with diameters of (A) 5.5 nm, (B) 3.2 nm, (C) 2.2 nm. The nanotubule diameter in (D) was too small to measure using the gas-permeation method (diameter < 0.6 nm). Only MV^{2+} was transported through this membrane. The data show more noise than A, B, and C because the flux is lower.

The data in Fig. 17 may be summarized as follows:

1. Even for the largest i.d. nanotubules investigated (5.5 nm), α_i was substantially greater than the ratio of the diffusion coefficients for these molecules in aqueous solution ($\alpha_i = 50$; ratio of diffusion coefficients = 1.5 [110,111]). Hence, size-based sieving occurred in these large-i.d. ($\gg$ molecular dimensions) nanotubules.*
2. As the nanotubule i.d. decreased, the fluxes for both molecules decreased; however, the flux of the larger $Ru(bpy)_3^{2+}$ decreased more rapidly. As a result, α_i increased with decreasing nanotubule i.d.† Values for the 5.5-, 3.2-, and 2.0-nm-i.d. nanotubule membranes are $\alpha_i = 50$, 88, and 172, respectively.
3. The smallest i.d. nanotubule membrane (Fig. 17D) showed a measurable flux for MV^{2+}, but the larger $Ru(bpy)_3^{2+}$ could not be detected in the permeate solution, even after a 2-week permeation experiment.

Although the results of these experiments were encouraging in terms of selectivity, the fluxes were low. As indicated above, higher fluxes should be obtained from the bottleneck nanotubule membranes (Fig. 15B). However, we first present flux data to illustrate the bottleneck character of these tubules. The MV^{2+} flux in a typical bottleneck membrane (Fig. 15B) is 19 nmol hr^{-1} cm^{-2}. When the surface gold layer containing the bottleneck (Fig. 15B) is removed [25], the flux increases by one order of magnitude to 180 nmol hr^{-1} cm^{-2}. However, the surface layer is only 150 nm in thickness. When the second surface layer is removed, the flux again increases by an order of magnitude. That these very thin (relative to the membrane) surface layers can have such a dramatic effect on the flux clearly shows that flux-limiting constrictions are present in the surface layers (Fig. 15B).

We have also shown that these bottleneck membranes can show the same selectivity but higher flux than the more conventional shape (Fig.

---

* Similar sieving was observed in radiotracer self-diffusion experiments on lightly etched films prepared via the track-etch process [112]. However, molecular filtration of the type described here was not observed.

† While α_i, in general, increased with decreasing tubule i.d., an interesting anomaly was observed for two membranes with tubule i.d.s between those in Fig. 17C and 17D. α_i values for those membranes were lower than the $\alpha_i = 172$ observed in Fig. 17C. We are currently exploring the genesis of this anomaly.

15A). To demonstrate this point, the rate and selectivity of transport across a conventional (Fig. 15A) and a bottleneck membrane were compared.* Both membranes were able to cleanly separate MV^{2+} from $Ru(bpy)_3^{2+}$ in the two-molecule permeation experiment (see below). Hence, these membranes showed comparable, excellent selectivity. However, as expected, the flux of MV^{2+} across the bottleneck nanotubule membrane was dramatically higher than for the conventional nanotubule membrane (14 vs. 0.07 nmol hr^{-1} cm^{-2}).

3. *Results of Dual-Molecule Permeation Experiments*

We now turn to the more interesting case of having both molecules of a pair in the feed solution together. These experiments were only done on bottleneck nanotubule membranes that showed $\alpha_i = \infty$. Typical results for the pyridine/quinine pair are shown in Fig. 18. Figure 18A shows the absorption spectra for 0.5 mM quinine and 0.5 mM pyridine. Pyridine shows a characteristic peak at ~252 nm. Quinine shows a much more intense band centered at 225 nm and two other bands at ~280 and ~330 nm.

Figure 18B shows the absorption spectrum for the feed solution used in the permeation experiment. Although both molecules are present in solution at the same concentration, the higher absorbance of the quinine nearly swamps out the 252-nm peak of the pyridine. Figure 18C shows the absorption spectrum of the permeate solution after a 72-hour permeation experiment. In spite of the higher absorbance of the quinine (larger molecule), only the peak for the pyridine (smaller molecule) is seen in this spectrum. Note, in particular, the complete absence of the very intense quinine band centered at 225 nm. Figure 18C shows that, to our ability to make the measurement, this bottleneck nanotubule membrane has filtered these two molecules on the basis of molecular size.

To verify this point, a much more sensitive analytical method, fluorescence† was used to search for traces of quinine in the permeate solution. The magnitude of the absorbance in Fig. 18C indicates that the pyridine concentration in the permeate is 7×10^{-5} M. With fluorescence analysis it is possible to detect 5×10^{-9} M quinine in the presence of 7×10^{-5} M pyridine. However, no quinine fluorescence could be detected from the

---

* Because of the nonuniform shape of the bottleneck tubules (Fig. 15B), it is difficult to extract an i.d. using the gas-flux method [71]. All bottleneck membranes were plated a pH 12 bath for a duration of 8 hours.

† Quinine was excited at $\lambda_{ex} = 308$ nm and detected at $\lambda_{em} = 403$ nm.

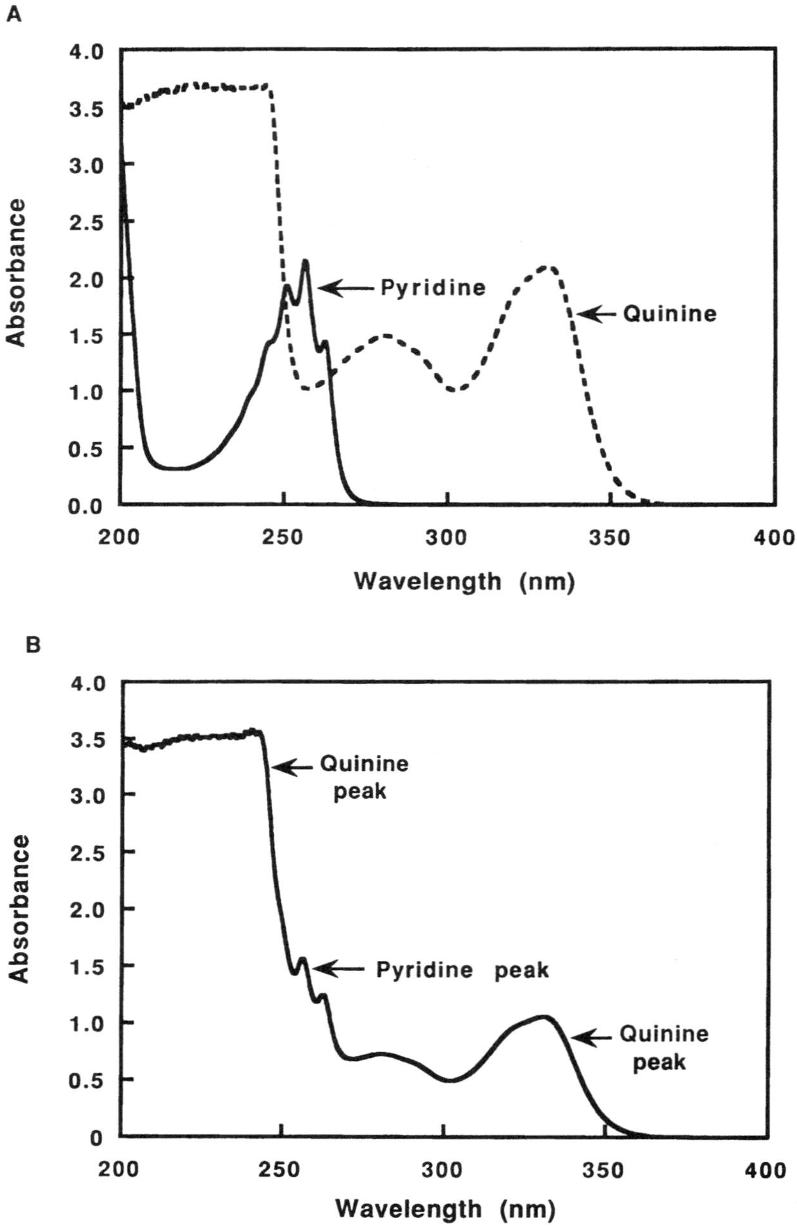

FIG. 18. Absorption spectra for (A) 0.5 mM pyridine and 0.5 mM quinine, (B) the pyridine/quinine feed solution (both molecules 0.25 mM), (C) permeate after transport through a bottleneck nanotubule membrane.

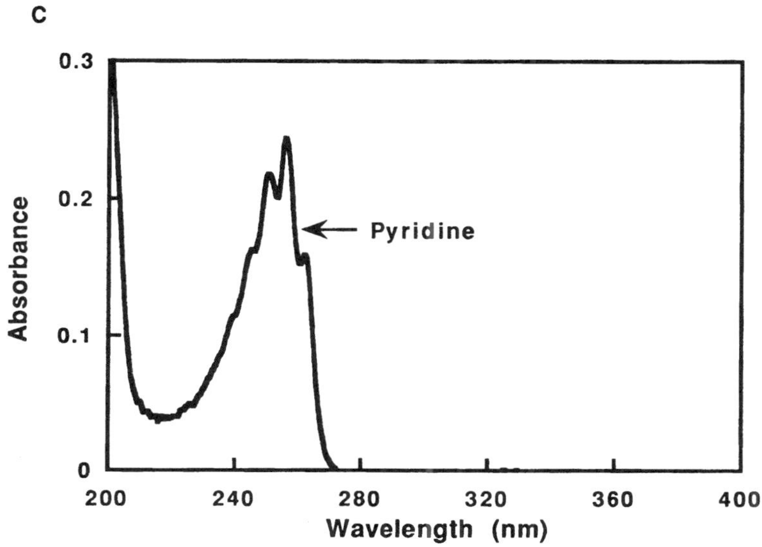

FIG. 18. Continued

permeate solution. Again, these data show that to our (now much more sensitive) ability to make the measurement, this membrane has cleanly separated these two molecules. This analysis also shows that if any quinine is present in the permeate solution, its concentration is less than 5×10^{-9} M.

These analytical data can be used to calculate a minimal selectivity coefficient, α_{min}. Because the concentration of the smaller molecule in the permeate solution was 7×10^{-5} M and the concentration of the larger molecule (if present at all) must be less than 5×10^{-9} M, the minimal selectivity coefficient for the pyridine/quinine pair is $\alpha_{min} = 15{,}000$. The minimal selectivity coefficients obtained in this way* [121] for the other

* The minimal quantity of $Ru(bpy)_3^{2+}$ that could be detected was determined via fluorescence; $\lambda_{ex} = 286$ nm; $\lambda_{em} = 594$ nm. The concentration of MV^{2+} was determined via UV absorbance (258 nm). The minimal quantity of rhodamine B that could be detected was determined via its extremely intense absorbance at 555 nm. The concentration of anilinium was determined via absorbance (254 nm).

TABLE 2
Minimal Selectivity Coefficients for Three Different Small Molecule/Large Molecule Pairs

Small molecule	Large molecule	Minimal selectivity coefficients
Methylviologen chloride	Ruthenium tris(2,2′-bipyridine) chloride	1,500
Pyridine	Quinine	15,000
Anilinium chloride	Rhodamine B	130,000

pairs are shown in Table 2. It is important to stress that, in all three cases, the larger molecule was undetectable in the permeate solution.

We demonstrated above that these Au nanotubule membranes can show charge-based transport selectivity, and we have now shown that these membranes can also have molecular size–based selectivity. The next step was to attempt to introduce chemical transport selectivity.

B. Chemical Transport Selectivity

In addition to the transport selectivities based on molecular charge or size described above, chemical interactions between the membrane material and the molecule to be transported can also strongly influence the rate and selectivity of transport. The introduction of chemically based transport selectivity was accomplished by chemisorbing thiols (RSH) to the Au tubule surfaces* [113]. Membranes derivatized with two different R groups—the hydrophobic R = $-C_{16}H_{33}$ and the more hydrophilic (2)R = $-C_2H_4$-OH—were prepared. The rate and selectivity of transport in these membranes is dramatically altered by the chemical identity of the R group.

1. Experimental

The electroless plating procedure described above was used to plate the Au nanotubules into the pores of commercially available polycarbonate track-etch filters [Osmonics, 6 μm thick, pore dia. = 50 nm (28 nm-dia. Au tubules) or 30 nm (all other Au tubules), 6×10^8 pores cm^{-2}]. The

* Chemisorption was accomplished by immersion for 12 hours in a 1 mM solution of either mercaptoethanol or hexadecylthiol; the solvent was absolute ethanol.

inside diameter of the nanotubule was varied by varying the plating time, and a gas-flux method was used to obtain approximate i.d. values for each membrane. As discussed above, these tubes are bottleneck in shape, thus the i.d. values are approximate and really only provide a relative measure of the effective tube diameter. In addition, the i.d.s reported here were measured before chemisorption of the thiol, and this will clearly result in further constriction of the pore. However, the decrease in the measured i.d. after incorporation of the thiol is not as dramatic as might be expected based on the length of the thiol. For example, a membrane that contained Au nanotubules with a measured i.d. value of 2.6 nm before chemisorption showed an i.d. value of 1.9 nm after chemisorption of the R = -$C_{16}H_{33}$ thiol. No change in the i.d. could be detected after chemisorption of the R = -C_2H_4-OH thiol.

Transport properties were determined by mounting the membrane between the two halves of a U-tube permeation cell [108]. The feed half-cell contained 5 ml of an aqueous solution (5 mM) of the molecule to be transported (the permeant molecule); the permeate half-cell initially contained 5 ml of pure water. The transport of the permeant molecule into the permeate half-cell was monitored by periodically assaying (via UV absorbance spectroscopy) the permeate solution. These membranes showed reproducible fluxes for periods of at least 10 days.

2. *Results*

We begin by comparing fluxes of the permeant molecule pyridine in untreated and thiol-treated nanotubule membranes. An untreated membrane that contained tubules with diameters of approximately 2.6 nm showed a pyridine flux of 1.8×10^{-7} mol cm^{-2} hr^{-1}. After chemisorption of the R = -C_2H_4-OH thiol the flux increased to 4.2×10^{-7} mol cm^{-2} hr^{-1}. In contrast, after chemisorption of the R = -$C_{16}H_{33}$ thiol, the pyridine flux dropped to 2.7×10^{-8} mol cm^{-2} hr^{-1}. These data clearly show that thiol chemisorption has a dramatic effect on permeant flux in these nanotubule membranes.

Au nanotubule membranes with the following approximate nanotubule diameters were used to obtain the majority of the data reported here: i.d. = 28 ± 1, 7.0 ± 0.1, 1.9 ± 0.1, and 1.5 ± 0.2 nm. Figure 19 shows permeation data for transport of pyridine through these various membranes. Data for membranes derivatized with both the R = -C_2H_4-OH (upper solid curve) and the R = -$C_{16}H_{33}$ (lower dashed curve) thiols are shown. The corresponding flux data are shown in Table 3. As would be expected [114], the flux of pyridine decreases with decreasing tubule diameter for both the

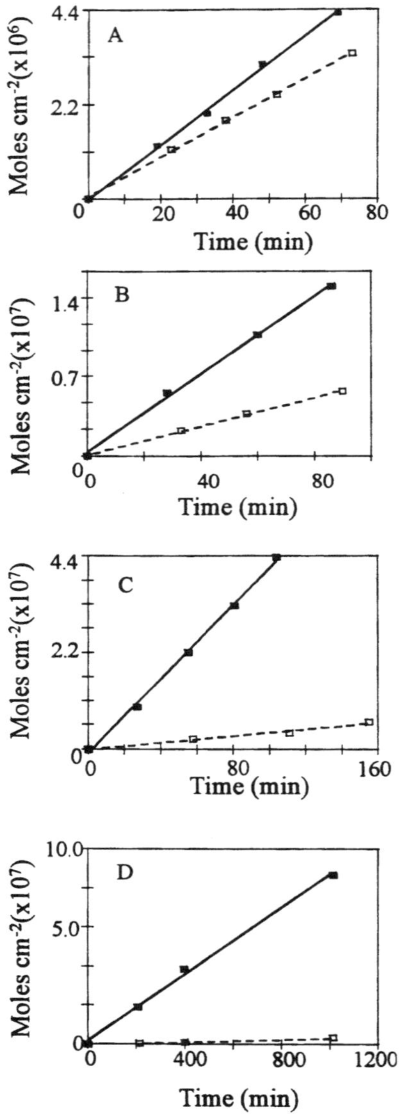

FIG. 19. Pyridine permeation data for membranes containing Au nanotubes with approximate inside diameters of (A) 28 ± 1 nm, (B) 7.0 ± 0.1 nm, (C) 1.9 ± 0.1 nm, and (D) 1.5 ± 0.2 nm. In each case, the upper solid line is for the R = $-C_2H_4-OH$ membrane and the lower dashed line is for the R = $-C_{16}H_{33}$ membrane.

TABLE 3

Pyridine and Toluene Flux and Selectivity Data

Permeant molecule	Nanotube diameter (nm)	Flux in R = -C_2H_4-OH membranes (mol cm^{-2} hr^{-1})	Flux in R = -$C_{16}H_{33}$ membranes (mol cm^{-2} hr^{-1})	$\alpha_{OH/C16}$	$\alpha_{C16/OH}$
Pyridine	7	9.7×10^{-7}	3.5×10^{-7}	2.8	—
Pyridine	1.9	2.5×10^{-7}	2.2×10^{-8}	11	—
Pyridine	1.5	1.2×10^{-7}	5.2×10^{-9}	23	—
Toluene	7	2.7×10^{-6}	5.5×10^{-6}	—	2.0
Toluene	1.9	1.3×10^{-6}	3.6×10^{-6}	—	2.8

R = -C_2H_4-OH and R = -$C_{16}H_{33}$ membranes. However, for any nanotube diameter, the pyridine flux in the R= -C_2H_4-OH membrane is greater than in the R = -$C_{16}H_{33}$ membrane. In addition, as the tubule diameter decreases, the difference in flux between the R = -C_2H_4-OH and the R = -$C_{16}H_{33}$ membranes becomes more dramatic.

This last point can be illustrated by defining a selectivity coefficient, $\alpha_{OH/C16}$, which is the flux of pyridine in the R = -C_2H_4-OH membrane divided by the flux of pyridine in the corresponding R = -$C_{16}H_{33}$ membrane. As shown in Table 3, this selectivity coefficient increases with decreasing tubule diameter. The smallest tubule-diameter R = -C_2H_4-OH membrane showed a factor of 23 higher selectivity for pyridine transport than the corresponding R = -$C_{16}H_{33}$ membrane. Similar large $\alpha_{OH/C16}$ values were obtained for two other relatively hydrophilic organic molecules—benzoic acid ($\alpha_{OH/C16}$ = 28) and phenol ($\alpha_{OH/C16}$ = 15).

Results of analogous permeation studies for the hydrophobic toluene molecule are shown in Fig. 20. Now the opposite selectivity pattern is observed; i.e., toluene is preferentially transported in the R = -$C_{16}H_{33}$ membranes. This can be illustrated by defining the alternative selectivity coefficient $\alpha_{C16/OH}$ (Table 3). As was the case for $\alpha_{OH/C16}$, the $\alpha_{C16/OH}$ values increase with decreasing tubule diameter. In addition to toluene, $\alpha_{C16/OH}$ values were determined for *p*-xylene and naphthalene in the i.d. = 1.9 nm membranes. The following $\alpha_{C16/OH}$ values were obtained: 2.8 for toluene, 6.2 for *p*-xylene, and 16 for naphthalene.

We suggest the following interpretation for these various data: Note first that of all the flux values reported in Table 3, the toluene fluxes in the R= -$C_{16}H_{33}$ membranes are, in general, the highest. This may, at first glance, seem surprising because the long C_{16} thiol might be expected to hinder diffusion in these membranes.* However, flux is proportional to both the diffusion coefficient and the partition coefficient for the permeant molecule in the membrane [115]. The comparison of $\alpha_{C16/OH}$ values for toluene, *p*-xylene, and naphthalene clearly shows that the hydrophobic ef-

---

* The mechanism of diffusion of these permeant molecules in these membranes is an issue that must be explored in detail. We have shown [71] that the R = -C_2H_4-OH–derivatized nanotubules flood when immersed in water. In contrast, permeation experiments with inorganic salts suggest that the R = -$C_{16}H_{33}$ nanotubules do not flood with water. Hence, in these membranes the permeate molecule is partitioned into and diffuses through the C_{16} phase within the tubes.

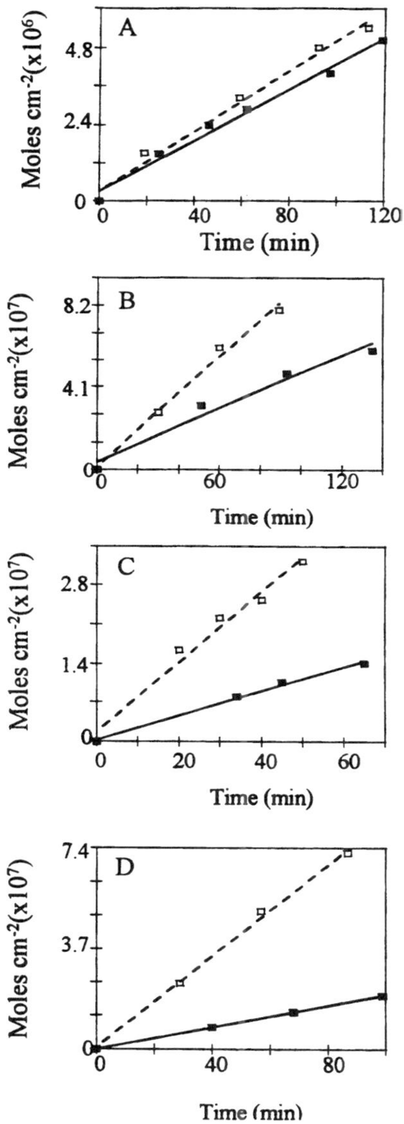

FIG. 20. Toluene permeation data (membranes as per Fig. 19). In each case, the upper dashed line is for the R = $-C_{16}H_{33}$ membrane and the lower solid line is for the R = $-C_2H_4$-OH membrane.

fect causes preferential partitioning of hydrophobic molecules into these hydrophobic membranes. Hence, we suggest that flux for hydrophobic molecules in the R = $-C_{16}H_{33}$ membranes is driven by favorable partitioning of such molecules from water (the feed solution) into the membrane.

This hypothesis is supported by the fact that the expected [114] decrease in flux with tubule diameter is not, in general, observed for toluene in the R = $-C_{16}H_{33}$ membranes (Table 3). This hypothesis is also supported by the fact that the next largest group of flux values in Table 3 is for toluene in the R = $-C_2H_4$-OH membranes. Water can still lower its free energy by partitioning the hydrophobic toluene molecule into these membranes, but much of the advantage is lost due to the lower hydrophobicity of the R = $-C_2H_4OH$ group relative to R = $-C_{16}H_{33}$.

The next highest set of fluxes is for pyridine in the R = $-C_2H_4$-OH membranes (Table 3). Clearly, the hydrophobically driven partitioning of this molecule is greatly diminished relative to toluene, and this accounts for the lower pyridine (vs. toluene) fluxes in the R = $-C_2H_4$-OH membranes. The lowest fluxes are for pyridine in the R = $-C_{16}H_{33}$ membranes. Now the relatively hydrophilic pyridine molecule pays an enthalpic penalty (loss of hydration) upon entering these hydrophobic membranes. We suggest that this results in a low partition coefficient and correspondingly low fluxes.

Finally, the ratio of the fluxes for toluene versus pyridine transport in the d = 1.5 nm R = $-C_{16}H_{33}$ membrane is greater than 400 (Table 3). This suggests that this membrane might be useful for separating mixtures containing hydrophobic and hydrophilic molecules, with the hydrophobic molecules being preferentially transported to the permeate. To explore this point we did a dual-molecule experiment in which the feed solution was 7 mM in toluene and 5 mM in pyridine. The ratio of the toluene to pyridine fluxes (corrected for the difference in feed concentrations) was 100. While not as high as predicted by the single-molecule permeation experiments, this datum does confirm that these membranes show promise for separating hydrophobic and hydrophilic molecules.

VI. NANOMATERIALS IN SECONDARY BATTERY RESEARCH AND DEVELOPMENT

The need to develop high rate performance electrodes [116] in advanced batteries, such as rocking-chair (lithium-ion) cells [117], has led to efforts to reduce the effects of mass transfer during discharge/charge [118–120]. Mass transfer effects are caused by the slow diffusion and long diffusion distances of Li^+ in the Li^+ intercalation materials used as the battery electrodes [118–120]. For example, Auborn and Barberio [120] showed that lithium ion battery performance is dramatically affected by the particle size of the Li^+ intercalation materials used to prepare the electrodes. Specifically, when large particles were used, the experimental capacity of the electrode was significantly lower than the theoretical capacity. Tipton et al. also discuss this effect of particle size on discharge capacity [118].

Capacity is lost when large particles are used to make the electrode because concentration polarization occurs within the particle before the entire capacity can be utilized [118–122]. Auborn and Barberio obviated this problem by decreasing the particle size. This lowered not only the current density, but also the diffusion distances of Li^+ within the electrode material [118,119], thus delaying the onset of concentration polarization. As a result, a larger fraction of the theoretical capacity of the electrode could be utilized [119–122]. Wittingham [123] also showed that utilization of the cathode capacity for a Li/TiS_2 battery was influenced by the size of TiS_2 particles. Theoretical analyses also point out the importance of particle size [121,122].

Given the importance of particle size to rate capabilities in Li^+ batteries, preparation of nanostructures of Li^+ insertion material for possible use as electrodes in Li^+ batteries seemed like an obvious extension of our work on nanomaterials. The fact that these nanostructures can be prepared as high-density ensembles that protrude from a surface like the bristles of a brush (Fig. 2A) seemed particularly useful for this proposed application because the substrate surface could then act as a current collector for the nanostructured battery electrode material.

We have proven that electrodes based on such ensembles of template-synthesized Li^+ insertion micro- and nanomaterials have superior rate capabilities relative to thin film electrodes composed of the same material. Specifically, these nanostructured electrodes can deliver higher capacities at high discharge currents than can the same amount of a thin film of the same material. We have proven this superior performance for nanostructures composed of two different Li^+ insertion materials—$LiMn_2O_4$ and TiS_2. This work demonstrates that the fundamental issues that make nanomaterials scientifically interesting can be applied to the area of electrochemical energy production leading to a new hybrid field of research/technology.

Results of our investigations of template-synthesized Li^+ battery electrode materials are reviewed here.

A. Investigations of Nanotubules of $LiMn_2O_4$

1. *Electrode Fabrication*

This work was done in collaboration with Professor Hiroshi Yoneyama of Osaka University [124]. The procedure used to prepare the $LiMn_2O_4$ tubules is shown schematically in Fig. 21. A commercially available alumina filtration membrane (Anopore, Whatman) was used as the template. Alumina is especially suited for this application because of its high porosity, monodispersity of pore size, and the fact that it can be heated to high temperature without degradation. This membrane contains ~200-nm-diameter pores, is 60 μm thick, and has a porosity of ~0.6. A 1.5 cm × 1.5 cm piece of this membrane was mounted on a Pt plate (2 cm × 2 cm) by applying a strip of plastic adhesive tape (also 2 cm × 2 cm; NICHIBAN VT-19) across the upper face of the membrane. The Pt plate will serve as the current collector for the $LiMn_2O_4$ battery electrode material. The strip of tape, which will be subsequently removed, had a 1.0 cm^2 circular hole punched in it, which defined the area of the membrane used for the template synthesis of the $LiMn_2O_4$.

The Pt current collector was first used to deposit short (~2 μm) Pt nanoposts [37,73] into the template membrane (Fig. 21A). These Pt nanoposts anchor the alumina membrane to the Pt surface and will serve to make electrical contact to the $LiMn_2O_4$ nanotubes. After Pt deposition, the pores in the membrane were filled with an aqueous solution that was 0.5 M in $LiNO_3$ and 1 M in $Mn(NO_3)_2$ (Fig. 21B). The excess solution was wiped from the membrane surface, and the solvent (water) was removed by heating (50°C) in vacuum for 1 hour. The assembly was then heated at 500°C in air for 5 hours. This burns away the plastic tape and also causes tubules of $LiMn_2O_4$ to form within the pores (Figs. 21C, 22).

The alumina template was then dissolved away using 2 M NaOH, and the resulting array of nanotubes (Fig. 22) was heated at 850°C in air for 24 hours. X-ray diffraction studies indicate that this yields the spinel $LiMn_2O_4$ [124]. The amount of $LiMn_2O_4$ deposited was determined by dissolving the nanotubes and using a visible absorption assay for MnO_4^- in the resulting solution [124]. This assay showed that 0.75 ± 0.03 mg of $LiMn_2O_4$ tubules were deposited into the 1 cm^2 portion of the template membrane.

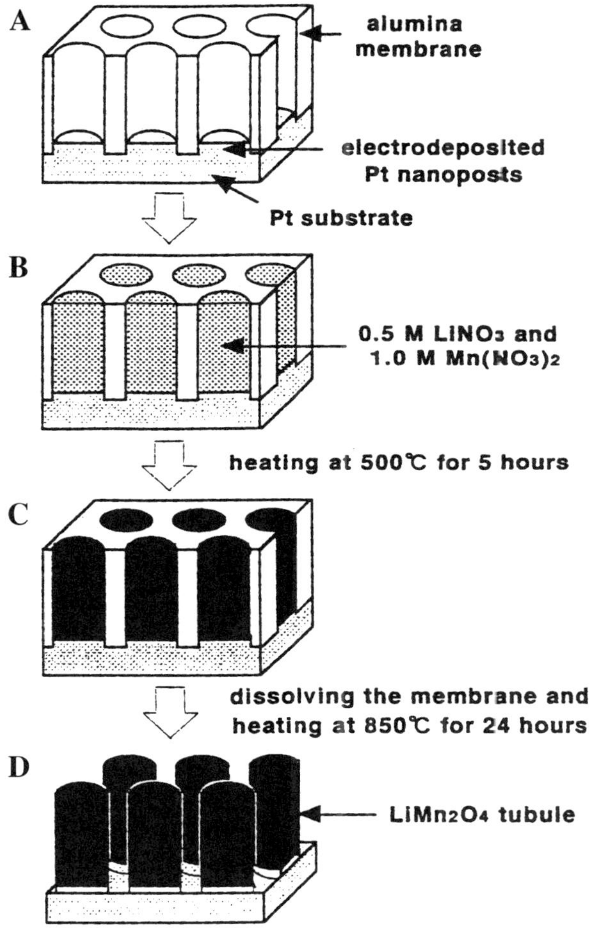

FIG. 21. Schematic of the procedure used to prepare the $LiMn_2O_4$ tubules.

The final step entailed coating both the inner and outer surfaces of the $LiMn_2O_4$ tubules with the conductive polymer polypyrrole [125]. It has been shown that such $LiMn_2O_4$/polypyrrole composite electrodes have lower resistance and higher capacity than electrodes prepared from $LiMn_2O_4$ alone [125]. Furthermore, the high porosity and fast (relative to

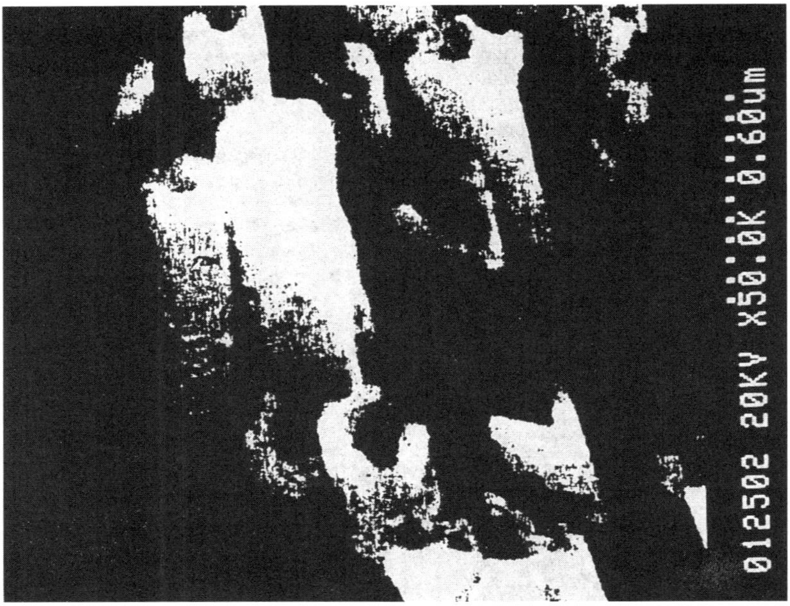

FIG. 22. Scanning electron micrograph of the $LiMn_2O_4$ tubules.

$LiMn_2O_4$) electrochemical kinetics ensure that all of the surfaces of the nanotubules remain accessible to solvent and electrolyte. The polypyrrole coat was deposited by simply applying 5 μl of a solution that was 1 M in $HClO_4$ and 0.2 M in pyrrole to the $LiMn_2O_4$ surface. This results in oxidative polymerization of all of the pyrrole, yielding 0.065 mg of polypyrrole per cm^2 of Pt substrate surface [125].

In order to determine whether the new nanotubule electrode shows improved performance, a control electrode composed of the same material but prepared via a more conventional method is required. This control $LiMn_2O_4$ electrode was prepared by applying the precursor solutions described above directly onto a 1 cm^2 Pt plate and thermally processing as before. Scanning electron micrographs showed that these films consisted of $LiMn_2O_4$ particles with diameters of ~500 nm [124]. Spectrophotometric assay showed that this control electrode also contained 0.75 mg of $LiMn_2O_4$ per cm^2. A polypyrrole coat identical to that applied to the tubular electrode (0.065 mg) was also applied to this control electrode.

2. *Electron Microscopy*

Figure 22 shows a scanning electron micrograph of a typical ensemble of $LiMn_2O_4$ nanotubules. The outer diameter of these tubes is determined by the diameter of the pores (200 nm) in the template membrane; the tubule wall thickness is ~50 nm. Tubules are obtained because the salts preferentially adsorb to the alumina pore wall after removal of the water from the precursor salt solutions. This tubular structure ensures that the electrolyte solution can access a large surface area of the $LiMn_2O_4$ electrode material. Furthermore, it is clear from the charge/discharge reactions for $LiMn_2O_4$ [see Eq. (2)] that Li^+ must diffuse into and out of this material. The extremely thin walls of these nanotubules ensure that the distance that Li^+ must diffuse within the $LiMn_2O_4$ phase is very small (~25 nm).

3. *Charge/Discharge Experiments*

Constant current charge/discharge experiments were carried out on both the nanotubular and control $LiMn_2O_4$ electrodes. The cell consisted of the $LiMn_2O_4$ working electrode, a Pt sheet counterelectrode, and an Ag/AgCl reference electrode (propylene carbonate saturated with LiCl; 2.9 V vs. Li/Li^+; potentials quoted here are vs. Li/Li^+). The electrolyte was 1 M $LiClO_4$ dissolved in a 1:1 (by volume) mixture of propylene carbonate and 1,2-dimethoxyethane. The electrodes were charged to an upper potential limit of 3.8 V and discharged to a lower limit of 2.2 V. The charge/discharge reactions for this material can be written as [126]:

$$LiMn_2O_4 + xLi^+ + x\,e^- \underset{\text{Charge}}{\xrightarrow{\text{Discharge}}} Li_{(x+1)}Mn_2O_4 \qquad (2)$$

The value of x for the potential region used here is approximately 1; i.e.; the fully charged material is $LiMn_2O_4$ and the fully discharged material has the approximate stoichiometry $Li_2Mn_2O_4$. All measurements were done in an Ar-filled glove box.

Figure 23 shows charge/discharge curves (at a current density of 0.1 mA cm^{-2} of Pt current collector surface area) for nanotubular and control $LiMn_2O_4$ electrodes. The key question to be addressed from this figure is: What charge and discharge capacity can be achieved (for each type of electrode) before the potential limits for charging (3.8 V) and discharging (2.2 V) are obtained? Figure 23 clearly shows that both the charge and discharge capacity are higher for the nanotubular electrode than for the control electrode. This higher capacity is obtained in spite of the fact that

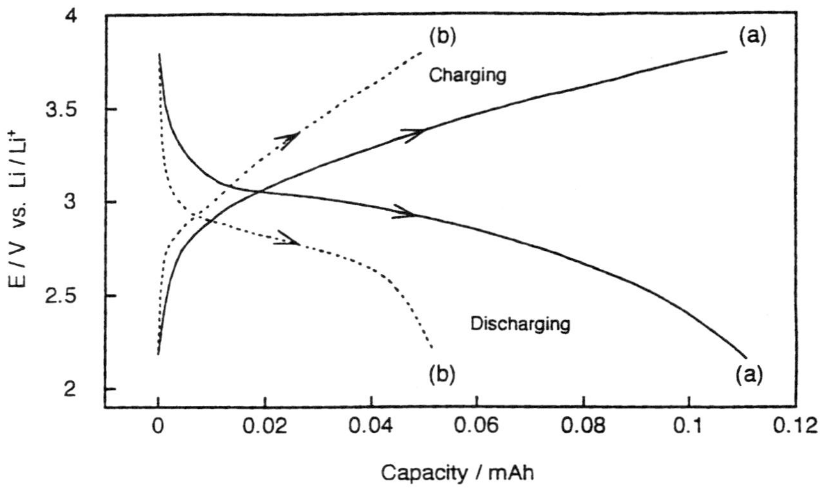

FIG. 23. Charge discharge curves for nanotubular (a) and thin film (b) $LiMn_2O_4$/polypyrrole electrodes. Current density = 0.1 mA cm^{-2}. Electrolyte was 1 M $LiClO_4$ in 1:1 (vol.) propylene carbonate:dimethoxyethane.

both electrodes contain the same quantities of both $LiMn_2O_4$ and polypyrrole.

It is easy to show [124] that the polypyrrole present in these electrodes contributes 8.85×10^{-3} mAh to the total experimental discharge capacity. Furthermore, it is known that over the potential window used here, $LiMn_2O_4$ has a theoretical (maximum) capacity of 148.3 mAh g^{-1} [124]. Correcting the experimental capacities in Fig. 23 for the polypyrrole contribution and dividing by the mass of the $LiMn_2O_4$ used shows that in the nanotubular electrode 90% (133.8 mAh g^{-1}) of the theoretical capacity is utilized, whereas in the control electrode only 37% (54.9 mAh g^{-1}) of the capacity is used. These data clearly show that the nanotubular electrode is superior to the control $LiMn_2O_4$ electrode.

It was noted above that a unique feature of the nanotubular electrode is the thin tubule walls, which ensure that the distance over which the Li^+ must diffuse within the $LiMn_2O_4$ is small. If this factor is partially responsible for the improved performance of the nanotubular electrode (vs. the control electrode), then the performance should be further improved at higher current densities. This is because higher current densities force the charge and discharge processes to occur at higher rates, and this will exacerbate the problem of slow Li^+ diffusion within the $LiMn_2O_4$. Figure 24 shows that the performance of the tubular electrode (relative to the control) is, indeed, improved at higher current densities. At a current density of 1 mA cm^{-2}, the experimental capacity of the tubular electrode is over an order of magnitude higher than that of the control electrode.

Finally, the other factor that could contribute to the improved performance of the nanotubule electrode is its higher surface area. This higher surface area would make the true current density at the nanotubular material lower than at the control material. In order to access the contribution of surface area, Brunauer-Emmett-Teller (BET) measurements were made on both the nanotubular and control electrodes [124]. The specific surface areas were found to be 40 $m^2\ g^{-1}$ (nanotubule electrode) and 13 $m^2\ g^{-1}$ (control electrode). This factor of 3 increase in surface area cannot account for the factor of 12 improvement in capacity observed at the highest current

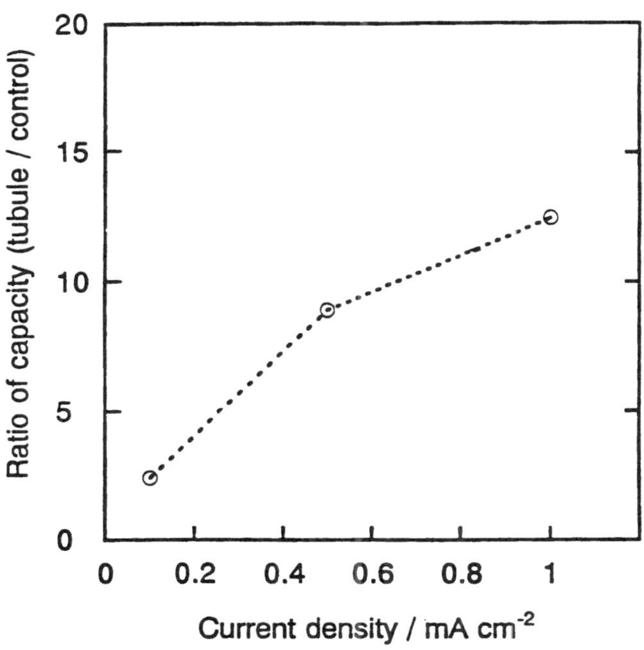

FIG. 24. Ratio of the specific capacities of the thin film and tubular electrodes.

density (Fig. 24). Hence, it is clear that the decreased diffusional distance in the nanotubular material plays a strong role in the improved performance of the nanotubular electrode.

B. TiS_2 Tubules Where Each Tube Has Its Own Built-In Current Collector

The $LiMn_2O_4$ example discussed above provides proof of concept that a nanostructured Li^+ battery electrode can provide superior rate capabilities relative to a thin film electrode. We will show in this section that this concept applies to TiS_2 also. In addition, we will demonstrate a second interesting feature of the template synthesis method. Specifically, we will show that by doing sequential template syntheses, it is possible to prepare composite structures in which outer tubules of the Li^+ intercalation material are coated onto concentric inner tubules of a metal. We have recently described a number of chemical strategies which can be used to prepare such concentric tubular composite micro- and nanostructures [31]. By coating tubes of TiS_2 over tubes of Au, each particle of the Li^+ intercalation material (the outer TiS_2 tube) has its own on-board current collector (the inner Au tube). This is important because, as illustrated by the $LiMn_2O_4$ example discussed above, the electronic conductivities of the Li^+ intercalation materials can often be low. This work was done in collaboration with Professor Ellen R. Fisher of Colorado State University [127].

1. *Electrode Fabrication*

A schematic diagram of the fabrication method used is shown in Fig. 25. The electroless plating procedure discussed above was used to deposit gold microtubules within the pores of a polyester template membrane. One gold surface layer was removed (Fig. 25C), and the membrane was placed (remaining Au surface layer down) on a piece of Al foil (Fig. 25D). The membrane was then dissolved by immersion of this assembly into hexafluoroisopropanol for 24 hours. This yielded an ensemble of Au tubules (0.9 μm o.d., 0.4 μm i.d.) protruding from the Au/Al substrate surface (Figs. 25E and 26A). Interestingly, the adhesion between the Au film that was grown on the membrane and the substrate Al foil is very strong. This is evidenced by the fact that the membrane can be dissolved away without loss of the Au tubes from the Al/Au substrate. Furthermore, this assembly can be immersed into an electrolyte solution and electrochemical experiments run (see below) without loss of the Au tubes.

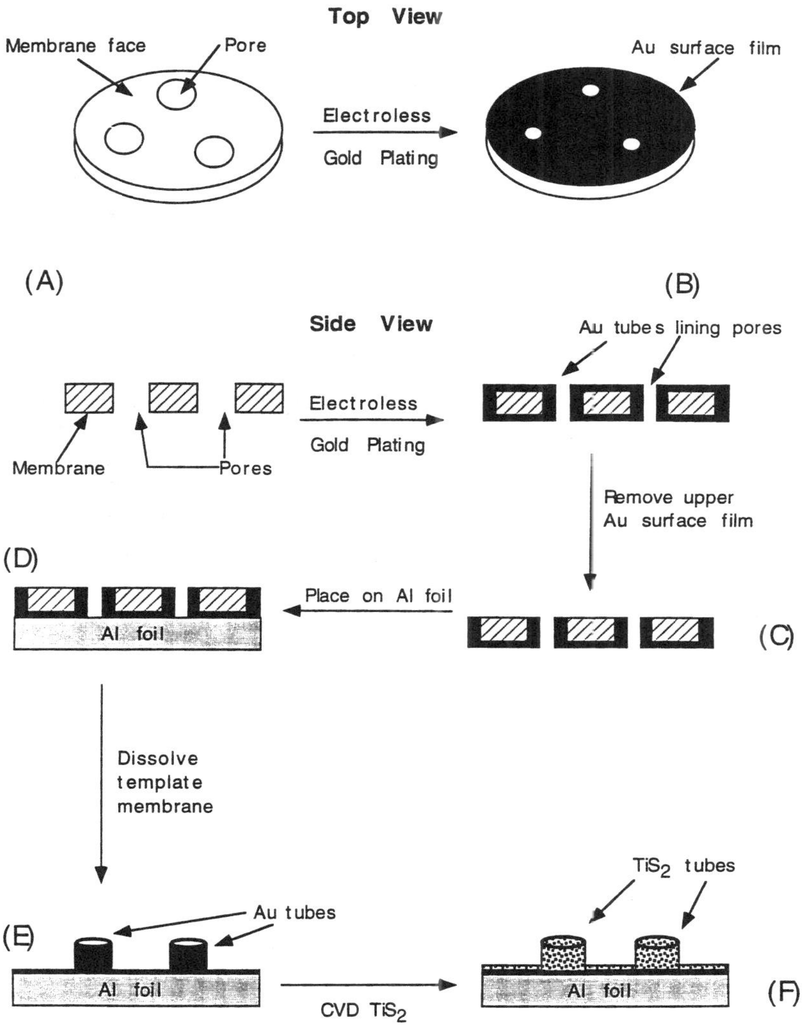

FIG. 25. Schematic diagram of the fabrication of the Au/TiS_2 concentric tubular electrode.

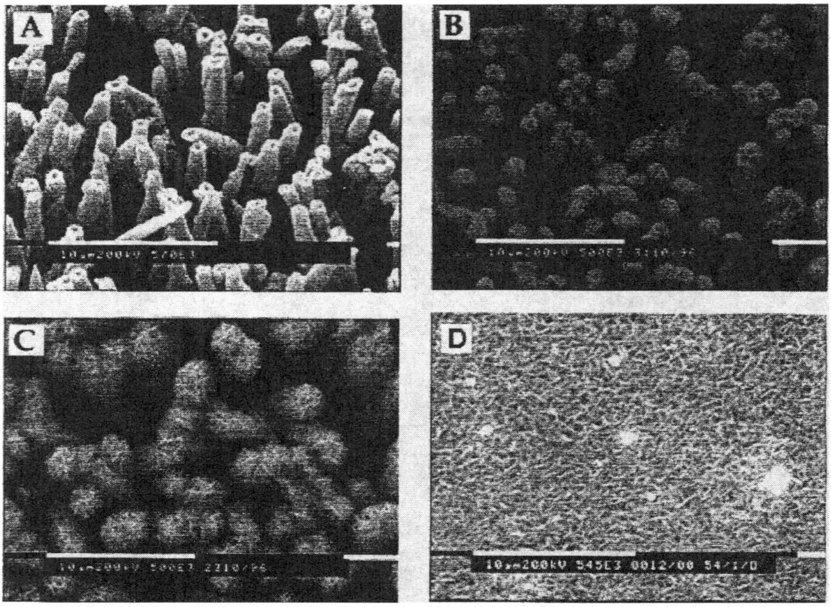

FIG. 26. Scanning electron micrographs: (A) the template-synthesized gold tubule ensemble obtained after dissolution of the polyester template membrane; (B) as per A, but after CVD synthesis of TiS_2 outer tubes on the Au inner tubes. These tubular microstructures contained 0.86 mg of TiS_2 cm^{-2} of substrate Al surface area; (C) as per B, but with a larger quantity (2.04 mg cm^{-2}) of TiS_2; (D) CVD TiS_2 film.

A CVD reactor was then used to coat the outer TiS_2 tubules onto the inner Au tubules; details of the equipment and the synthesis have been described previously [127,128]. TiS_2 was found to deposit as thin tubular skins on the outer surfaces of the Au microtubules; the thickness of the TiS_2 skin could be controlled by varying the deposition time (Fig. 26B and 26C). The quantity of TiS_2 deposited was determined by dissolving the TiS_2 from the substrate and using inductively coupled plasma atomic emission analysis to determine the amount of Ti^{4+} in the resulting solution [127]. Thin film TiS_2 electrodes (control electrodes) were prepared by CVD of TiS_2 onto Al foil.

2. *Electron Microscopy*

Figure 26A shows an SEM image of the Au microtubules used as the current collector for the microstructured TiS_2 electrode. Note that a high-density ensemble of monodisperse tubules is obtained. Figure 26B shows an analogous SEM image after CVD of TiS_2 onto the Au tubules (deposition time = 3 min). The Au tubules have been coated with outer tubules of TiS_2. That the outer material is, indeed, TiS_2 was confirmed via energy-dispersive x-ray analysis and by x-ray diffraction [127,128]. For this deposition time, the thickness of the walls of the outer TiS_2 tubes is ~360 nm. Figure 26C shows an SEM image after a longer TiS_2 deposition time (5 min). Note that the outer TiS_2 tubes now have thicker walls (~910 nm).

Figure 26D shows an SEM image of a CVD TiS_2 film that had been deposited on Au foil (deposition time = 2 min). The morphology of this TiS_2 is similar to that of the TiS_2 deposited on the Au microtubules. It is of interest to note, however, that when placed into electrolyte solution, such TiS_2 films deposited on Au foil detached from the Au foil surface. In contrast, no detachment of TiS_2 was observed from the Au microtubular substrate. The enhanced roughness of the microtubular Au surface (Fig. 26A) is likely responsible for this improved adhesion. Because TiS_2 detached from Au foil surfaces, electrochemistry on CVD TiS_2 films (the control electrodes) was done by depositing TiS_2 onto Al foil surfaces, where good adhesion was observed.

3. *Electrochemical Experiments*

A three-electrode cell—with microtubular or control TiS_2 working electrode and Li foils as both the counter- and reference electrodes—was used. The electrolyte was 1 M $LiClO_4$ in a 30:70 (v/v) mixture of ethylene carbonate and diethyl carbonate. All electrochemical measurements were made in a glove box filled with argon. Au/TiS_2 composite microstructures with two different wall thicknesses for the outer TiS_2 tubes were investigated. These two different composite microstructures correspond to the images shown in Fig. 26B and 26C. The thin-walled (~360 nm) TiS_2 tubules were obtained by using a 3-minute deposition time (Fig. 26B). These tubules contained 0.86 mg of TiS_2 per cm^2 of geometric surface area. (Geometric surface area is the area of the planar Al foil substrate that supports the Au microtubules.) The thick-walled (~910 nm) TiS_2 tubules were obtained by using a 5-minute deposition time (Fig. 26C). These tubules contained 2.04 mg of TiS_2 per cm^2 of geometric surface area.

Cyclic voltammograms associated with the reversible intercalation of Li^+ [Eq. (3)] into the thick-walled TiS_2 tubules are shown as the solid curves in Fig. 27. Voltammograms for two different scan rates are shown.

$$TiS_2 + xe^- + xLi^+ \rightarrow Li_xTiS_2 \qquad (3)$$

The analogous voltammograms for a thin-film control electrode containing an essentially identical quantity of TiS_2 are shown as the dotted curves

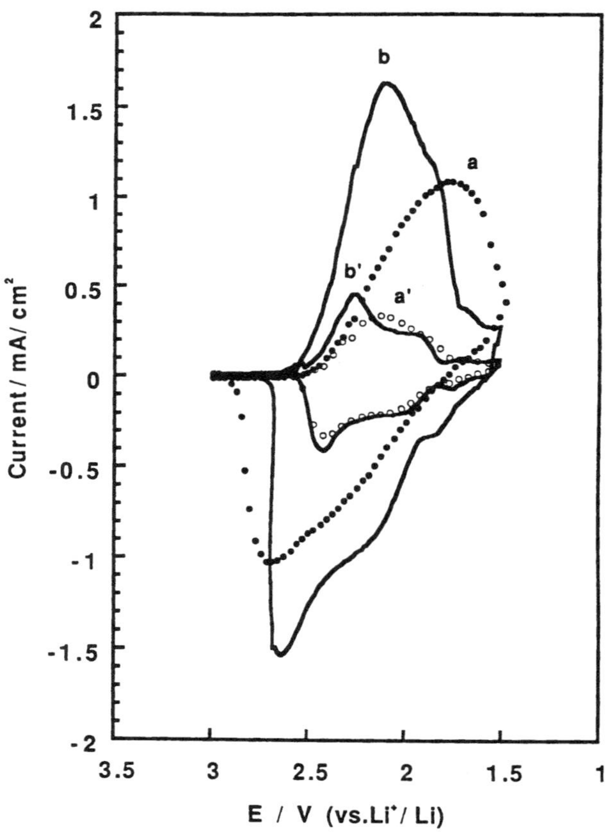

FIG. 27. Lithium intercalation cyclic voltammograms of control TiS_2 film sample (2.13 mg TiS_2 cm^{-2}) (a,a′) and microtubular TiS_2 sample (2.04 mg cm^{-2}) (b,b′). Scan rate: (a,b) = 0.5 mV s^{-1} and (a′,b′) = 0.1 mV s^{-1}.

in Fig. 27. Considering, first, the thin-film control electrode, note that the difference in peak potential (ΔE_{pk}) is enormous ($\sim$1 V) at the 0.5 mV sec^{-1} scan rate. The ideal peak potential separation for a thin-film electrode is 0 mV [129]. However, if the diffusion coefficient for the electroactive species is low (and/or the scan rate is high), ΔE_{pk} can increase to 60 mV (for a one-electron transfer) due to the onset of semi-infinite linear diffusion in the film. Clearly this is not the major contribution to the $\sim$1 V ΔE_{pk} observed here. Other factors that can contribute to peak separation are potential drop in the solution, potential drop due to either low electronic or ionic conduction in the film, and slow electron transfer kinetics. Since TiS_2 has high electronic conductivity [130,131], the film resistance term (if significant) would be due to low ionic (Li^+) conductivity.

A number of studies were done in order to determine which of these various factors contribute to the large peak separations observed here. First, it is well known that the effects of resistive elements can be obviated by applying positive feedback [132]. When positive feedback was applied to a thin-film control electrode similar to that described in Fig. 27, the peak separation decreased from $\sim$0.8 to $\sim$0.35 V (Fig. 28). These data show that resistance does, indeed, contribute to the large ΔE_{pk} values observed here. However, the fact that $\sim$0.35 V of this peak splitting cannot be removed by applying positive feedback clearly indicates that slow electron transfer kinetics also contribute to ΔE_{pk}.

One final issue remains to be resolved: Of the portion of the ΔE_{pk} that is due to resistance, what part is caused by solution resistance and what part is caused by film resistance? To explore this issue we examined the electrochemistry of a reversible redox couple (ferrocene/ferricinium) at a polished glassy carbon electrode in the electrolyte used for the TiS_2 electrochemistry. At a peak current density essentially identical to the peak current density for the thin film electrode in Fig. 27 (0.5 mV sec^{-1}), this reversible redox couple showed a ΔE_{pk} of 0.32 V (without application of positive feedback). Since this is a reversible couple (no contribution to the peak separation due to slow kinetics) and since there is no film on the electrode (no contribution to the peak separation due to film resistance), the largest portion of this 0.32 V is due to solution resistance. However, the reversible peak separation for a diffusional one-electron redox process is $\sim$0.06 V. This analysis indicates that we can anticipate a contribution of 0.32 V $-$ 0.06 V $=$ 0.26 V from solution resistance in the 0.5 mV sec^{-1} control TiS_2 voltammogram in Fig. 27.

Adding these various contributions to the peak splitting together, we

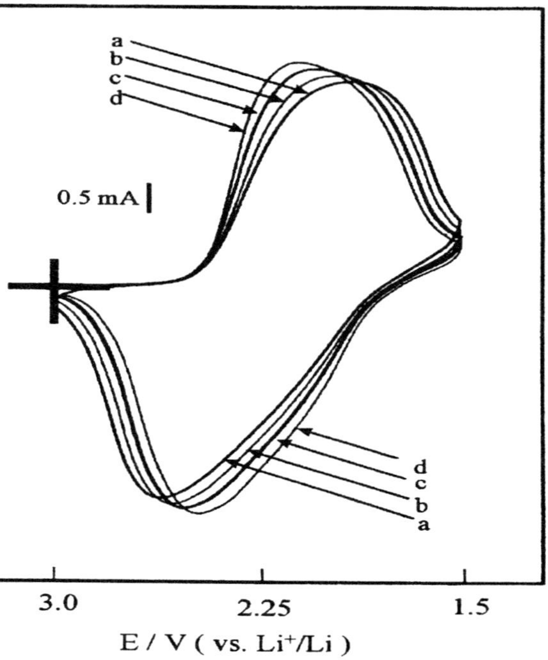

FIG. 28. Cyclic voltammograms of TiS_2 film electrode with increasing amounts of applied positive feedback. Scan rate = 1 mV s^{-1}. The level of applied positive feedback increased from a (none applied) to d (maximum applied).

find that solution resistance can contribute 0.26 V, the onset of semi-infinite linear diffusion in the film can contribute 0.06 V, and slow electron transfer can contribute 0.35 V, for a total of 0.67 V. However, as indicated above, the observed peak separation is ~1 V. This analysis suggests that film resistance also contributes to the observed peak separation in the 0.5 mV sec^{-1} voltammogram for the thin film control electrode shown in Fig. 27. This nonzero film resistance term results from the low solid-state diffusion coefficient associated with Li^+ diffusion in TiS_2 (10^{-10}–10^{-12} cm^2 sec^{-1}) [133]. This low diffusion coefficient creates ionic resistance in the film.

The final point to make concerning this control electrode is that when scanned at lower rates, the observed ΔE_{pk} decreases (Fig. 27, lower dotted-curve voltammogram). This is not surprising because all of the possible contributions to peak splitting discussed above will be decreased at lower scan rates.

Consider, now, the 0.5 mV sec^{-1} microtubular TiS_2 voltammogram in Fig. 27 (upper solid-curve voltammogram). The key differences between the microtubular and control voltammograms are the higher peak current densities and dramatically smaller ΔE_{pk} values for the microtubular electrode. Considering ΔE_{pk} first, we note that the experimental ΔE_{pk} for the microtubular electrode is ~0.5 V, so we must account for a 0.5 V change (decrease) between this value and the 1 V ΔE_{pk} observed for the comparable control electrode. A change in the solution-resistance term cannot be responsible for this decrease in ΔE_{pk} because the current density for the microtubular electrode is, in fact, higher than for the control electrode. This clearly shows that the decrease in ΔE_{pk} for the microtubular electrode is due to decreases in the film resistance and electron transfer kinetic contributions to ΔE_{pk}.

The decreased contribution of film resistance for the microtubular electrode makes sense because the effective film thickness for the microtubular system is less than for the thin film control electrode. This is because the surface area of the microtubular current collector is eight times higher than the surface area of the planar current collector. (This factor is calculated from the membrane thickness and the density and diameter of the pores in the membrane.) Since the control and microtubular electrodes contain the same amount of TiS_2, the eight times higher underlying surface area of the microtubular electrode means that the TiS_2 film is effectively a factor of 8 thinner, relative to the control electrode.

The decreased contribution due to slow electron transfer kinetics for the microtubular electrode is also attributable to the higher underlying surface area of the tubular current collector. Because the surface area is higher, the effective current density for the microtubular TiS_2 is less than for the thin film TiS_2, which has a conventional planar current collector. The decreased contributions of film resistance and slow electron transfer kinetics also account for the higher peak current density of the microtubular electrodes (Fig. 27).

It is important to point out that the effects of both film resistance and slow electron transfer kinetics are undesirable from a battery discharge point of view. Resistance is undesirable because it causes a portion of the energy produced by the battery to be lost as resistive heating. Kinetic limitations are undesirable because such limitations will cause the electrode to discharge at potentials lower than the theoretical values. We will prove these points below.

Figure 29 shows analogous Li^+ intercalation voltammograms for the thin-walled TiS_2 tubular electrode and for a control electrode containing approximately the same amount of TiS_2. As would be expected, the magni-

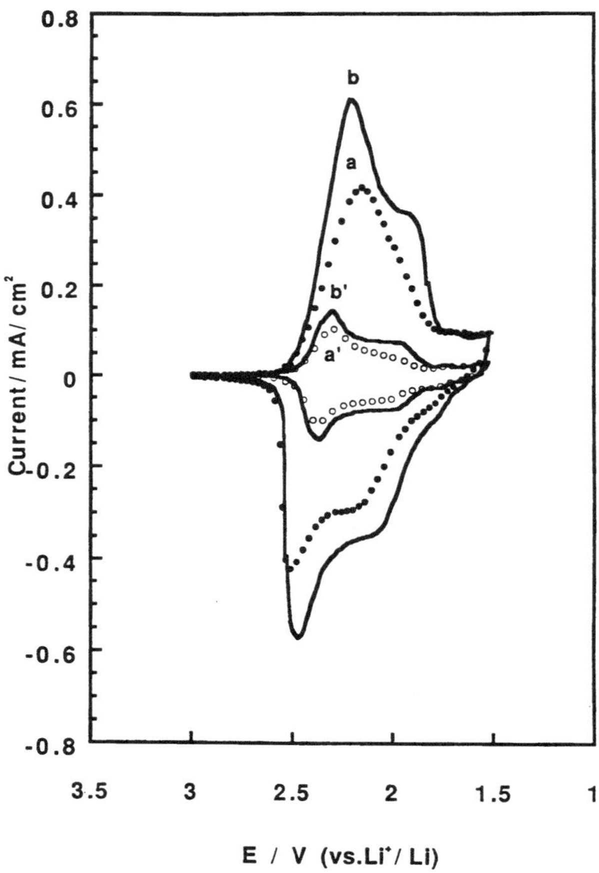

FIG. 29. Lithium intercalation cyclic voltammograms of control TiS_2 film sample (0.60 mg TiS_2 cm^{-2}) (a,a′) and microtubular TiS_2 sample (0.86 mg TiS_2 cm^{-2}) (b,b′). Scan rate: (a,b) = 0.5 mV s^{-1} and (a′,b′) = 0.1 mV s^{-1}.

tude of ΔE_{pk} for these thinner films is less than that for the thicker films shown in Fig. 27. This decrease in ΔE_{pk} is anticipated for two reasons: (1) the current densities are lower, and this will decrease the effects of any resistive element; (2) the films are thinner, and this will decrease any film resistance contribution to ΔE_{pk}.

We turn now to the issue of electrode capacity, which was explored using both voltammetric and constant current discharge experiments. In keeping with prior work on TiS_2, the constant current experiments were done between potential limits of 3.0 and 1.5 V versus Li^+/Li [116]. These potential limits were also used in the voltammetric evaluation of capacity. This was accomplished by scanning the potential of the TiS_2 electrode between these limits. The dependence of the discharge capacity on the rate of discharge was evaluated voltammetrically by scanning between these limits at various scan rates.

Typical voltammograms used to obtain such capacity data are shown in Fig. 30. The discharge capacity was obtained by integrating the forward (cathodic) wave. Figure 31 shows the discharge capacities as function of scan (discharge) rate for both the tubular and thin film TiS_2 electrodes that contain the larger amount of TiS_2. The theoretical capacity for TiS_2 (assuming a maximum intercalation level of 1 mole of Li^+ per mole of TiS_2) is 239 mA hr g^{-1}. At the lowest scan rate employed, the experimental capacity for the control electrode is below this theoretical capacity, and capacity falls off sharply with increasing scan rate.

The reason for this loss in capacity with increasing scan can be clearly seen in the voltammograms in Fig. 30A and 30A′. The peak separation, discussed in detail above, becomes larger as the scan rate is increased. The result of this enhanced distortion of the voltammetric wave is the inability to utilize the capacity of the electrode over the useful potential window of the electrode (3.0 to 1.5 V). As would be expected (see above), this distortion is less for the microtubular electrode, and this should result in higher capacities for this electrode.

At the lowest scan rate employed, the microtubular electrode delivers an experimental capacity (256 $\pm$ 21 mA hr g^{-1}) that is identical to the theoretical capacity. As scan rate is increased, capacity does fall off; however, at any scan rate, the experimental capacity obtained from the microtubular electrode is greater than the capacity obtained at the control electrode. At the highest scan rate employed, the microtubular electrode delivers almost seven times the experimental capacity of the control electrode, even though both electrodes contain the same amount of TiS_2.

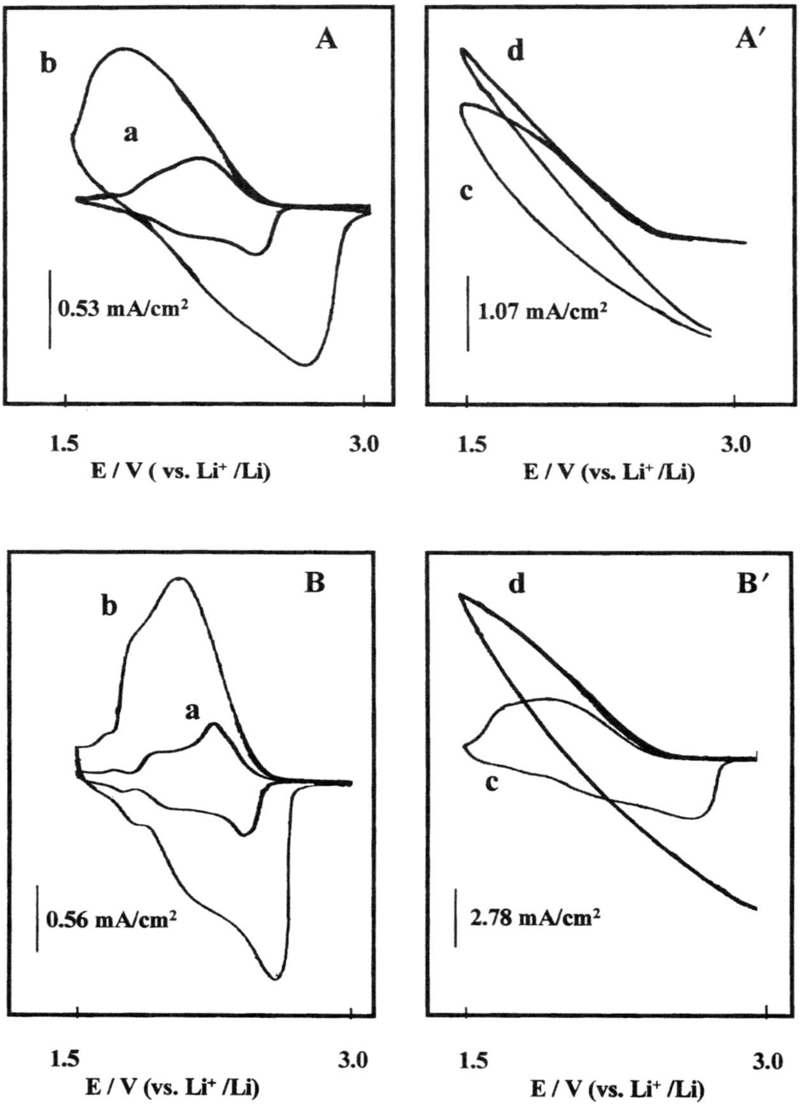

FIG. 30. Lithium intercalation cyclic voltammograms of control TiS_2 film sample (A,A′) and microtubular TiS_2 sample (B,B′). Scan rate: a = 0.1 mV s^{-1}; b = 0.5 mV s^{-1}; c = 1 mV s^{-1}; and d = 5 mV s^{-1}.

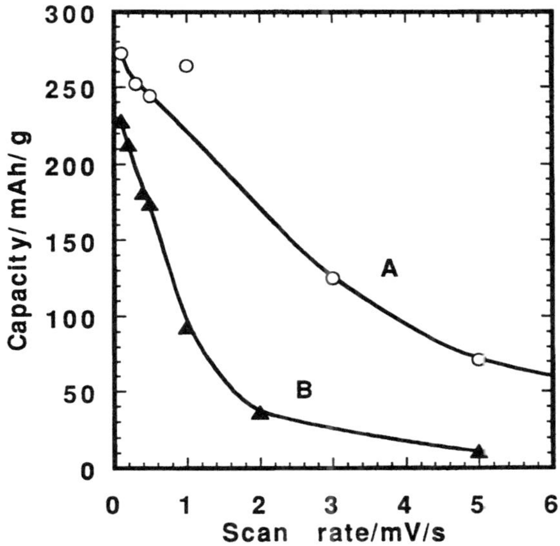

FIG. 31. Discharge capacity versus scan rate for (A) TiS_2 microtubular electrode (2.04 mg cm^{-2}) and (B) TiS_2 film control electrode (2.13 mg cm^{-2}). (Data from Fig. 30.)

Figure 32 shows analogous plots of capacity versus scan rate for the control and microtubular electrodes containing the smaller amount of TiS_2. Because the distortion of the voltammograms is less for the smaller amount of TiS_2 (Fig. 29), the capacity falls off less sharply with scan rate than for the electrodes containing the larger amount of TiS_2. However, the microtubular electrode, again, shows higher capacity at any scan rate than does the control electrode.

Finally, Fig. 33 shows the results of constant current discharge experiments at a microtubular electrode and a control electrode containing the same amount of TiS_2. Note that at this discharge current density, the microtubular electrode delivers 90% of its theoretical capacity. In contrast, as would be expected, the control electrode delivers significantly less capacity.

4. Conclusions

A new approach for preparing microstructured Li ion battery electrodes was demonstrated here. This approach entails using the template method

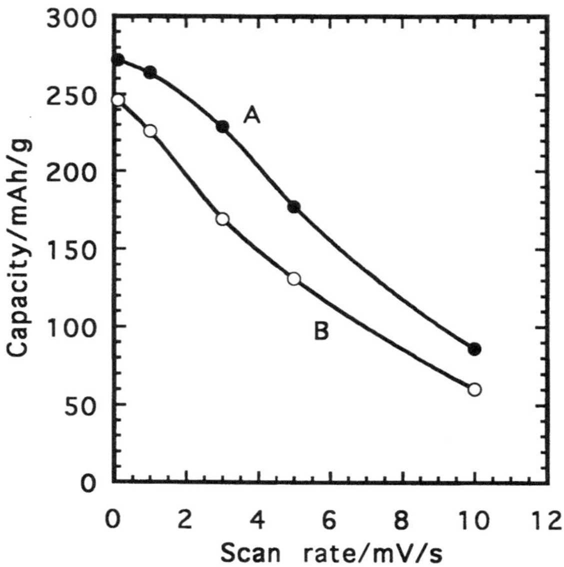

FIG. 32. Discharge capacity versus scan rate for (A) TiS_2 microtubular electrode (0.86 mg TiS_2 cm^{-2}) and (B) TiS_2 film control electrode (0.60 mg TiS_2 cm^{-2}).

[1–3] to prepare a microtubular current collector and then doing CVD synthesis of the Li^+ intercalation material over this microstructured current collector. We have demonstrated that this microstructured electrode can deliver higher discharge capacities at any discharge rate than a thin film electrode containing the same amount of TiS_2. This enhanced capacity results because the distance over which Li^+ must diffuse within the TiS_2 is effectively decreased and because the effective current is smaller for the microtubular electrode.

The microtubular electrode concept described here also offers another possible advantage. In these concentric tubular electrodes, each particle of the Li^+ intercalation material (the outer tube) has its own current collector (the inner metal microtubule). This could be an important advantage for Li^+ intercalation materials with low electrical conductivity. This advantage was not demonstrated here because TiS_2 has relatively high electronic conductivity. We have recently shown that electrochemical synthesis can be used to coat the gold microtubular current collector with outer tubes of a

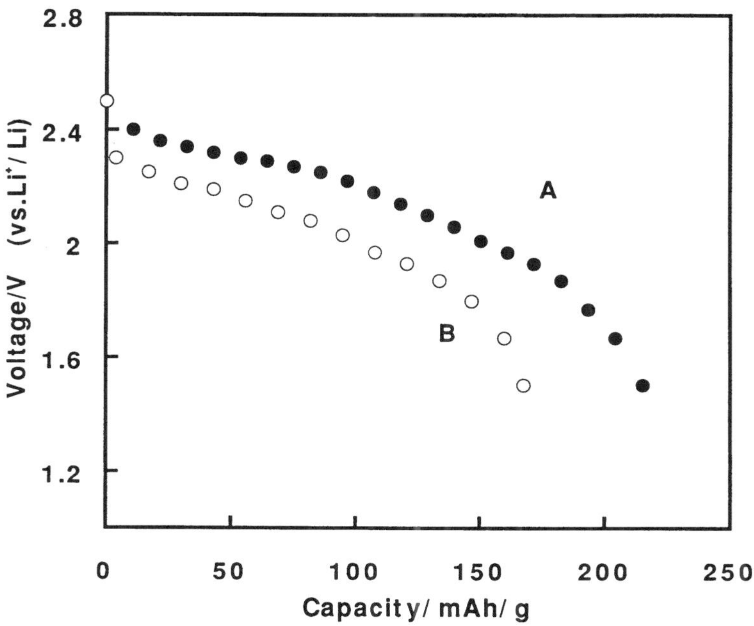

FIG. 33. Constant current (1 mA cm^{-2}) discharge curves: (A) A microtubular electrode (1.5 mg TiS_2 cm^{-2}); (B) a control electrode (1.3 mg TiS_2 cm^{-2}).

Li^+ intercalation material (V. M. Cepak and C. R. Martin, unpublished). These results, which will be the subject of a future paper, show that other synthetic methodologies, in addition to CVD, can be used to make microstructured battery electrodes like those described here. In addition, the underlying microtubular current collector does not have to be Au. Microtubules composed of graphite [35] or other metals [1,3] (e.g., Ni) could be used. Finally, for the advantages noted above to be realized in practical cells, large-scale template-fabrication methods would have to be developed.

ACKNOWLEDGMENTS

First, we would like to acknowledge the contributions of our professional colleagues, Professor Hiroshi Yoneyama of Osaka University and Professor Ellen R. Fisher of Colorado State University. Second, this work would not have been possible without the efforts of a talented group of postdoctoral research associates and graduate students. These include Dr. Vinod P. Menon, Dr. Ranjani V. Parthasarathy, Dr. Matsuhiko Nishizawa, Dr. Susumu Kuwabata, Mr. Kiyoshi Mukai, Dr. Guangli Che, Ms. Kshama B. Jirage, Ms. Brinda B. Lakshmi, Dr. John C. Hulteen, and Dr. G. Louis Hornyak. The various aspects of this work have been supported by the Department of Energy, Office of Energy Research, Grant DE-FG03-95ER14576 (joint grant with Prof. Fisher), the Office of Naval Research, and the National Science Foundation. Collaboration between CRM and Professor Yoneyama was made possible by Grant-in-Aid for International Scientific Research Program: Joint Research, No. 0744150, from the Ministry of Education, Science, Culture, and Sports, Japan.

REFERENCES

1. C. R. Martin, Science *266*:1961 (1994).
2. C. R. Martin, Acc. Chem. Res. *28*:61 (1995).
3. C. R. Martin, Chem. Mater. *8*:1739 (1996).
4. Science (special issue on nanomaterials) *254*:1300 (1991).
5. R. T. Bate, Sci. Am. *258*:96 (1988).
6. E. A. Medcalf, D. J. Newman, E. G. Gorman, and C. P. Price, Clin. Chem. *36*:446 (1990).
7. J. M. Simo, J. Joven, X. Cliville, and T. Sans, Clin. Chem. *40*:625 (1994).
8. L. B. Bangs, J. Clin. Immunoassay *13*:127 (1990).
9. L. B. Bangs, Pure Appl. Chem. *68*:1873 (1996).
10. G. A. Ozin, Adv. Mater. *4*:612 (1992).
11. R. Gref et al., Science *263*:1600 (1994).
12. R. Parthasarathy and C. R. Martin, J. Appl. Poly. Sci. *62*:875 (1996).
13. S. DeVito-Luebben and C. R. Martin, Chem. Mater *10*:1738 (1998).
14. A. Imhof and D. J. Pine, Nature *389*:948 (1997).
15. C. J. O'Connor et al., J. Appl. Phys. *81*:4741 (1997).
16. P. M. Ajayan, O. Stephan, P. Redlich, and C. Colliex, Nature *375*:564 (1995).
17. J. M. Schnur, Science *262*:1669 (1993).
18. J. S. Beck et al., J. Am. Chem. Soc. *114*:10834 (1992).
19. E. Kim, Y. Xia, and G. M. Whitesides, J. Am. Chem. Soc. *118*:5722 (1996).
20. R. L. Fleischer, P. B. Price, and R. M. Walker, *Nuclear Tracks in Solids*, University of California Press, Berkeley, 1975.
21. R. L. Fleischer, P. B. Price, and R. M. Walker, J. Appl. Phys. *34*:510 (1963).
22. P. B. Price and R. M. Walker, J. Appl. Phys. *33*:3407 (1962).
23. G. E. Possin, Rev. Sci. Instrum. *41*:772 (1970).
24. W. D. Williams and N. Giordano, Rev. Sci. Instrum. *55*:410 (1984).

25. V. P. Menon and C. R. Martin, Anal. Chem. *67*:1920 (1995).
26. C. Schonenberger et al., J. Phys. Chem. B *101*:5497 (1997).
27. A. Blondel, J. P. Meier, B. Doudin, and J.-P. Ansermet, Appl. Phys. Lett. *65*:3019 (1994).
28. L. Piraux et al., Appl. Phys. Lett. *65*:2484 (1994).
29. L. Wang, K. Yu-Zhang, A. Metrot, P. Bonhomme, and M. Troyon, Thin Solid Films *288*:86 (1996).
30. L. Piraux, S. Dubois, and S. Demoustier-Champagne, Nucl. Instr. Meth. Phys. Res. B *131*:357 (1997).
31. V. M. Cepak et al., Chem. Mater. *9*:1065 (1997).
32. D. Routkevitch et al., IEEE Trans. Elect. Dev. *43*:1645 (1996).
33. C. M. Zelenski and P. K. Dorhout, J. Am. Chem. Soc. *120*:734 (1998).
34. G. Che, B. B. Lakshmi, C. R. Martin, and E. R. Fisher, Chem. Mater. *10*: 260 (1998).
35. R. Parthasarathy, K. L. N. Phani, and C. R. Martin, Adv. Mater. *7*:896 (1995).
36. G. Hornyak et al., Chem. Eur. J. *3*:1951 (1997).
37. G. L. Hornyak, C. J. Patrissi, and C. R. Martin, J. Phys. Chem. B *101*:1548 (1997).
38. R. C. Furneaux, W. R. Rigby, and A. P. Davidson, Nature *337*:147 (1989).
39. T. Asada, Japanese patent No. 310,401 (1963).
40. S. Kawai and R. Ueda, J. Electrochem. Soc. *122*:32 (1975).
41. A. Andersson, O. Hunderi, and G. C. Granqvist, J. Appl. Phys. *51*:754 (1980).
42. R. D. Patel, M. G. Takwale, V. K. Nagar, and V. G. Bhide, Thin Solid Films *115*:169 (1984).
43. S. M. Marinakos, I. L. C. Brousseau, A. Jones, and D. L. Feldheim, Chem. Mater. *10*:1214 (1998).
44. R. J. Tonucci, B. L. Justus, A. J. Campillo, and C. E. Ford, Science *258*: 783 (1992).
45. P. P. Nguyen, D. H. Pearson, R. J. Tonucci, and K. Babcock, J. Electrochem. Soc. *145*:247 (1998).
46. O. D. Velev, K. Furusawa, and K. Nagayama, Langmuir *12*:2374 (1996).
47. L. Qi, J. Ma, H. Cheng, and Z. Zhao, J. Phys. Chem. B *101*:3460 (1997).
48. C. N. R. Rao, B. C. Satishkumar, and A. Govindaraj, Chem. Commun. *16*: 1581 (1997).
49. B. C. Satishkumar, A. Govindaraj, E. M. Vogl, L. Basumallick, and C. N. R. Rao, J. Mater. Res. *12*:604 (1997).
50. W. Han, S. Fan, Q. Li, and Y. Hu, Science *277*:1287 (1997).
51. J. M. Schnur and R. Shashidhar, Adv. Mater. *6*:971 (1994).
52. J. D. Hartgerink, J. R. Granja, R. A. Milligan, and M. R. Ghadiri, J. Am. Chem. Soc. *118*:43 (1996).
53. D. D. Archibald and S. Mann, Nature *364*:430 (1993).

54. S. L. Beckett and S. Mann, Chem. Comm. *3*:321 (1996).
55. G. M. Chow et al., Mater. Sci. Eng. *A158*:1 (1992).
56. Z. Qi and R. B. Lennox, Electrochem. Soc. Proc. *97-5*:173 (1997).
57. R. Kirsch et al., Thin Solid Films *305*:248 (1997).
58. C.-G. Wu, and T. Bein, Science *266*:1013 (1994).
59. C.-G. Wu and T. Bein, Science *264*:1757 (1994).
60. S. Y. Chou, P. R. Krauss, and P. J. Renstrom, Science *272*:85 (1996).
61. E. Kim, G. M. Whitesides, L. K. Lee, S. P. Smith, and M. Prentiss, Adv. Mater. *8*:139 (1996).
62. P. Cahill, Q. D. Walker, J. M. Finnegan, G. E. Mickelson, and R. M. Wightman, Anal. Chem. *68*:3180 (1996).
63. F. F. Fan and A. J. Bard, Science *277*:1791 (1997).
64. R. S. Ingram et al., J. Am. Chem. Soc. *119*:9279 (1997).
65. F. F. Fan and A. J. Bard, Science *267*:871 (1995).
66. A. L. Crumbliss et al., Biotechnol. Bioeng. *40*:483 (1992).
67. K. R. Brown, A. P. Fox, and M. J. Natan, J. Am. Chem. Soc. *118*:1154 (1996).
68. R. M. Penner and C. R. Martin, Anal. Chem. *59*:2625 (1987).
69. I. F. Cheng, L. D. Whiteley, and C. R. Martin, Anal. Chem. *61*:762 (1989).
70. J. C. Hulteen, V. P. Menon, and C. R. Martin, J. Chem. Soc. Far. Trans. 1 *92*:4029 (1996).
71. M. Nishizawa, V. P. Menon, and C. R. Martin, Science *268*:700 (1995).
72. K. B. Jirage, J. C. Hulteen, and C. R. Martin, Science *278*:655 (1997).
73. C. A. Foss, G. L. Hornyak, J. A. Stockart, and C. R. Martin, J. Phys. Chem. *98*:2963 (1994).
74. H. C. Van de Hulst, *Light Scattering by Small Particles*, Dover Publications, New York, 1981.
75. K. R. Wehmeyer and R. M. Wightman, J. Electroanal. Chem. *196*:417 (1985).
76. B. R. Scharifker, in *Microelectrodes: Theory and Applications* (M. I. Montenegro, M. A. Querios, J. L. Daschbach, eds.), Kluwer, Dordrecht, 1991, p. 227.
77. R. M. Penner, M. J. Heben, T. L. Longin, and N. S. Lewis, Science *250*: 1118 (1990).
78. E. Sabatani and I. Rubinstein, J. Phys. Chem. *91*:6663 (1987).
79. R. Bilewicz and M. Majda, J. Am. Chem. Soc. *113*:5464 (1991).
80. R. S. Nicholson, Anal. Chem. *37*:1351 (1965).
81. C. Amatore, J. M. Saveant, and D. Tessier, J. Electroanal. Chem. *147*:39 (1983).
82. B. Ballarin et al., Anal. Chem. *64*:2647 (1992).
83. V. P. Menon, Ph.D. dissertation, Colorado State University, 1996.

84. H. S. White, A. J. Bard, and F.-R. Fan, personal communication, 1994.
85. J. C. Hulteen, K. B. Jirage, and C. R. Martin, J. Am. Chem. Soc. *120*:6603 (1998).
86. W. J. Petzny, J. A. Quinn, Science *166*:751 (1969).
87. C. Liu, Ph.D. dissertation, Texas A & M University, 1991.
88. N. Lakshminarayanaiah, *Transport Phenomena in Membranes*, Academic Press, New York, 1969.
89. J. Koryta, *Ion-Selective Electrodes*, Cambridge University Press, Cambridge, 1975, p. 21.
90. S. Biggs, P. Mulvaney, C. F. Zukoski, and F. Grieser, J. Am. Chem. Soc. *116*:9150 (1994).
91. A. J. Bard and L. R. Faulkner, *Electrochemical Methods*, Wiley, New York, 1980, p. 500.
92. C. Liu and C. R. Martin, J. Electrochem. Soc. *137*:3114 (1990).
93. R. L. Dotson and K. E. Woodard, eds., *Perfluorinated Ionomer Membranes*, Vol. 180, American Chemical Society, Washington, DC, 1982.
94. D. L. Rath and W. N. Hansen, Surf. Sci. *136*:195 (1984).
95. J. Clavilier and C. N. V. Huong, J. Electroanal. Chem *80*:101 (1977).
96. M. R. Deakin, T. T. Li, and O. R. Melroy, J. Electroanal. Chem. *243*:343 (1988).
97. C. D. Bain et al., J. Am. Chem. Soc. *111*:321 (1989).
98. A. Ulman, *An Introduction to Ultrathin Organic Films from Langmuir-Blodgett to Self-Assembly*, Academic Press, San Diego, 1991.
99. P. Burgmayer and R. W. Murray, J. Phys. Chem. *88*:2515 (1984).
100. J. M. S. Hennis and M. K. Tripodi, Science *220*:11 (1983).
101. P. H. Abelson, Science *224*:1421 (1989).
102. C. Liu and C. R. Martin, Nature *352*:50 (1991).
103. U.S. Dept. of Energy, *Membrane Separation Systems—A Research Needs Assessment*, DE90-01170, U.S. Department of Energy, Washington, D.C., 1990.
104. R. W. Spillman, Chem. Eng. Prog. *85*:41 (1989).
105. M. Mulder, *Basic Principles of Membrane Technology*, Kluwer Academic Press, Dordrecht, The Netherlands, 1990, p. 260.
106. Y. Kobayashi and C. R. Martin, J. Electroanal. Chem. *431*:29 (1997).
107. R. J. Peterson, J. Membrane Sci. *83*:81 (1993).
108. B. B. Lakshmi and C. R. Martin, Nature *388*:758 (1997).
109. R. V. Parthasarathy, V. P. Menon, and C. R. Martin, Chem. Mater. *9*:560 (1997).
110. Z. Prat, Y.-M. Tricot, and I. Rubinstein, J. Electroanal. Chem. *315*:225 (1991).
111. C. R. Martin, I. Rubinstein, and A. J. Bard, J. Electroanal. Chem. *115*:267 (1983).

112. T. K. Rostovtseva, et al., J. Membrane Biol. *151*:29 (1996).
113. M. D. Porter, T. B. Bright, D. L. Allara, and C. E. D. Chidsey, J. Am. Chem. Soc. *109*:3559 (1987).
114. V. M. Deen, A. I. Ch. E. J. *33*:1409 (1987).
115. W. R. Veith, *Diffusion In and Through Polymers*, Hanser Verlag, Munich, 1991, p. 19.
116. S. Hossain, in *Handbook of Batteries*, 2nd ed. (D. Linden, Ed.), Wiley, New York, 1995, p. 36.
117. S. Megahed and B. Scrosati, J. Power Sources *51*:79 (1994).
118. A. L. Tipton, S. Passerini, B. B. Owens, and W. H. Smyrl, J. Electrochem. Soc. *143*:3473 (1996).
119. T. D. Tran, J. H. Feikert, R. W. Pekala, and K. Kinoshita, J. Appl. Electrochem. *26*:1161 (1996).
120. J. J. Auborn and Y. L. Barberio, J. Electrochem. Soc. *134*:638 (1987).
121. T. F. Fuller, M. Doyle, and J. Newman, J. Electrochem. Soc. *141*:1 (1994).
122. M. Doyle, T. F. Fuller, and J. Newman, J. Electrochem. Soc. *140*:1526 (1993).
123. M. S. Wittingham, Prog. Solid State Chem. *12*:41 (1978).
124. M. Nishizawa, K. Mukai, S. Kuwabata, C. R. Martin, and H. Yoneyama, J. Electrochem. Soc. *114*:1923 (1997).
125. A. H. Gemeay, H. Nishiyama, S. Kuwabata, and H. Yoneyama, J. Electrochem. Soc. *142*:4190 (1995).
126. M. M. Thackeray, J. Electrochem. Soc. *142*:2558 (1995).
127. G. L. Che, K. B. Jirage, E. R. Fisher, and C. R. Martin, J. Electrochem. Soc. *144*:4296 (1997).
128. J. R. Bottin, P. R. McCurdy, and E. R. Fisher, Rev. Sci. Instrum. *68*:2149 (1997).
129. E. Laviron, J. Electroanal. Chem. *52*:395 (1974).
130. K. Kanehori, F. Kirano, Y. Ito, K. Miyauchi, and T. Kudo, J. Electrochem. Soc. *136*:1265 (1989).
131. S. Kikkawa, M. Miyazaki, and M. Koizumi, J. Mater. Res. *5*:2894 (1990).
132. D. K. Roe, in *Laboratory Techniques in Electroanalytical Chemistry* (P. T. Kissinger, W. R. Heineman, eds.), Marcel Dekker, New York, 1996, p. 195.
133. B. V. Ratnakumar, G. Nagasubramanian, S. D. Stefano, and C. P. Bankston, J. Electrochem. Soc. *139*:1531 (1992).

ELECTROCHEMICAL ATOMIC LAYER EPITAXY

John L. Stickney

Department of Chemistry
University of Georgia
Athens, Georgia

I. INTRODUCTION

The formation of electronic grade compound semiconductor thin films is generally performed using techniques based on vacuum or gas phase reactors. Such methods include chemical vapor deposition (CVD) [1–4] and molecular beam epitaxy (MBE) [5–11]. In addition, methods based on sputtering or evaporation are frequently used. The work being performed in this author's laboratory seeks to answer the question: Can electronic grade compound semiconductor thin films be formed electrochemically? The work described in this chapter is directed towards understanding the basic limits to electrodeposition as a method for forming high-quality compound thin films.

Electrodeposition has a number of possible advantages, including low-temperature deposition (near room temperature), relatively low-cost hardware, coverage measurements via coulometry, uniform coverage on odd shapes, more tractable waste issues [12], and increased selectivity, thus lower impurity levels in some systems [12–15]. For the formation of large-area photovoltaics, the cost factors are inviting, while, in the formation of electro-optical devices, low-temperature deposition is generally desirable. Possible drawbacks to electrodeposition in the formation of compound semiconductors include the need for a conductive substrate, the need for some level of conductivity in the deposits themselves, and the fact that a condensed deposition medium is used and may result in increased levels of some impurities.

This chapter describes the current state and progress in the development of electrochemical ALE (ECALE), an electrodeposition methodology that may provide the control necessary to produce electronic grade thin films. ALE stands for atomic layer epitaxy, a technique involving the formation of thin films one atomic layer at a time [7–11,16–21]. Generally, a cycle is used where atomic layers of the individual elements are deposited independently and alternately. The idea is that if surface-limited reactions are used to form each atomic layer, three-dimensional growth modes will never be initiated and epitaxial deposits will result. ECALE is the application of electrochemical surface-limited reactions, underpotential deposition (UPD) [22,23], to ALE. Ideally, in a solution containing an ion of an element that can be electrodeposited, there is a potential beyond which a bulk deposit of the element remains stable. That a monolayer or so of the element frequently deposits on a second element at a potential prior to (under) that potential has been known for more than 80 years [24,25], and is generally referred to as UPD:

$$M^{2+} + 2e^- \Leftrightarrow M_{(UPD)} \tag{1}$$

One way to view UPD is as formation of a surface compound. In other words, deposition of the first atomic layer of an element on a second element involves a larger deposition driving force than subsequent layers, as it benefits from the ΔG of compound formation. For deposits formed at underpotential, once the substrate is covered the deposition stops because the reaction is surface limited. No more of the substrate element is available to react, unless it can quickly diffuse to the surface through or around the initially deposited monolayer (an example would be amalgam formation at a mercury electrode surface). Subsequent deposition is then only observed when the bulk deposition potential has been exceeded.

UPD generally involves the reductive deposition of an atomic layer of one element on a second. It has clearly been shown, however, that if there is a preadsorbed atomic layer of a third element, the UPD process may involve reactions with both the substrate and the preadsorbed layer.

For example, the voltammogram in Fig. 1 depicts Ag UPD on an I-coated Pt(111) electrode [26]. Three features can be attributed to the UPD of Ag, each of which results in the formation of a new structure on the surface, as indicated by the LEED patterns diagrammed in the circles. It was concluded in that work that UPD involved more than a single monolayer of Ag. Ag depositing at underpotential reacted with the Pt substrate as well as with the adsorbed I atom layer. It is also interesting to note that Ag underpotentially deposited in Fig. 1 reacted with the adsorbed atomic layer of I atoms to form a monolayer of the I-VII compound AgI on the Pt surface.

ECALE is then ALE where the elements are deposited by controlling the substrate's electrochemical potential, so that atomic layers are formed at underpotentials [Eq. (1)]. The underpotentials are used in order to obtain surface-limited deposition reactions. Compounds are deposited using a cycle where a first solution containing a precursor to one of the elements is introduced to the substrate and an atomic layer is electrodeposited at its underpotential. The cell is then rinsed, a solution containing a precursor to

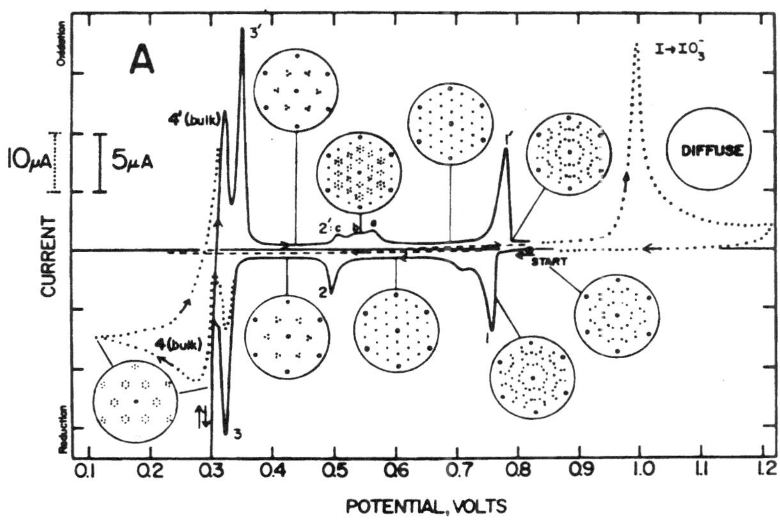

FIG. 1. Cyclic current-potential curves and LEED patterns: 10^{-4} M Ag^+ in 1 M $HClO_4$ at the Pt(111)($\sqrt{7} \times \sqrt{7}$)R19.1°-I surface. Scan rate = 2 m V/s. (From Ref. 26.)

the second element is introduced and an atomic layer is electrodeposited at its underpotential. An ECALE cycle thus involves the use of separate solutions and potentials to deposit atomic layers of each of the elements and to sequentially form a monolayer of the compound.

Electrodeposition is by its nature a condensed phase process, whereas most studies of ALE have been performed using gas phase or vacuum methodologies, CVD or MBE. A solution phase deposition methodology related to ALE has been developed in France by Nicolau et al. [27–32] (Fig. 2), in which adsorbed layers of elements are formed by rinsing a substrate in aqueous solutions containing ionic precursor for the desired elements, sequentially, in a cycle. After exposure to each precursor, the substrate is copiously rinsed and then transferred to a solution containing the precursor for the next element. The method is referred to as successive ionic layer adsorption and reaction (SILAR). Reactivity in SILAR appears to be controlled by the rinsing procedure, solution composition, pH, and specifically

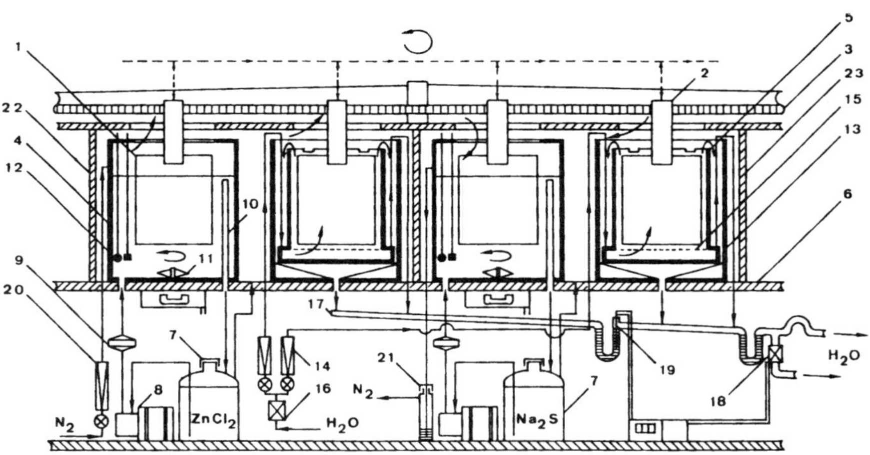

FIG. 2. Detailed drawing of the circulation of fluids and of the immersion and rinsing vessels for a SILAR deposition system: (1) substrates, (2) tweezers, (3) moving crown, (4) immersion beakers, (5) rinsing beakers, (6) circulatory tray, (7) bottles, (8) pumps, (9) filters, (10) overflows, (11) magnetic stirrers, (12) electrodes, (13) rinsing vessels, (14) rotameters, (15) sieves, (16) electrogate, (17) draining pipe, (18) electrogate, (19) conductivity cell, (20) rotameter, (21) bubbler, (22) zinc (or cadmium) compartments, (23) sulfur compartments. (From Ref. 30.)

by the activities of the reactant species. It might be thought of as a kind of chemical bath [33–37] ALE.

A number of different methods have been and are being used to form compound semiconductor thin films electrochemically (see Table 1). Most of the previous work has involved formation of various II-VI and related compounds. In general, other electrodeposition methodologies appear to be inherently faster then ECALE but result in deposits with less than ideal structure and morphology. Most compound electrodeposition methodologies lack sufficient control over deposit structure to form electronic grade deposits and require postdeposition annealing before reasonable x-ray diffraction patterns can be obtained. However, some recent advances have been very encouraging [38,39].

One of the major benefits of the ECALE methodology is that it breaks compound electrodeposition into a series of identical cycles and each cycle into a set of individual steps. Each step is examined and optimized independently, resulting in increased control over deposit structure, composition, and morphology. Better understanding of the individual steps in the deposition mechanism should allow the electrochemical formation of high-quality thin films of compound semiconductors.

The remainder of this introduction will involve a brief discussion of previous work in the area of semiconductor electrodeposition. The focus will be on the strengths and limitations of the various electrodeposition methods with regard to controlling deposit structure, composition, and morphology. Most of this work has been well reviewed by others [13,40–44].

The methodology most practiced is referred to here as codeposition, where a single solution contains precursors for all the elements being deposited and is reduced at a fixed potential or current density. The earliest report appears to be that by Gobrecht et al., which was published in 1963 [45]. Two anodes were used in the study, one of Se and one of Cd (or Ag), to form selenite and cadmium ions, respectively. CdSe was then formed by co-reduction of both species at the cathode. Reports of the formation of GaP in 1968 [46] and ZnSe in 1975 [47] via codeposition were subsequently published, and both involved molten salt electrolysis.

In 1976 a paper published by Hodes et al. described the codeposition of CdSe using a solution made by dissolving $CdSO_4$ and SeO_2 in sulfuric acid [129]. This appears to be the first of a large number of similar studies where II-VI compounds were formed from aqueous solutions by co-reduction of Cd^{+2} and $HSeO_3^-$ (or $HTeO_2^+$, as performed in the classic study by Panicker et al. [218].

TABLE 1
Compound Electrodeposition Studies

Compound	Mechanism	Solvent	Characterization	Substrates	Ref.
	Codeposition				41
	Codeposition		Theory		48
	Codeposition		Theory		49
	Codeposition		Review		40
			Review		50
	Codeposition		Review		51
	Codeposition		Review		52
	Codeposition, pulse plate		Theory		53
Bi_2S_3	Codeposition	DMSO, DMF, EG	Con, XRD	Pt, Au, SS, Ni, Zn	54
Bi_2S_3	Precipitation	1 M KOH	RRDE	Cd, Bi	55
Bi_2S_3	Precipitation	Base	RDE, PEC	Cd, Bi	56
CdBiS	Codeposition			ITO	57
CdS			OS, Raman, XRD		58
CdS	ECALE		Raman	Au, on Si	59
CdS	ECALE		STM, EC	Au(111)	60
CdS	ECALE	pH 5.9, pH 9	TLE	Au	61
CdS	ECALE	pH 9–10	EC, STM	Au(100)	62
CdS	Codeposition		J-V	SnO_2	63
CdS	Codeposition		J-V	ITO, CdS	64
CdS	Codeposition	1 M NaOH	SEM, XRD, EDX, XPS, DP, PEC	Cd	65
CdS	Codeposition	DEG	EC	Pt, Au	66
CdS	Codeposition	DEG:H_2O, 10:1	Hall, XRD, SEM, Con	Mo	67
CdS	Codeposition	DMSO	EC	Pt	68
CdS	Codeposition	DMSO	EC,DME	Hg, Pt, Au	69

CdS	Codeposition	DMSO	EC, PEC	Au	70
CdS	Codeposition	DMSO	PEC, OS	Pt	71
CdS	Codeposition	DMSO	SEM, XRD, RBS	Pt, Au	72
CdS	Codeposition	DMSO	XRD, SEM, AES, Photocon	ITO	73
CdS	Codeposition	DMSO, DMF, EG	Con, XRD	Pt, Au, SS, Ni, Zn	54
CdS	Codeposition	DMSO, EG	TEM, XRD, PEC, OS, ED	Au	43
CdS	Codeposition	DMSO, PC	XRD, EDX, OS	SS, SO	74
CdS	Codeposition	EG, LiClO4	SEM, OS	ITO	75
CdS	Codeposition	LiCl-KCl eutectic	XRD, SEM	GC	76
CdS	Codeposition	LiCl-KCl eutectic	XRD, SEM, RHEED	Ag, Cu	77
CdS	Codeposition	NH_3 buffer	PEC, J-V	Ti	78
CdS	Codeposition	PC	EC	Au, Pt	79
CdS	Codeposition	PC	EDX, EC	Ti	80
CdS	Codeposition	PC	PEC, SEM, XRD, OS	Ti, CdS, ITO	81
CdS	Codeposition	Aqueous	Con	ITO	82
CdS	Codeposition	pH 1, H_2SO_4	SEM, OS	ITO	57
CdS	Codeposition	pH 1.6	EC, AA, OS, J-V, XRD	ITO, ITO/CdS	27
CdS	Codeposition	pH 1.6	SEM, EDX, J-V	ITO/CdS	83
CdS	Codeposition	pH 2	I-V, SEM, Cap	ITO	84
CdS	Codeposition	pH 2, pH 1.8	i-T, i-V	ITO,	85
CdS	Codeposition	pH 2.3	XRD, SEM, EC, Cap	Pt, Mo, Al	86
CdS	Codeposition	pH 2.3	XRD, SEM, EDX, EC, ED, OS	Al	87
CdS	Codeposition	pH 2.5, pH 2	I-V, CAP	ITO	88
CdS	Codeposition	pH 2.5, pH 2	XRD, SEM, I-V, CAP	ITO	89
CdS	Codeposition	pH 2.8	XRD, SEM, EC, Cap, EPMA, RDE	Pt	90
CdS	Codeposition	pH 4	CV, EC	Pt, ITO	91
CdS	Codeposition	pH 8, DEG, PC	Hall, Con	ITO	92
CdS	Codeposition, pulse plate		OS, Con	ITO	93
CdS	Codeposition, pulse plate	pH 2	XRD, J-V, CAP	FTO	15

TABLE 1 Continued

Compound	Mechanism	Solvent	Characterization	Substrates	Ref.
CdS	Precipitation		PEC, elip., SNMS, EC	HgCdTe	94
CdS	Precipitation		Raman, EC	Cd	95
CdS	Precipitation	1 M KOH	RRDE	Cd, Bi	55
CdS	Precipitation	1 M $NaHCO_3$	AC, modeling	Hg, Cd(Hg)	96
CdS	Precipitation	1 M $NaHCO_3$	EC, Mott, AC	Cd	97
CdS	Precipitation	1 M $NaHCO_3$	PEC, SEM	Cd	98
CdS	Precipitation	1 M NaOH	Raman,	Cd	99
CdS	Precipitation	Base	RDE, PEC	Cd, Bi	56
CdS	Precipitation	pH 14	EC, modeling	Cd	100
CdS	Precipitation	pH 7–13	RRDE, Cap, AES-DP	Cd	101
CdS	Precipitation	pH 9	EC		102
CdS	Precipitation	pH 9, carbonate buffer	PEC, Mott, CAP, RRDE	Cd	103
CdSSe	Codeposition	DMSO	SEM, XRD, OS	ITO	104
CdSSe	Codeposition	DMSO, 0.1 M H_2SO_4	XRD, SEM, RBS, PEC	Pt on glass	105
CdSSe	Codeposition	LiCl-KCl eutectic	XRD, SEM	GC	76
CdSe	ECALE		LEED, AES, STM	Au(111)	106
CdSe	ECALE		LEED, AES, STM	Au, single crys	107
CdSe	ECALE	Sulfite	TLEC	Au	108
CdSe	SMD	0.5 M HCl, 0.25 M H_2SO_4	XRD, SEM, EC, EDX	Ni, Ti	109
CdSe	SMD	0.25 H_2SO_4	SEM, EDX map, J-V, XRD	Ni	110
CdSe	SMD	0.25 M H_2SO_4	PEC, EC	ITO	111
CdSe	Two stage, selenization	pH 9	SEM, EDAX		112
CdSe	Codeposition				42

CdSe	Codeposition		Carrier type determination	ITO	113
CdSe	Codeposition		RHEED, XRD, MBE	InAs	114
CdSe	Codeposition		TEM, ED	Au	115
CdSe	Codeposition		TEM, ED	Ti	116
CdSe	Codeposition	0.5 M H_2SO_4	EC, PEC	GC	117
CdSe	Codeposition	0.5 M H_2SO_4	EC, XRD	Ti	118
CdSe	Codeposition	0.5 M H_2SO_4	PEC, SEM, AES	Ti	119
CdSe	Codeposition	0.5 M HCl	PEC	Ti	120
CdSe	Codeposition	0.5 M H_2SO_4	SEM, PL, Raman, AES	Ti	121
CdSe	Codeposition	1 M H_2SO_4	RRDE, PEC	Ti	122
CdSe	Codeposition	1 M NH_4Cl	PEC	Ti, Ni	123
CdSe	Codeposition	1 M NaOH	SEM, XRD, EDX, XPS, DP, PEC	Cd	65
CdSe	Codeposition	DEG, PC	P, PEC, OS, J-V, SEM	Ti, ITO	124
CdSe	Codeposition	DMSO	TEM, ED	Au(111)	125
CdSe	Codeposition	DMSO	TEM, ED	Au(111)	115
CdSe	Codeposition	DMSO	TEM, ED	Au(111)	125
CdSe	Codeposition	DMSO, 0.1 M H_2SO_4	XRD, SEM, RBS, PEC	Pt on glass	105
CdSe	Codeposition	DMSO, DMF, EG	Con, XRD	Pt, Au, SS, Ni, Zn	54
CdSe	Codeposition	DMSO, EG	TEM, XRD, PEC, OS, ED	Au	43
CdSe	Codeposition	H2SO4	PL, EL	Ti	126
CdSe	Codeposition	KCN	EC, PEC, J-V, XRD	Ti	127
CdSe	Codeposition	Acid, base(NaCN)	SEM, XRD, PEC,	Ti	128
CdSe	Codeposition	Acidic sulfate	PEC, XRD, EDX?, J-V	Ti	129
CdSe	Codeposition	Aqueous	XPS-DP	Ti	130
CdSe	Codeposition	pH 0.7	PEC	Ti	131
CdSe	Codeposition	pH 1–2			45
CdSe	Codeposition	pH 10	RRDE, J-V, PEC	Ti, Au, Cd	132
CdSe	Codeposition	pH 2	J-V, OS	ITO	133
CdSe	Codeposition	pH 2–7	EC, Pourbaix, XPS-DP, PEC	Ti, Pt, GC	134

TABLE 1 Continued

Compound	Mechanism	Solvent	Characterization	Substrates	Ref.
CdSe	Codeposition	pH 2.2	XRD, EDAX, SEM, TEM, OS	Ti	135
CdSe	Codeposition	pH 2.2	XRD, SEM, EDAX, PEC, OS	Tin oxide	136
CdSe	Codeposition	pH 2.2	XRD, SEM, EDX, RDE, OS, PEC	Ni	137
CdSe	Codeposition	pH 2.7	AES DP, PEC		138
CdSe	Codeposition	pH 3–4	QCM, EC	Au	139
CdSe	Codeposition	pH 4	SEM, XRD, OS, I-V	Ti, ITO?	140
CdSe	Codeposition	pH 7?	Hall, resistivity	ITO, + Se, Cd	141
CdSe	Codeposition	pH 8	EC, PEC	Ti	142
CdSe	Codeposition	pH 8	EC, RBS, SEM, XRD, P	Ti	143
CdSe	Codeposition	pH 8	SEM, EDX, PEC	Ti	144
CdSe	Codeposition	pH 8, DEG, PC	Hall, Con	ITO	92
CdSe	Codeposition	pH 9	RBS, SEM, PEC, EC	Ti	145
CdSe	Codeposition	pH 9	Reflectance	Ti, SS, Si	146
CdSe	Codeposition	pH 9, 1 M H_2SO_4	SEM, Reflectance	Ti on Si wafers	147
CdSe	Codeposition, pulse plate	Aqueous	XRD, SEM, EDX, EC	Ti, GC	148
CdSe	Precipitation	1 M KOH	PEC	Cd, Cd on Fe, SS	149
CdSe	Precipitation	1 M KOH	PEC	Cd	150
CdSeTe	Codeposition	1 M H_2SO_4	PL, XRD, PEC	Ti	151
CdSeTe	Codeposition	1 m H_2SO_4	PEC, Con, J-V	Ti	152
CdSeTe	Codeposition	KCN	EC, PEC, J-V, XRD	Ti	127
CdSeTe	Codeposition	Acid, base(NaCN)	SEM, XRD, PEC	Ti	128
CdSeTe	Codeposition	pH 1–2.5	PEC, XRD, EDX, AES-DP, OS	Ti	153
CdSeTe	Codeposition	pH 2.2	XRD, EDAX, SEM, TEM, OS	Ti	135
CdSeTe	Codeposition	pH 2.2	XRD, SEM, EDX, RDE, OS, PEC	Ni	137

CdSeTe	Codeposition	pH 9.6	PEC, RRDE, XRD	Au, Ti	154
CdSeTe	Codeposition, pulse plate	Aqueous	XRD, SEM, EDX, EC	Ti, GC	148
CdTe	ECALE	20 mM H_2SO_4	EC, Pourbaix, AES, LEED, STM	Au-SC	155
CdTe	ECALE	Aqueous		Au	156
CdTe	ECALE	Aqueous	EC, AES, LEED	Au-SC	157
CdTe	ECALE	Aqueous	EC, EPMA, ICP-AES, SEM	Au	158
CdTe	ECALE	Aqueous	STM, EC	Au-SC	159
CdTe	ECALE	pH 2.9 and pH 1.3	TLE, review	Au-SC	160
CdTe	ECALE	pH 4.6 and pH 2.2	TLE, LEED, AES	Au	161
CdTe	ECALE	pH 4.7 and pH 2.1	AES, LEED, EC	Au-SC	162
CdTe	ECALE	pH 5.2 and pH 2.9	EC, TLE, EPMA, SEM, STM	Au	44
CdTe	SMD	0.25 H2SO4	SEM, EDX map, J-V, XRD	Ni	110
CdTe	Two-stage	Aqueous	XRD, EPMA	Mo	163
CdTe	Codeposition		AES-DP, J-V	ITO/CdS	164
CdTe	Codeposition		Con	ITO	165
CdTe	Codeposition		J-V	SnO_2	63
CdTe	Codeposition		J-V	ITO, CdS	64
CdTe	Codeposition		PEC, SEM, EDS	Si	166
CdTe	Codeposition		Theory		167
CdTe	Codeposition	0.1 M H_2SO_4	EC	VC	168
CdTe	Codeposition	0.1 M H_2SO_4	QCM, EC	Au	169
CdTe	Codeposition	0.5 M H_2SO_4	EC	GC	170
CdTe	Codeposition	0.5 M H_2SO_4	EC, PEC	GC	117
CdTe	Codeposition	0.5 M H_2SO_4	RDE, DSC	Pt, Ti	171
CdTe	Codeposition	0.5 M H_2SO_4	RRDE	Pt, Te on Pt	172
CdTe	Codeposition	1 M H_2SO_4	RDE, AA	Ti	173
CdTe	Codeposition	1 M NaOH	Pourbaix, XPS, AES, SEM	Cd	174
CdTe	Codeposition	1 M NaOH	SEM, XRD, EDX, XPS, DP, PEC	Cd	65
CdTe	Codeposition	DMSO	ED, EDX, SEM	SnO_2	175

TABLE 1 Continued

Compound	Mechanism	Solvent	Characterization	Substrates	Ref.
CdTe	Codeposition	DMSO	Con, TEM, SEM, ED	ITO	176
CdTe	Codeposition	EG	SEM, EC	Ni	177
CdTe	Codeposition	EG	XPS-DP, SEM, PEC, XRD	Ni	178
CdTe	Codeposition	EG	XPS-DP, SEM, XRD	Ni	179
CdTe	Codeposition	EG, $LiClO_4$	SEM, OS	ITO	75
CdTe	Codeposition	H_2SO_4	AES-DP, XRD, EPMA	Cu, Ti, SS, ITO	180
CdTe	Codeposition	H_2SO_4	EDS, PEC, PL, Raman, SEM, XRD	ITO	181
CdTe	Codeposition	H_2SO_4	PEC	Ti	182
CdTe	Codeposition	H_2SO_4	PL	ITO-CdS	183
CdTe	Codeposition	H_2SO_4	XRD, SEM, EC, EPMA	Ni, Te on Ni	184
CdTe	Codeposition	LiCl-KCl molten salt	XRD, SEM	GC	185
CdTe	Codeposition	PC	Con, Hall, SEM	ITO	186
CdTe	Codeposition	PC	Con, SEM	ITO	187
CdTe	Codeposition	PC	EC, XRD, SEM, PEC	Ti	188
CdTe	Codeposition	PC	P, EC, SEM, PEC	ITO, Ti	189
CdTe	Codeposition	PC	PEC, SEM, XRD, OS	Ti, CdS, ITO	81
CdTe	Codeposition	PC	Con, depth profil, Hall, XPS	ITO	190
CdTe	Codeposition	PC, $LiClO_4$	Hall	ITO	191
CdTe	Codeposition	Aqueous	Cap, Con	Ti	192
CdTe	Codeposition	Aqueous	Con	ITO	82
CdTe	Codeposition	Aqueous	EC, XPS, optical microscopy	GC	193
CdTe	Codeposition	pH 0	PEC	Ti	194
CdTe	Codeposition	pH 0.5	EC, XRD, SEM, EDX,	Si n-type	195
CdTe	Codeposition	pH 1.4	AES-DP, XRD	Ti	196
CdTe	Codeposition	pH 1.4	EC, PEC, XRD	Ti, Ni	197
CdTe	Codeposition	pH 1.4	PEC, XRD, AES	Ti, Ni	198

CdTe	Codeposition	pH 1.4	XRD, AES, Con	Ti, Ni	199
CdTe	Codeposition	pH 1.5	RDE, Pourbaix	SnO_2	200
CdTe	Codeposition	pH 1.5–2.0	RDE, EC, Theory	VC, SS	201
CdTe	Codeposition	pH 1.6	EC, AA, OS, J-V, XRD	ITO, ITO/CdS	202
CdTe	Codeposition	pH 1.6–2	SEM, EDAX, XRD,	Ni	203
CdTe	Codeposition	pH 1.6–2	SEM, XRD, EDAX, OS	SnO_2	204
CdTe	Codeposition	pH 1.7	XRD, SEM	GaAs	205
CdTe	Codeposition	pH 1.8	XRD, DSC, XPS, EC, ellipsometry, RBS	n- p- Si	206
CdTe	Codeposition	pH 1.8	In situ ellipsometry, raman, EDAX, DSC	MCT	207
CdTe	Codeposition	pH 2	I-V, SEM, Cap	ITO	84
CdTe	Codeposition	pH 2	J-V, OS	ITO	133
CdTe	Codeposition	pH 2	SEM, XRD	CdS/SnO_2/glass	208
CdTe	Codeposition	pH 2	SIMS, AA, SEM, XRD, NAA, PIXE, AES	ITO, CdS, Ni, Ti	13
CdTe	Codeposition	pH 2	XRD, RHEED	InP-CdS	39
CdTe	Codeposition	pH 2, pH 1.8	i-T, i-V	ITO	85
CdTe	Codeposition	pH 2–3	XRD, V-J	Ni	209
CdTe	Codeposition	pH 2.0–3.5	Con, XRD	SS, ITO	210
CdTe	Codeposition	pH 2.2	AC, RDE	Ni	211
CdTe	Codeposition	pH 2.2	EC, XRD, SEM, OS, PEC	SnO_2	212
CdTe	Codeposition	pH 2.2	SEM, J-V, OS	Cu, Steel, Ni, Cd	213
CdTe	Codeposition	pH 2.2	SEM, XRD, XPS, OS, PEC	SnO_2	214
CdTe	Codeposition	pH 2.2	XRD, EDAX, SEM, TEM, OS	Ti	35
CdTe	Codeposition	pH 2.2	XRD, SEM, EDAX, PEC, OS	Tin oxide	136
CdTe	Codeposition	pH 2.2	XRD, SEM, EDX, RDE, OS, PEC	Ni	137
CdTe	Codeposition	pH 2.2	XRD, SEM, RHEED, EDS	Ni, Ti	215
CdTe	Codeposition	pH 2.5	EPMA, AES-DP, V-J, XRD	ITO, Mo	216

TABLE 1 Continued

Compound	Mechanism	Solvent	Characterization	Substrates	Ref.
CdTe	Codeposition	pH 2.5	Theory	C, Pt, etc.	217
CdTe	Codeposition	pH 2.5	XRD, OS, PEC	TO-Cds	38
CdTe	Codeposition	pH 2.5, pH 2	I-V, CAP	ITO	88
CdTe	Codeposition	pH 2.5, pH 2	XRD, SEM, I-V, CAP	ITO	89
CdTe	Codeposition	pH 2.5–3	XRD, SEM, EDX, EC, Pourbaix, Con-type	Ni, SnO_2:Sb	218
CdTe	Codeposition	pH 2.5–3.0	PEC, XRD, OS, SEM, EC, EDX	Ti, Nesatron	219
CdTe	Codeposition	pH 2.5–3.0	SEM, AES-DP, EPMA, J-V, Mott	ITO, Mo	220
CdTe	Codeposition	pH 8, DEG, PC	Hall, Con	ITO	92
CdTe	Codeposition, pulse plate		Theory		221
CdTe	Codeposition, pulse plate	0.3 M H_2SO_4	SEM, RDE, PEC	Cd	222
CdTe	Codeposition, pulse plate	Aqueous	XRD, SEM, EDX, EC	Ti, GC	148
CdTe	Codeposition, pulse plate	pH 2	XRD, J-V, CAP	FTO	15
CdTe	Precipitation	1 M $CdSO_4$, 1 M $ZnSO_4$	PEC, EC	Te	223
CdTe	Precipitation	pH 4.5	EC, AA	Te, Cd	224
CdZnS					57
CdZnS	Codeposition				57
CdZnS	Codeposition	DMSO	SEM, XRD, OS	ITO	104
CdZnS	Codeposition	pH 2	XPS, XRD, SEM, AES depth profile	Ti, Pt, ITO	225
CdZnSe	Codeposition	pH 2–7	EC, Pourbaix, XPS-DP, PEC	Ti, Pt, GC	134
CdZnSe	Codeposition	pH ?	AC, I-V	Ti	226

CdZnTe	Two-stage	1 M H_2SO_4	SEM, XRD, J-V, OS	ITO, SnO_2	227
CoS	Codeposition	DMSO, DMF, EG	Con, XRD	Pt, Au, SS, Ni, Zn	54
Cu_2S	Codeposition	DMSO, DMF, EG	Con, XRD	Pt, Au, SS, Ni, Zn	54
Cu_2Se	Codeposition	pH 2.9, 3.28, 3.7	XRD, SEM, EDX, OS	Ti, Ni	228
Cu_9In_4	Codeposition	pH 2.9, 3.28, 3.7	XRD, SEM, EDX, OS	Ti, Ni	228
$CuISe_3$	Codeposition	4 M HI	J-V	ITO	229
$CuIn_5S_8$	Codeposition, two-stage	pH 2	SEM, PEC, EPMA, XRD, AES	Ti	230
$CuInS_2$	Codeposition, two-stage	pH 2	SEM, PEC, EPMA, XRD, AES	Ti	230
$CuInS_2$	Codeposition, two-stage				12
$CuInSe_2$	Two-stage				231
$CuInSe_2$	Two-stage		SEM, XRD, depth profil	Mo	232
$CuInSe_2$	Two-stage		XRD, XPS	Mo, ITO	233
$CuInSe_2$	Codeposition		AC, Con	Mo	234
$CuInSe_2$	Codeposition		J-V	Mo	235
$CuInSe_2$	Codeposition		J-V, XRD	Mo	236
$CuInSe_2$	Codeposition		PEC, Mott		237
$CuInSe_2$	Codeposition		SEM, XRD, J-V, SIMS, EPMA	Mo, W	238
$CuInSe_2$	Codeposition	0.4 citric acid	AA, XRD, EDX	Mo	239
$CuInSe_2$	Codeposition	4 M HI	J-V	ITO	229
$CuInSe_2$	Codeposition	pH 1	RRDE, PEC, Pourbaix	GC	240
$CuInSe_2$	Codeposition	pH 1	SEM, ED, OS, Pourbaix, EDX, J-V	Ti	241
$CuInSe_2$	Codeposition	pH 1	XRD, PEC	SnO_2:F	242
$CuInSe_2$	Codeposition	pH 1.6–1.7	RRDE, SEM, XRD, EDX, PEC	Ti, Ni, Ml, Cu	243
$CuInSe_2$	Codeposition	pH 2.9, 3.28, 3.7	XRD, SEM, EDX, OS	Ti, Ni	228
$CuInSe_2$	Codeposition, two-stage	pH 2	SEM, PEC, EPMA, XRD, AES	Ti	230

TABLE 1 Continued

Compound	Mechanism	Solvent	Characterization	Substrates	Ref.
$CuInSe_2$	Codeposition, pulse plate	pH 2.3	SEM, XRD, EC	Ti	244
$CuInSe_2$	Codeposition, pulse plate		XRD, SEM		245
$CuInSe_2$	Codeposition, two-stage				12
$CuInSe_2$	Two-stage	H_2SO_4	XRD, SEM, EPMA	W/C	246
$CuInTe_2$	Codeposition	pH 2.0, pH 8–9	EC, EDX, XRD	Ti	247
$CuIn_xSe_yI_z$	Codeposition	4 M HI	J-V	ITO	229
CuS	?		RHEED, OS, J-V	Ag-Cr/CdS	248
CuS	Codeposition	pH 3–5	OS	CdS	249
CuSe	Two-stage, selenization	pH 9	SEM, EDAX		112
CuSe	Codeposition		RRDE, PEC, Pourbaix	GC	240
CuSe	Codeposition	Melt, LiCl added		Al oxide	250
FeS_2	Two-stage	pH 10, pH ?	EQCM, EC	Au	251
GaAs	ECALE	Aqueous	EC, AES, LEED	Au-SC	157
GaAs	ECALE	pH 1.9 and pH 4.0	TLE, AES, LEED, Pourbaix	Au-SC	252
GaAs	ECALE	pH 2.7 and pH 3.2	AES, EC, LEED, STM	Au(100),(110)	253
GaAs	Codeposition		PEC		254
GaAs	Codeposition	B_2O_3, NaF, molten salt	XRD	GaAs, Ni	14
GaAs	Codeposition	$KGaCl_4$ molten salt	XRD, EC	Au on Ni	255
GaAs	Codeposition	$pH < 3$, $pH > 12$	SIMS, AES depth profile, EDAX, XRD	Ti, Si, Pb, Sn, C	256
GaAsSb	Two-stage	5 M KOH, 7 M HCl	XRD, EDX, SEM	Ti, Ni/Cu, VC	257
GaP	Codeposition	$NaPO_3$, NaF	SEM, XRD, J-V, Laue, ES	GrC, Si-SC	46
GaP	Codeposition	$NaPO_3$, NaF	SEM, Laue	GrC, Si, GaP	258

GaSb	Two-stage	5 M KOH, 1.6 M, H_2SO_4-HCl	XRD, EC, SIMS, EDX	Ni-plated Cu	259
HgCdTe	Codeposition	H_2SO_4	I-V, PEC	Ti	182
HgCdTe	Codeposition	PC	XRD, OS, AA, PIXE	ITO	260
HgCdTe	Codeposition	pH 1.6	EC, AA, OS, J-V, XRD	ITO, ITO/CdS	202
HgCdTe	Codeposition	pH 1.6	SEM, EDX, J-V	ITO/CdS	83
HgCdTe	Codeposition	pH 1.6	XRD, OS, PEC	ITO	261
HgCdTe	Codeposition	pH 1.6	XRD, SEM, EDX, PEC	Ti	262
HgCdTe	Codeposition	pH 2, H_2SO_4	EDS, OS, PEC, XRD	Ti	263
HgCdTe	Codeposition	H_2SO_4	PEC	Ti	182
HgS	Codeposition	DMSO, DMF, EG	Con, XRD	Pt, Au, SS, Ni, Zn	54
HgS	Precipitation	1 M $NaHCO_3$	EC	DME	264
HgSe	Precipitation	0.1 M $HClO_4$	EC, PEC	DME, Hg pool	265
HgTe	Precipitation	pH 4.5	EC, AA	Te, Cd	224
InAs	Two-stage	7M HCl, pH 2	XRD, EDX, EC	Ni	266
InAs	Codeposition, two-stage	Citric acid	XRD, EC, PEC, AC, Mott	Ti	267
InGaSb	Two-stage	5M, KOH, pH 1.5–2, pH 0	XRD, SEM, EDX, SIMS, EC	Ni, Pt, Sb, InSb	268
InP	Codeposition	DMF	XRD, EDS, Con	Ti	269
InP	Codeposition	$NaPO_3$/NaF molten salt	EC, SEM	CdS, InP(111)	270
InP	Codeposition	pH 2	XRD, EDS, OS	Ti	271
InP	Codeposition, two-stage	Citric acid	XRD, EC, PEC, AC, Mott	Ti	267
InSb	Two-stage	1 M H_2SO_4	EC, XRD	Sb	272
InSb	Two-stage	5M KOH, pH 1.5–2, pH 0	XRD, SEM, EDX, SIMS, EC	Ni, Pt, Sb, InSb	268
InSb	Codeposition, two-stage	Citric acid	XRD, EC, PEC, AC, Mott	Ti	267
InSe	Two-stage, selenization	pH 9	SEM, EDAX		112

TABLE 1 Continued

Compound	Mechanism	Solvent	Characterization	Substrates	Ref.
In_2Se_3	Codeposition	pH 3.45	XRD, CV, OS	SnO_2, Mo on glass	273
InSe	Codeposition	pH 1	SEM, EPMA, XRD	Ti	274
NiS	Codeposition	DMSO, DMF, EG	Con, XRD	Pt, Au, SS, Ni, Zn	54
PbS	Codeposition	DMSO	EC, DME	Hg, Pt, Au	69
PbS	Codeposition	DMSO, DMF, EG	Con, XRD	Pt, Au, SS, Ni, Zn	54
PbS	Precipitation	0.1 M Na_2S	EC	Hg(Pb)	275
PbS	Precipitation	pH 9–14	EC	Pb	276
SnS	Codeposition/ pulse	EG	SEM, XRD, SPX, EPMA, AES, PEC, Pourbaix	ITO	277
SnSe	Codeposition	pH 3, DMF	XRD, O, Con, EMPA	ITO	278
Tl_2S	Codeposition	DMSO, DMF, EG	Con, XRD	Pt, Au, SS, Ni, Zn	54
ZnMgSeTe	Codeposition		RHEED, XRD, MBE	InAs	114
ZnS	ECALE		TLEC, EC	Au	279
ZnS	Codeposition	pH 2.5	XRD, OS	ITO	280
ZnS	Codeposition	pH 8–10	SEM, EC, OS	Ti, SS, SnO_2	281
ZnS	Codeposition	DMSO	EC, DME	Hg, Pt, Au	69
ZnSe		pH 2	EC, XRD	Ti, SS, SnO_2	282
ZnSe	ECALE		TLEC, EC	Au	279
ZnSe	Codeposition		AC,PEC	Ti	283
ZnSe	Codeposition	0.5 M H_2SO_4	EC, PEC	GC	117
ZnSe	Codeposition	KCl + LiCl, eutectic	XRD, ED, SEM	Ge(III), Si(111)	47
ZnSe	Codeposition	Aqueous	XPS-DP	Ti	130
ZnSe	Codeposition	pH 2–7	EC, Pourbaix, XPS-DP, PEC	Ti, Pt, GC	134
ZnSe	Codeposition	pH 2–6	AC, PEC	Ti	284

ZnSeTe	Codeposition	pH 2.3–3	PEC, EC	Ti	285
ZnTe	ECALE		TLEC, EC	Au	279
ZnTe	ECALE	pH 5.2 and pH 2.9	EC, TLE, EPMA, SEM, STM	Au	44
ZnTe	Codeposition	pH 4.5	XRD, SEM, EPMA, OS	Ti, Ni, SnO_2	286
ZnTe	Codeposition		RHEED, XRD, MBE		114
ZnTe	Codeposition	pH 4	XPS, XPS depth profile, EC, PEC, Cap	MCT	287
ZnTe	Precipitation	1 M $CdSO_4$, 1 M $ZnSO_4$	PEC, EC	Te	223
ZnTe	Precipitation	pH 4.5 or less	XRD, gravimetry	Homogeneous	288

Analytical techniques: AA = atomic adsorption; AC = ac impedance; AES = Auger electron spectroscopy; Cap = capacitance; Con = conductivity measurements; DME = dropping mercury electrode; DP = depth profiling; DSC = differential scanning calorimetry; EC = electrochemical analysis; ED = electron diffraction; EDX = energy dispersive x-ray analysis; EL = electroluminescence; EPMA = electron probe microanalysis; ES = emission spectroscopy; ICP-AES = inductively coupled plasma–atomic emission spectroscopy; J-V = current voltage curves; LEED = low-energy electron diffraction; Mott = Mott-Schottky; NAA = neutron activation analysis; OS = optical spectroscopy; P = polarography; PEC = photoelectrochemical cells; PIXE = photon-induced x-ray emission; PL = photoluminescence; QCM = quartz crystal microbalance; RBS = Rutherford backscattering; RDE = rotating disk electrode; RHEED = reflection high-energy electron diffraction; RRDE = rotating ring disk electrode; SEM = scanning electron microscopy; STM = scanning tunneling microscopy; TEM = transmission electron microscopy; TLE = thin layer electrochemistry; VPD = van der Pauw; XPS = x-ray photoelectron spectroscopy; XRD = x-ray diffraction.
Reactants: BPS = tri(*n*-butyl)phosphine selenide; BPT = tri(*n*-butyl)phosphine teluride; NTA = nitrilotriacetate complex ion; SeCN = sellenocyanide; SOS = selenosulfite; STS = sodium thiosulfate; TeCN = tellurocyanide; TFMS = trifluoromethane sulfanate; TSS = triphenylstibine sulfide.
Solvents: aq = aqueous; DEG = diethylene glycol; EG = ethylene glycol; PC = propylene carbonate.
Substrates: DME = dropping mercury electrode; FTO = fluorine-doped tin oxide; G = graphite; GC = glassy carbon; GrC = graphic carbon; ITO = indium tin oxide–coated glass; SC = single crystals; SS = stainless steel; TCO = transparent conducting oxide; VC = vitrious carbon.
Miscellaneous: ECALE = electrochemical atomic layer epitaxy; ED = electrodeposition; ML = monolayer; RT = room temperature; SMD = sequential monolayer deposition; V = vacuum.

Codeposition produces some of the better II-VI electrodeposits and, as can be seen in Table 1, has been used and studied extensively. Aqueous codeposition of CdTe serves as a good example of the method. The deposition is usually performed at an underpotential for Cd, at a potential where the Cd deposits exclusively on previously deposited Te. Te, on the other hand, is more noble than Cd and is thus deposited at an overpotential. The tellurite concentration, however, is kept far below that of the Cd^{+2}, so there is a large excess of Cd^{+2}. As soon as Te deposits, Cd quantitatively underpotentially deposits on top, providing control over deposit stoichiometry.

One of the exciting aspects of the early work in this area was the observation that both n- and p-type semiconductors could be formed from the same bath simply by adjusting the deposition potential (Fig. 3) [218]. At the more positive deposition potentials, where p-type behavior was observed, Cd is less reactive and small amounts of excess Te may be present.

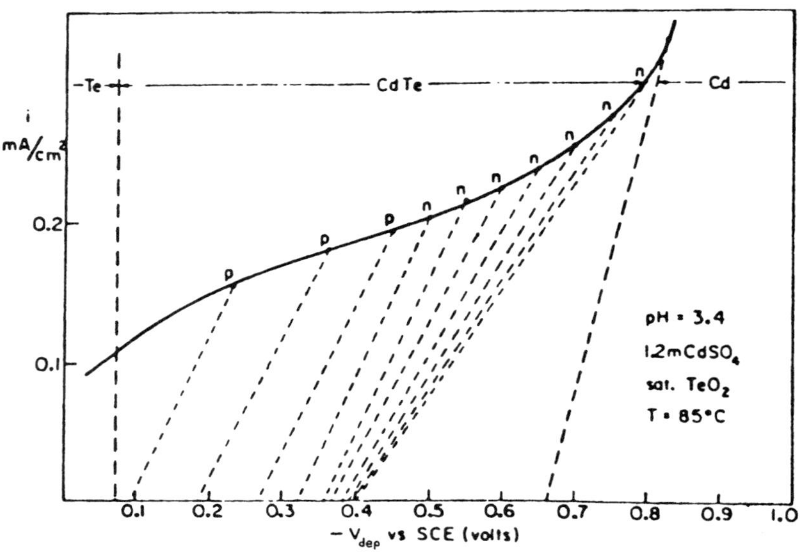

FIG. 3. Current density vs. cathode voltage for deposition at 85°C from a 1.2 M $CdSO_4$ solution saturated with TeO_2; pH = 3.4, stirring rate 160 rpm. (From Ref. 218.)

On the other hand, n-type behavior was observed at more negative potentials, where excess Cd may have been present. Type conversion, from n- to p-, was also frequently achieved by annealing deposits near 350°C [218].

The downside to the codeposition methodology is that the chalcogenide is generally deposited at an overpotential, and it is not clear that the chalcogenide atoms necessarily deposit in optimal sites. It is expected that significant surface diffusion of the chalcogenide is required in order to avoid high defect densities. The use of higher deposition temperatures has been observed to improve deposit structure significantly (Fig. 4), in line with the need for adequate surface diffusion. In addition, postdeposition annealing generally improved deposit crystallinity.

The case in which both elements are deposited at underpotentials, simultaneously, from a single bath has been considered by Engelken [49]. A deposition potential that did not exceed the reversible potential for either element could be used if both elements have similar UPD potentials. The elements could induce the UPD of each other, possibly forming a higher-quality deposit than those where one element is deposited at an overpotential.

In the case of CdSe formation using the codeposition methodology, a problem was encountered early on and studied by Skyllas-Kazacos and Miller [122]. It concerned the formation of selenide ions and their reaction with the selenite starting material to form elemental Se:

$$HSeO_3^- + 2HSe^- + 3H^+ \Leftrightarrow 3Se + 3H_2O \tag{2}$$

This is an example of conproportionation. The net results were CdSe deposits that required thermal treatments to obtain optimal photoactivity due to the presence of elemental Se [129]. The initial solution to this problem was to search for a Se precursor in a lower oxidation state, such as selenosulfite, $SeSO_3^{2-}$ [132]. The nominal oxidation state of Se in this species is zero, as it is formed by the reaction of elemental Se with sulfite:

$$Se + SO_3^{2-} \Leftrightarrow SeSO_3^{2-} \tag{3}$$

Use of selenosulfite in combination with EDTA complexed Cd^{+2}, eliminated the elemental Se contamination, and improved the photoresponse of the as-formed deposits [132]. A second method for avoiding conproportionation, also suggested by Skyllas-Kazacos, was to use a cyanide solution to dissolve elemental Se (or Te) and high concentrations of $CdCl_2$ [127]. Again, the Se was felt to be in the zero oxidation state.

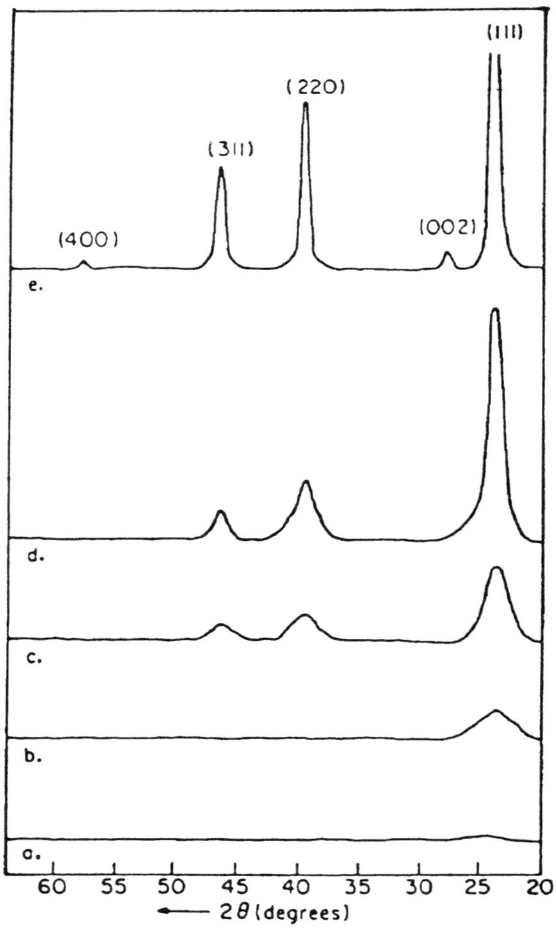

FIG. 4. X-ray diffraction patterns of films deposited on SnO_2-covered glass from a solution of 1 M $CdSO_4$, saturated with TeO_2, pH ≈ 2.7, with current density 0.5 mA/cm^2 (E_{rest} = 0.18V) at 22°C (a), 35°C (b), 65°C (c), and 90°C (d), and the pattern (e) of film (a) after annealing for 3.5 hours at 350°C in an argon atmosphere. (From Ref. 218.)

A major modification of the codeposition methodology was the introduction of nonaqueous solvents by Baranski and Fawcett [54,72]. The nonaqueous solvent allowed the use of elemental chalcogenides, again avoiding conproportionation. High-quality films were observed in those studies, although some were prone to cracking. Another advantage of using the nonaqueous solvents was the ability to deposit films at higher temperatures, which in general improved deposit quality.

As indicated previously, the state of the art in codeposition appears to be the work of Lincot et al. [38,39]. They have shown that the optimal deposition potential is around 5 mV positive of the reversible potential for Cd^{+2}/Cd. In addition, they have incorporated a feedback step in the deposition to correct for the increasing resistance of the films as they grow [289]. The resulting deposits have shown a remarkable degree of epitaxy, observed with reflection high-energy electron diffraction (RHEED) and XRD pole patterns. Their results offer great promise for the formation of high-quality compound semiconductor thin films.

A second major methodology for electrochemically forming compound semiconductors, investigated by a number of workers (Table 1), involves the dissolution of a substrate in a solution containing a species that will react with the dissolving ions. This methodology is referred to here as the precipitation method, because it can be thought of as electrochemically initiated precipitation at the electrode surface. The first example appears to be work performed by Panson [288], where ZnTe was deposited by forming Te^{2-} ions electrochemically and then precipitating them using a zinc salt. Similar studies were performed a few years later by Miles and McEwan [224] in the formation of CdTe and HgTe, where a tellurium cathode was used to form telluride ions and a cadmium anode was used to form Cd^{2+} ions. Attempts were made to use a Hg anode to form Hg^{2+} ions in solution, but much better results were obtained using a mercury (II) salt. Use of a solution containing a salt of one of the ions proved important, as it localized the precipitation to one of the electrodes, resulting in much better films.

In 1976 Miller and Heller used a solution of sulfide ions and formed Cd^{2+} ions from a Cd anode and Bi^{3+} ions from a Bi anode [56], precipitating the corresponding sulfides. Reasonably photoactive deposits of the sulfides were formed on both the Cd and the Bi anodes. Subsequently, extensive studies of the growth of CdS by this method were published by Miller et al. [55] and by Peter [97,98] in 1978.

Advantages of the precipitation methodology were simplicity and good adhesion of the deposits to the substrate. There are inherent deficiencies, however, such as the lack of control over deposit structure. Deposit growth by the precipitation mechanism requires the transport of ions from solution to the substrate-compound interface, or from the substrate to the compound-solution interface. The process is analogous to formation of passive films during corrosion. It is not clear how control over the structure and morphology of the depositing material could be significantly improved.

Another group of methods associated with the electrodeposition of compounds are referred to here as two-stage methods. This designation covers deposits where the elements are deposited in a first stage after which a reaction is initiated to form the compound in a second stage. Both stages are not necessarily electrochemical in nature, and thus these techniques are not strictly electrodeposition methodologies. Several examples have been included in Table 1, however, for completeness. A good example of a two-stage method is delineated in the patent by Kapur et al. [290], where thin films of the two component elements of the compound to be formed are first electrodeposited individually, from separate solutions, one on the other. The second stage is then a heat treatment, or anneal, resulting in interdiffusion and reaction of the component elements. A variation on this scenario was developed by Hodes and Cahen in 1985 [12], where $CuInS_2$ and $CuInSe_2$ were formed by first electrodepositing a Cu-In alloy and then heating the deposits in the presence of H_2S or H_2Se gases to form the respective ternary compounds. Other scenarios include formation of a layer of Cd on an inert substrate (via electrodeposition, vapor deposition, etc.) and then use of the precipitation methodology, oxidation in a sulfide, selenide or telluride solution to form the corresponding II-VI compounds electrochemically [112]. A completely electrochemical two-stage method has recently been described by Rajeshwar et al. [251], where a layer of sulfur was first electrodeposited and then transferred to a solution containing cations of the desired metal, in that case Fe^{2+}. The metal was then reduced into the sulfur layer forming a thin film of the corresponding metal sulfide.

One advantage of the two-stage methods is control over the deposited amounts. If all the elements are electrodeposited prior to thermal annealing, for instance, coulometry can be used to account for the amounts deposited and thus help control the stoichiometry of the resulting film. The annealing step can be a problem, as mentioned previously, when interdiffusion at heterojunctions is not desirable. It is interesting to note that the annealing temperatures used to form compounds by the two-stage method [290] are similar to those used to postanneal deposits formed by other electrodeposition methodologies, possibly indicating that the quality of many of the as-deposited films listed in Table 1 was not much better than a deposit consisting of segregated domains of the component elements.

There are a number of electrodeposition methods described in the literature based on potential programs, of which pulse plating is a good example [53,222,244,277,290]. One of these methodologies, sequential monolayer electrodeposition (SMED), developed by Sailor et al., is a more atomic-level approach to the formation of compounds electrochemically and may provide the control needed to form high-quality deposits [109,110]. The method was developed in order to avoid the traces of elemental Se that are generally incorporated in deposits of CdSe formed by codeposition from a solution containing selenite species [Eq. (2)]. SMED involves a potential program, which starts low in order to deposit both submonolayer quantities of CdSe and several monolayers of bulk Cd. Under those low potential conditions, it is assumed that all the available Se reacts with Cd. The next step in the cycle involves oxidative removal of the bulk Cd using a potential where only the CdSe is stable and remains behind. In the original studies, a fast cyclic potential program was used to form the CdSe deposits (Fig. 5).

Given the diversity of results described in the papers listed in Table 1, the future importance of electrodeposition in the formation of compound semiconductor thin films is not clear. ECALE is suggested as a means for examining limits to compound semiconductor electrodeposition. The rest of this chapter, then, describes studies of ECALE and has been organized as follows: Sec. II describes the use of manual thin layer electrochemical cells (TLECs) to investigate the potentials and solutions needed for an ECALE cycle. Section III describes the formation of thin films of a number of compounds using an automated flow cell electrodeposition system. Section IV describes studies of the nucleation and growth of the individual atomic layers, while Sec. V describes the inverse of electrochemical ALE, electrochemical digital etching (where a cycle is used to remove atomic layers of the elements constituting a compound, one at a time). Finally, Sec. VI describes possible new and future directions for ECALE-related work.

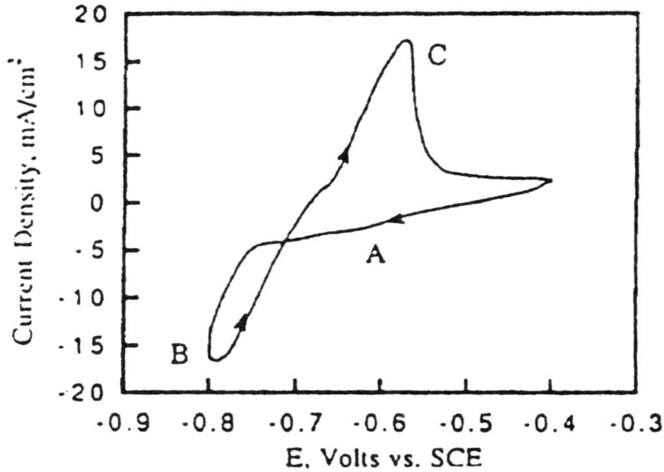

FIG. 5. Current-voltage trace obtained during cyclic electrodeposition of CdSe onto a Ni rotating disk electrode (1000 rpm): (A) potential region where Se and CdSe are deposited; (B) deposition of bulk Cd; (C) stripping wave of excess Cd. Voltage scale is referenced to SCE. Negative potentials are to the left; cathodic (negative) currents are in the downward direction. Scan rate is 10 V/s. (From Ref. 109.)

II. THIN LAYER ELECTROCHEMICAL CELL STUDIES

The development of ECALE in our group was directly stimulated by work and discussions with M. L. Norton (Marshall University, Huntington, WV) [291]. Those studies involved the use of a TLEC to investigate the deposition of Cd and Te on a series of metal electrodes (Cu, Au, and Pt) and of Cd and Te on atomic layers of each other [291]. TLECs were used for a number of reasons, such as that they provide a clean defined environment for the electrode. In the TLECs used in these studies, the electrodes and deposits were exposed to only a few µl of solution instead of the tens of ml used in corresponding thick layer cells. The decreased amount of solution results in a corresponding decrease in the amounts of solution-borne contaminates exposed to the electrode. Another advantage of these TLECs is that they allowed careful control of the amounts of reactant exposed to the electrode. The volumes of the TLECs used in these studies were easily determined to within 1–2%, using a standard solution of an electroactive species (Fig. 6). Thus if the concentration of the reactant solution is known, the total number of moles exposed to the electrode surface is easily determined. Additionally, accurate coulometry can be performed with TLECs [292,293] because of minimal and reproducible background currents. The most important reason for using a TLEC in these studies, however, was the ease with which solutions could be exchanged without exposing the surface to air. Application of a small overpressure of N_2 gas on the inside of the TLEC (Fig. 6) results in expulsion of the solution aliquot through the two pinholes at the tip of the cell. The TLEC can then be dipped into a new solution, and after relief of the N_2 overpressure, an aliquot of the solution in contact with the tip wicks in via capillary action.

Figures 7B and C display voltammetry for the deposition of Cd and Te, respectively, on a polycrystalline Au electrode. In Fig. 7B, Cd UPD is clearly evident as a broad peak centered at −0.15 V [all potentials are listed vs. Ag/AgCl (1 N NaCl)]. The resulting Cd coverage corresponded to about 1/2 monolayer (ML), where 1 ML corresponds to the deposition of one atom for each substrate surface atom. Figure 7C displays the deposition of Te from a tellurite solution ($HTeO_2^+$). Two features are evident: a UPD peak at 0.32 V followed by a small bulk deposition feature at 0.0 V, accounting for the remainder of the tellurite aliquot.

The next step was to alternate the deposition of Cd and Te. As Te is more noble than Cd, various amounts of Te were first deposited and then exposed to a Cd^{+2} ion solution at underpotential. Figure 8 is a graph of the Cd coverages observed to form on a Cu electrode initially coated with various amounts of Te. The slope of the graph is 0.95 (the Cd/Te ratio), which is consistent with the Cd reacting 1 : 1 with the Te. Similar results were observed for deposits formed on Pt and Au electrodes. The graph indicates that the Cd reacted at underpotential quantitatively with the Te, even when multiple atomic layers of Te were present.

Deposition of the first ML of CdTe was straightforward; the problem with development of an electrochemical ALE cycle for CdTe was forming the second ML of CdTe. UPD of a less noble element on the UPD of a more noble element appears to work well. However, in the second cycle, use of the same conditions for UPD of the more noble element is a thermodynamically untenable arrangement, as it must form on the deposit of the less noble element (Fig. 9). Figures 9A and B are Pourbaix diagrams for Cd and Te, respectively, and the nobility of Te relative to Cd is evident.

The Pourbaix diagrams in Figs. 9A and B do not, however, take into account the fact that the two elements are forming a compound. When the

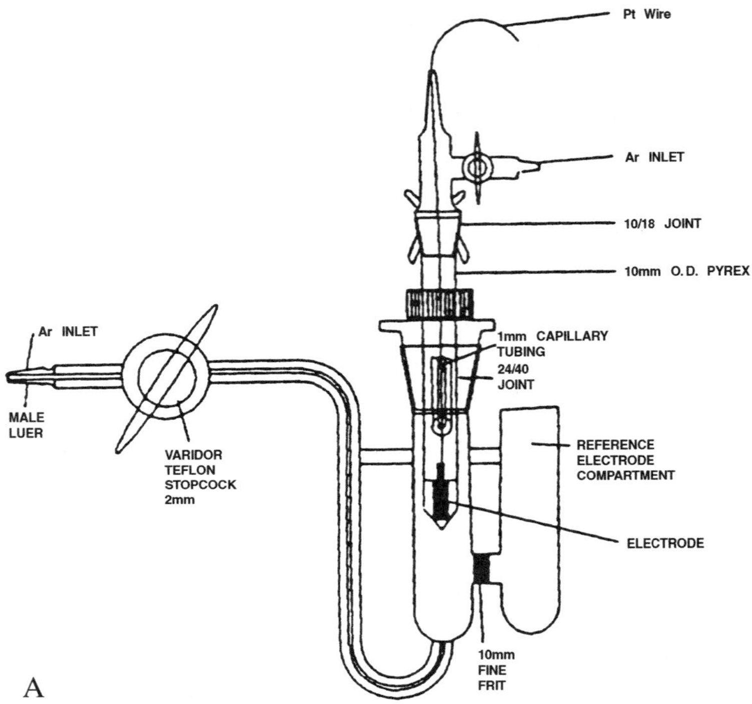

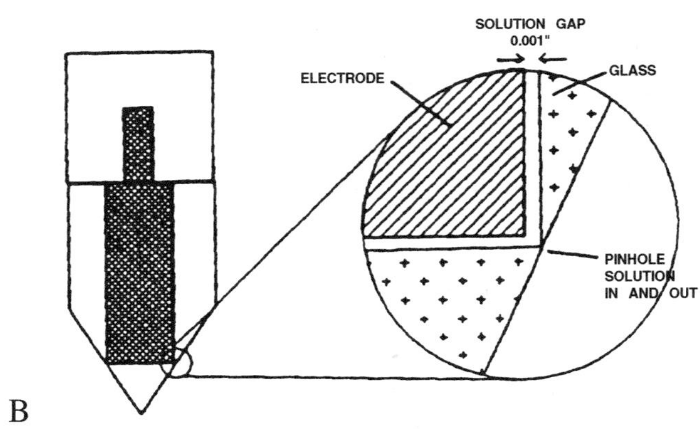

FIG. 6. Diagram of thin-layer electrochemical cell (TLEC): (A) TLEC in conjunction with electrochemical H-cell; (B) enlarged diagram showing pinhole region. (From Ref. 161.)

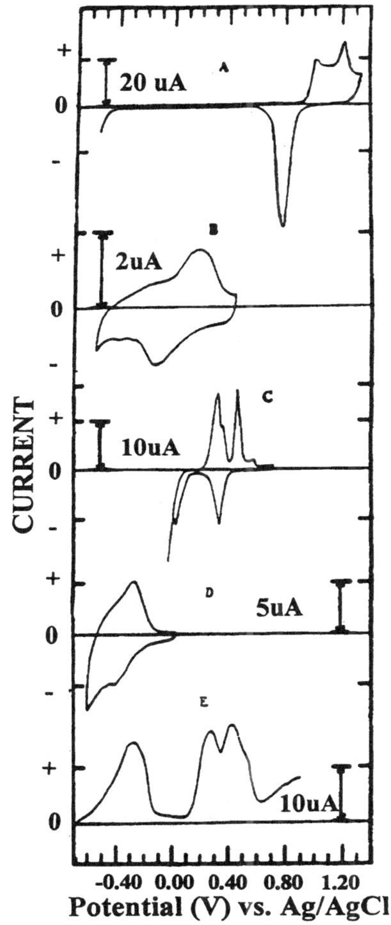

FIG. 7. Current-potential curves of Au tricrystal: (A) clean Au in 10 mM H_2SO_4 (pH = 1.7); (B) in 1.0 mM $CdSO_4$ + a buffer consisting of 2.0 mM acetic acid and 1.0 mM CsOH (pH = 4.6), sweep rate = 2 mV/s; (C) in 0.40 mM TeO_2 + 10 mM H_2SO_4 (pH = 2.2), sweep rate = 1 mV/s; (D) Te UPD-covered Au in 1.0 mM $CdSO_4$ + a buffer consisting of 2.0 mM acetic acid and 1.0 mM CsOH (pH = 4.6), sweep rate = 2 mV/s; (E) stripping of two ECALE layers of CdTe from polycrystalline Au in a TLEC. (From Ref. 161.)

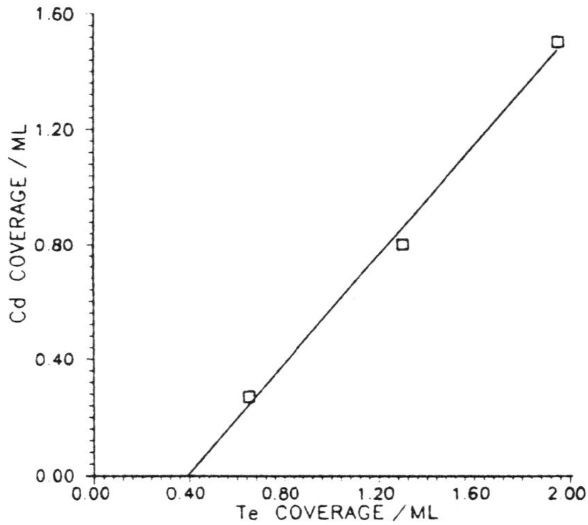

FIG. 8. Plot of Cd coverage vs. Te coverage on a polycrystalline Cu electrode. Slope = 0.95; x-intercept = 0.40. (From Ref. 291.)

free energy of compound formation is considered, the origins of UPD become clear [155]. Figure 9C is a Pourbaix diagram describing the stability of CdTe, where the free energy for CdTe formation has been included in the calculations. Figure 9D is a combination of Figs. 9A, B, and C. The differences in stability between the elements and the compound are clearly evident and can be equated to underpotentials.

Also evident from examination of the Pourbaix diagram shown in Fig. 9B is the equilibrium between Te and Te^{2-}:

$$Te_{(s)} + 2e^- \Leftrightarrow Te^{2-} \tag{4}$$

It was the realization that the equilibrium in Eq. (4) might be used to form atomic layers of Te that allowed development of the first workable ECALE cycle for CdTe. From Fig. 9D it can be seen that Te UPD can be performed oxidatively from a telluride solution:

$$Te^{2-} \Leftrightarrow Te_{(UPD)} + 2e^- \tag{5}$$

The vast majority of previous UPD studies [22,23] have involved "reductive" UPD of a less noble metal on a more noble one "Oxidative" UPD is not unprecedented, however. Some evidence for oxidative UPD has been shown for S [61,294], Se [295], Te [160,161,291], and even O

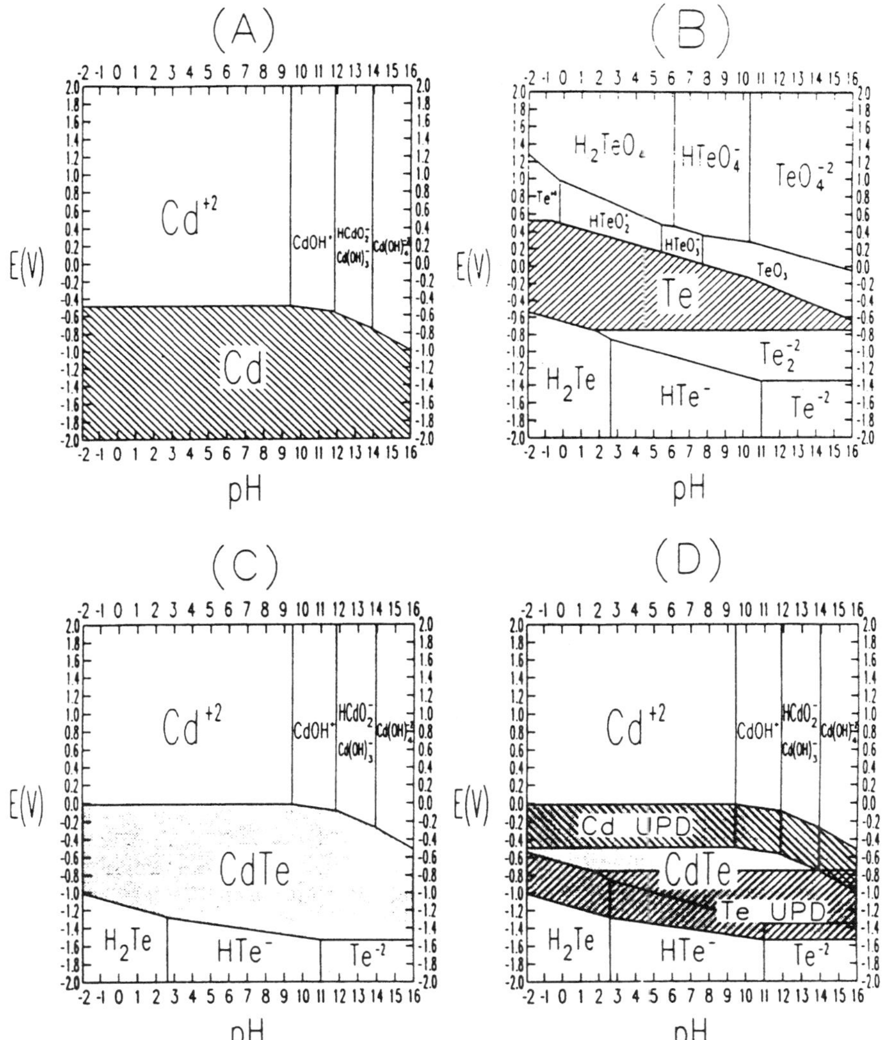

FIG. 9. Pourbaix diagrams describing (A) Cd, (B) Te, (C) CdTe, and (D) the underpotential deposition of Cd and Te on CdTe in water. The diagrams were calculated using an activity of 10^{-3} M for all soluble species. The hatched areas in (D) represent the differences in potentials, UPD, associated with deposition on CdTe as opposed to deposition on the pure elements. (From Ref. 155.)

[296]. In addition, the adsorption of halides from solution on metal surfaces can be thought of as UPD: Cl [297], Br [298–300], and I [299,301–304]. Normally, UPD is considered a precursor to the formation of a bulk deposit of an element. Bulk deposits of the halides are generally soluble, but the first atomic layer is formed at an underpotential. Recent studies have indicated that oxidative UPD of As can be performed as well [252,253].

At present, it appears that one of the criteria for construction of an ECALE cycle for a compound is whether atomic layers of one of the compound's elements can be formed using an equilibrium analogous to Eq. (5): oxidative UPD. As indicated above, the elements that can be deposited using oxidative UPD are mostly nonmetals such as the halides, chalcogenides, and pnictides.

Some telluride compounds, such as K_2Te, are available commercially, however, black specks and/or a deep purple color (dependent on the pH of the solution) are frequently observed when they are dissolved in the presence of traces of oxygen. The black specks are elemental Te, while the purple color has been attributed to Te_2^{2-} ions [305]. The apparent lack of stability of such solutions makes them undesirable for use in an ECALE cycle. The equilibrium shown in Eq. (5) can still be used to form Te atomic layers, however, the deposition is generally run as a stripping experiment. A small amount of bulk Te is first formed from a solution of $HTeO_2^+$, which is more tractable:

$$2HTeO_2^+ + 6H^+ + 8e^- \Leftrightarrow Te_{(UPD)} + Te_{(Bulk)} + 4H_2O \tag{6}$$

The excess (bulk) Te is then reduced off the surface as Te^{2-} ions at a potential just below that where bulk Te is stable:

$$Te_{(UPD)} + Te_{(bulk)} + 2e^- \Leftrightarrow Te_{(UPD)} + Te^{2-} \tag{7}$$

These two steps can then be used to form Te atomic layers in an ECALE cycle that works for the formation of more than one monolayer.

In the first TLEC studies of CdTe formation, a one-step variation on the reactions in Eqs. (6) and (7) was used to form the Te atomic layers [160,161]. In those studies, an aliquot of a 0.3 mM $HTeO_2^+$ solution was first rinsed into the TLEC. Given the low concentration and low volume of the cell, a total of a little over 1 ML of Te was initially present in the cell. A potential of −1.1 V was then applied, and used to reduce most of the $HTeO_2^+$ aliquot all the way to Te^{2-}, while leaving about 1/2 ML of Te on the surface:

$$2HTeO_2^+ + 6H^+ + 10e^- \Leftrightarrow Te_{(UPD)} + Te^{2-} + 4H_2O \tag{8}$$

The resulting Te^{2-} was then flushed from the cell, leaving the Te atomic layer.

Deposition of atomic layers of Te in this way allowed multiple cycles of alternated Cd and Te deposition to be performed, with each cycle producing another monolayer of CdTe. The total amounts of deposited Cd and Te were subsequently determined using stripping coulometry. Figure 7E displays a typical stripping voltammogram after two cycles of CdTe deposition. The peak centered at -0.3 V corresponds to oxidative stripping of the Cd from the CdTe deposit, while the twin peaks at 0.3 and 0.5 V correspond to oxidation of the remaining Te. As the electrodes were cylindrical and difficult to take apart after each deposition in these early studies, the only analysis performed on the deposits was coulometric stripping.

At present, TLEC studies of the ECALE deposition of a number of II-VI compounds have been performed, including CdS [61], CdSe [106], CdTe [155–157,160,161], ZnTe [279], ZnSe [279], and ZnS [279]. In addition, ECALE deposition of the III-V compound GaAs [157,252,253] has been reported. Other compounds that should be amenable to deposition by ECALE include IV-VI compounds such as PbS and the family of ternary compounds exemplified by $CuInSe_2$ [306].

In the rest of this section, studies of the deposition of Zn-based II-VI compounds will be discussed in more depth as an example of how TLECs have been used to develop initial ECALE cycles for a number of compounds. Figure 10 is a series of thin layer voltammograms showing the deposition of the chalcogenides (S, Se, and Te) on polycrystalline Au. Prior to each voltammogram, a series of oxidation and reduction cycles were performed in H_2SO_4 to clean the electrode, followed by recording a voltammogram. Clean electrodes produced voltammetry equivalent to that shown in Fig. 7A. The TLEC was then immersed into the chalcogenide solution and an aliquot was rinsed in. Figure 10A is the voltammogram for an aliquot of $HTeO_2^+$ and displays two reduction peaks, at 0.2 and -0.1 V, corresponding to Te UPD and bulk Te deposition, respectively. The peaks are fairly broad for several reasons, one of which is that the Au electrode is polycrystalline and the oxidation-reduction cycles used to clean the electrode also cause surface roughening [307–309]. In addition, the peaks are broadened by significant changes in the $HTeO_2^+$ activity in the cell during the deposition.

At potentials just below -1.0 V, a reversible couple corresponding to reduction of bulk Te to Te^{2-} [Eq. (7)] is clearly visible. As the Te^{2-} was formed inside the TLEC, it was trapped and quantitatively reoxidized to

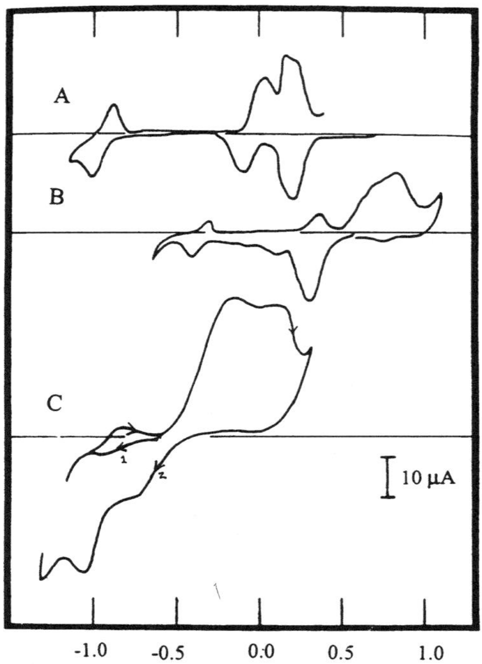

FIG. 10. (A) Clean Au electrode in 0.25 mM TeO_2 + 20 mM H_2SO_4 + 0.5 M Na_2SO_4, pH = 9.2. (B) Clean Au electrode in 1.0 mM SeO_2 + 10 mM $Na_2B_4O_7$ + 1.0 M $NaClO_4$, pH = 8.6. (C) Clean Au electrode in 2.5 mM Na_2S + 0.5 m $NaClO_4$, pH = 11.

Te during the subsequent anodic scan. No corresponding reduction feature for the atomic layer of Te on the Au was observed under any of the conditions investigated, indicating the dramatic stability of the first atomic layer.

An atomic layer of Te can be formed by deposition through the UPD peak in the $HTeO_2^+$ solution (at 0.0 V). The excess $HTeO_2^+$ can then be rinsed from the cell with an aliquot of pure electrolyte. Alternatively, the whole $HTeO_2^+$ aliquot can first be deposited (at -0.5 V), and then the potential can be shifted to -1.0 V to reduce the bulk Te. The resulting

Te^{2-} can be rinsed out of the cell with an aliquot of fresh electrolyte to leave an equivalent Te atomic layer. This latter case is more consistent with the procedure used in an ECALE cycle.

Figure 10B is a corresponding voltammogram for the deposition of Se from a solution of $HSeO_3^-$. The voltammograms in Figs. 10A and B are very similar, with the main differences being that the potentials for Se features are shifted positively from the corresponding features for Te. The largest shift is for bulk Se reduction, which occurs at −0.4 V, over half a volt positive of the corresponding Te feature [Eq. (7)].

As mentioned previously, Te^{2-} solutions are not generally stable, and Se^{2-} solutions are similarly hard to work with. Na_2S solutions, however, are stable and in principle should make the oxidative UPD of S a straightforward process. Figure 10C is the voltammetry of an aliquot of a Na_2S solution. Two cycles are displayed, the first initiated at −1.2 V and scanned up to −0.6 V and reversed. The oxidative UPD of S on Au is evident at −0.8 V. On the second cycle, the potential was scanned to 0.3 V and reveals a very large feature for the oxidation of the remaining S^{2-} in the aliquot, forming bulk S. Above 0.3 V, the current began to increase again as the deposited S converts to sulfate [61].

Voltammetry such as that in Fig. 10 was used to determine potentials for formation of chalcogenide atomic layers as a first step in the development of ECALE cycles for forming the zinc chalcogenides. For instance, the first atomic layer of Te was easily produced by depositing between 0.0 and 0.1 V, while that for Se was formed at 0.2 V, and the S atomic layers were formed oxidatively between −0.6 and −0.7 V. Generally, deposits were given 2 minutes to form before the reactant solutions were flushed from the cell. Several rinses with the corresponding blank solutions were performed after which the $ZnSO_4$ solution was introduced.

Figure 11 is a series of voltammograms for the deposition of Zn on atomic layers of Te, Se, and S. A definite trend in the Zn UPD peak potentials is evident, going up the periodic table. Zn is hardest to deposit on the Te atomic layer, where deposition is not initiated until −0.7 V. A well-defined Zn UPD peak is evident on the Se layer, initiated near −0.5 V, while Zn deposition on the S atomic layer begins near −0.3 V. These numbers are consistent with differences in the free energies of formation of the three compounds: −115.2, −173.6, and −200.0 kJ/mole for ZnTe, ZnSe, and ZnS respectively [310]. For a two-electron process, these differences in the stabilities of the compounds correspond to 0.30 V and 0.14 V, respectively, in line with the shifts observed in Fig. 11.

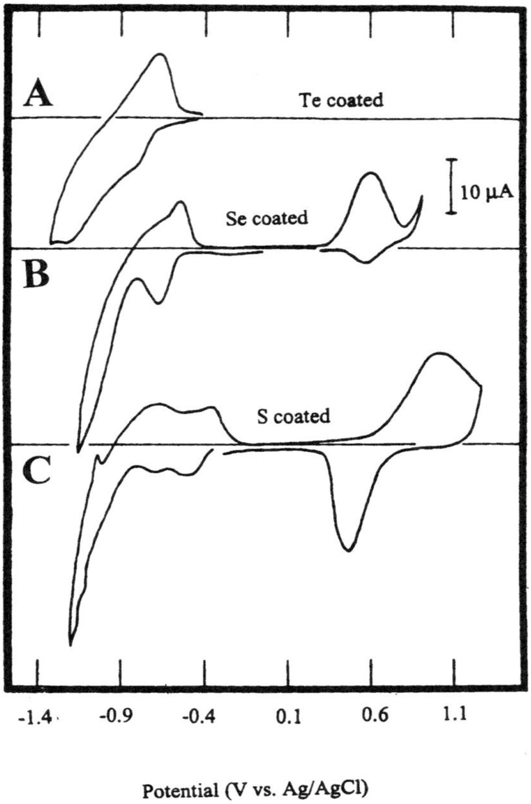

FIG. 11. Zn deposition on a Au TLEC coated with an atomic layer of (A) Te, (B) Se, and (C) S.

Subsequent studies of Zn chalcogenide deposition, using the TLEC, involved coulometric stripping of deposits to characterize elemental coverages/cycle as a function of cycle conditions, specifically deposition potentials and solution compositions. Those experiments proved tedious, each cycle requiring about 12 steps (Fig. 12), but worthwhile. Figure 13 displays stripping curves for deposits of ZnTe, ZnSe, and ZnS, each formed from four ECALE cycles. The same trend in Zn stability observed in the deposition scans (Fig. 11) is seen in Fig. 13; i.e., Zn is easier to strip from

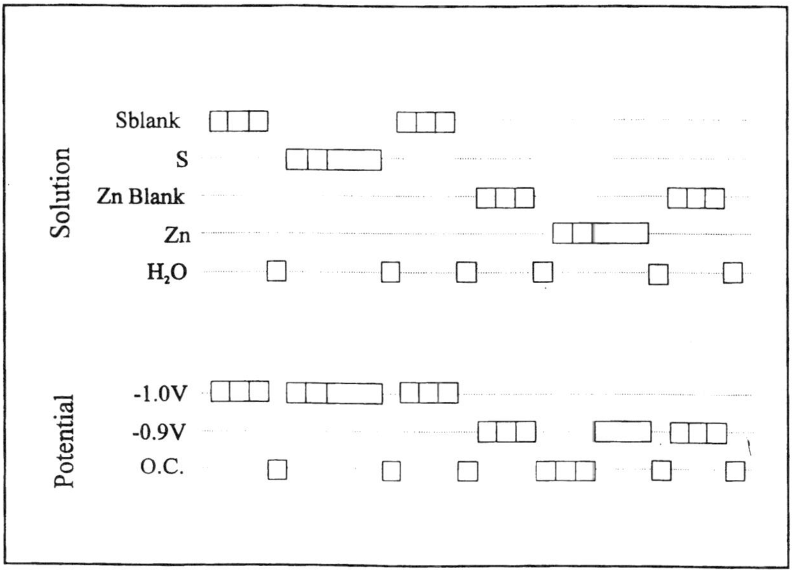

FIG. 12. Diagram of steps involved in a single ECALE cycle using a TLEC for the deposition of ZnS.

ZnTe than from ZnSe than from ZnS, in line with the free energies of formation. Examination of Fig. 13 also reveals that there are generally two Zn-stripping features in each scan. This is seen most clearly for ZnTe (Fig. 13A), where the low potential peak (−0.7 V) grows in first as a function of the number of cycles performed but does not significantly increase in size after the second cycle. Instead, the higher potential stripping feature (−0.3 V) starts to grow. As the number of cycles is increased further, this second peak grows and shifts to still higher potentials. A simple explanation would be that the first peak represents the top layer(s) of Zn atoms, those not completely coordinated to Te, while the second peak results from stripping the more highly coordinated interior Zn atoms. Studies where the first peak is stripped and then the potential is held for 5 minutes, just prior to the potential needed to strip the second peak, evidenced the full second stripping feature, when the stripping scan was resumed. This may suggest a thermodynamic effect, as opposed to a problem with the kinetics of transporting interior Zn atoms or ions out of the Te matrix.

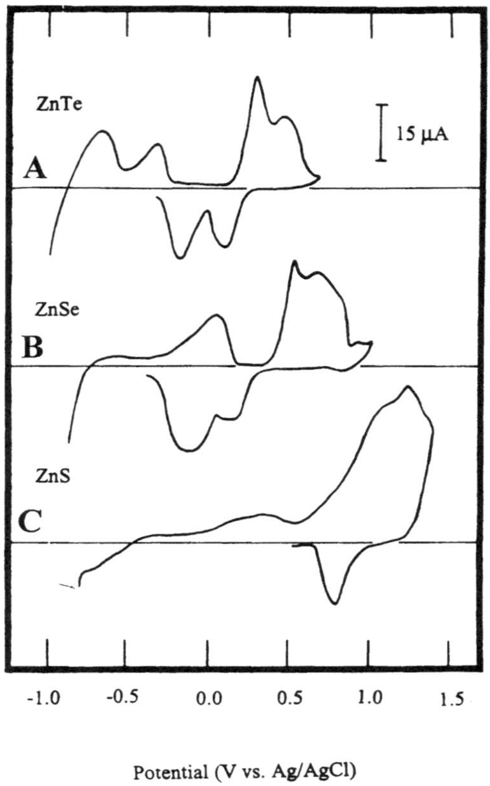

FIG. 13. Voltammetry showing the stripping of deposits of (A) ZnTe, (B) ZnSe, (C) ZnS. Each deposit was the result of four ECALE cycles.

Another trend in Fig. 13 involves the size of the valley between the Zn and chalcogenide stripping features, where that for ZnTe > ZnSe > ZnS. The width of the valley in each case shrinks as the number of cycles is increased due to the tendency of the Zn-stripping feature to broaden to higher potentials as the coverage increases. As the Zn feature continues to broaden, it eventually runs into the chalcogenide stripping feature, and it becomes difficult to independently determine the coverages of the two elements coulometrically. In the present studies, clear problems were encountered after 16 cycles of ZnTe, after 10 cycles of ZnSe, and after only 4 cycles of ZnS (Fig. 13).

In the next series of graphs, the number of monolayers of an element deposited per cycle is plotted as a function of the potential used to deposit one of the compound's elements, with other cycle conditions held constant. The coverage nomenclature used here can be a little confusing. The discussion is based on a simple model for the deposition of the elements on a polycrystalline Au substrate. The model states that it takes about 1/2 ML of each element to form one monolayer of the compound. That 1/2 ML of each element should be deposited each cycle is based on the idea that a compound monolayer is equivalent to a half unit cell thick slab of the compound in the zinc blende structure. This model is very simplistic and is based on a large number of assumptions. In addition, the coverages should not be taken too literally either, as their measurements were made with a number of assumptions, such as the actual number of Au surface atoms and the background correction used in the coulometry. Given the above statements, the data provide an idea of the trends in growth that can be expected for a given variable and of the conditions that might be used for subsequent flow cell studies, such as are described in the next section.

Figure 14 is a graph of the Zn and S coverages per cycle resulting from ZnS formation. Each point was determined after execution of three ECALE cycles, as a function of the S deposition potential. The Zn-deposition potential was held constant at −0.9 V. The Zn and S curves clearly parallel each other but do not coincide well. One possible explanation for the deviation is that there was a problem calculating the elemental coverages. Recall Fig. 13C, where the Zn-stripping feature overlapped significantly with the S oxidation feature after only four cycles. If some of the Zn was, in fact, stripping with the S feature, the Zn coverages would be seen as lower than expected. There would, however, be little change in the calculated S coverage because S oxidation to sulfate is a six-electron process:

$$S_{(UPD)} + 4H_2O \Rightarrow SO_4^{2-} + 8H^+ 6e^- \tag{9}$$

while Zn oxidation is a two-electron process. A second explanation for the discrepancy would be that there is something about the structure of the initial few monolayers that is not adequately explained as being 1/2 ML of S on 1/2 ML of Zn. Regardless of the discrepancy between the Zn and S coverages in these initial TLEC studies, the trends appear valid. Ideal ALE behavior is frequently represented by an S curve [17–21], i.e., for deposits formed as a function of a given variable, at one extreme the coverage will be too high, while at the other extreme the coverage will be too

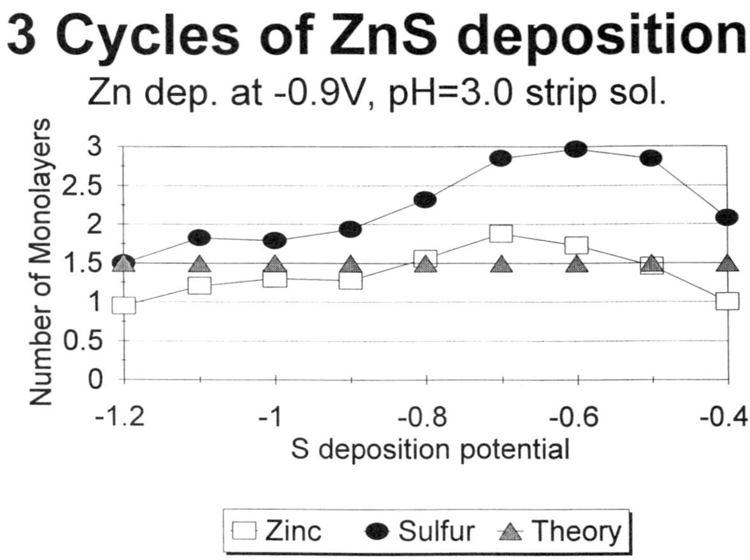

FIG. 14. Coverages of zinc and sulfur per cycle, after three ECALE cycles, as a function of the potential used to deposit sulfur. The zinc-deposition potential was held constant at −0.9 V.

low. In between, ALE behavior is frequently signified by the presence of a plateau, where the coverages are independent of the particular variable and the growth is truly controlled by surface-limited reactions alone. In Fig. 14, at the most negative potentials, below −1.1 V, insufficient S is being deposited with each cycle, and because the Zn requires the presence of S to deposit, both coverages begin to drop. At potentials above −0.9 V, both coverages start to increase as some bulk S begins to deposit. Those increases, however, start to drop off at potentials above −0.7 V because it becomes difficult to hold the Zn on the surface while the S is depositing. The conclusion drawn from this study is that S should be deposited at potentials between −1.1 and −0.9 V, where there is a small plateau corresponding to surface-limited control over the deposition.

Given the results shown in Fig. 14, the Zn potential dependence was investigated while holding the S deposition potential at −1.0 V (Fig. 15). The dependence on the Zn potential is not dramatic, but it again shows some S-curve behavior. At low potentials the Zn coverage builds up as the

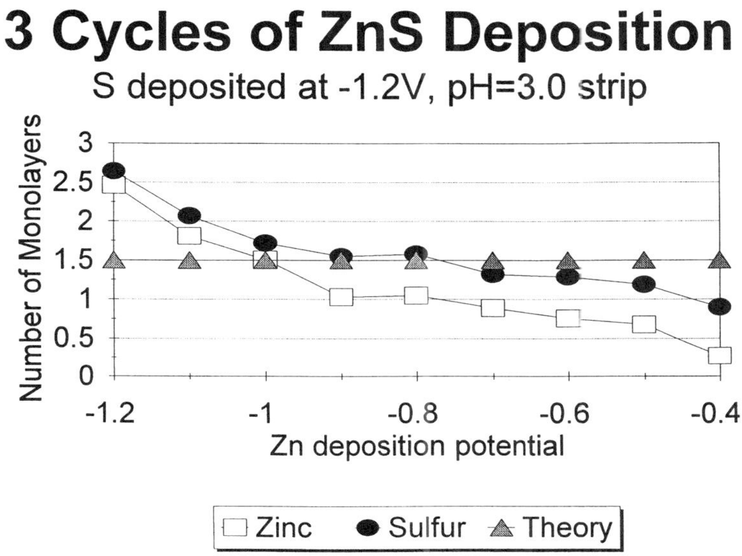

FIG. 15. Zinc and sulfur coverages per cycle, after three ECALE cycles, as a function of the potential used to deposit zinc. The sulfur-deposition potential was held constant at −1.00 V.

bulk deposition potential is exceeded, and the corresponding S coverage stays one for one. At higher potentials, the Zn coverage falls off, followed by the S coverage. A Zn potential of −0.9 V was used in Fig. 14, resulting in significant differences between the Zn and S coverages at all potentials. It appears from Fig. 15 that use of −1.0 V for Zn deposition may result in deposits closer to the desired 1/2 ML per cycle of each element.

Figure 16 is a graph of Zn and Se coverages per cycle for ZnSe deposits as a function of the Zn potential. The Se deposition was carried out by first depositing two monolayers at −0.9 V and then reducing off the excess at −0.9 V. The drop in coverage above −0.8 V is due to decreased stability of the Zn (Fig. 11). A plateau in both the Zn and Se coverages is evident between −1.2 and −0.9 V, however, the Zn coverage per cycle is nearly 3/4 ML, while the Se remains at 1/2 ML. The standard potential for Zn deposition is about −1.0 V (vs. Ag/AgCl), and given that a mM solution of $ZnSO_4$ was used, bulk deposition would not be expected until −1.1 V. The reason for the disparity between the Zn and Se in the

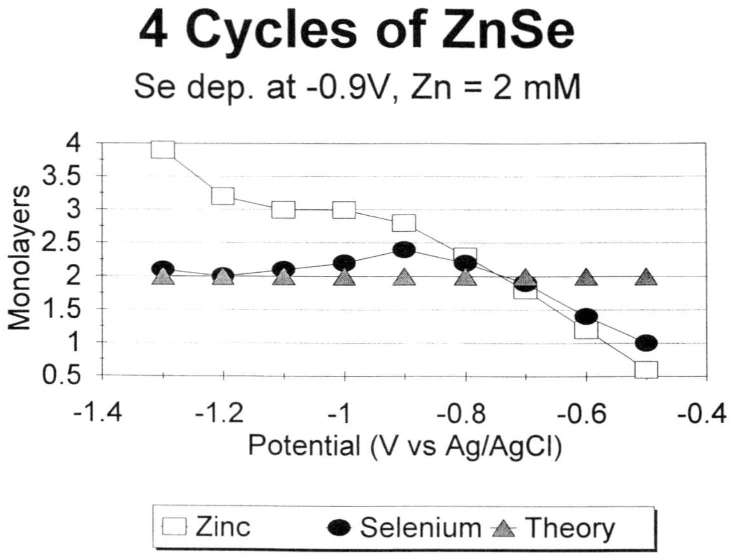

FIG. 16. Zinc and selenium coverages, per cycle, after four ECALE cycles, as a function of the potential used to deposit zinc. The Se atomic layers were formed by first depositing a couple of monolayers of Se at −0.9 V and then reducing off the excess at −0.9 V in the corresponding blank electrolyte solution.

plateau region is not clear. It is interesting to note that in Fig. 16 it is the metal coverage that is high, not the chalcogenide, as was observed for ZnS (Fig. 14). It is probable that the conditions used to form the Se atomic layers were not optimized and needed more study. Given the data in Fig. 16, however, the optimal Zn deposition potential appears to be between −0.9 and −0.7 V, with −0.8 V looking best.

Figure 17 is a graph of the Zn and Te coverages per cycle as a function of the Zn deposition potential. Atomic layers of Te were formed by first depositing a little more than a monolayer of Te at −0.8 V and reducing off the excess Te at −1.1 V in a blank electrolyte solution. The Te coverage per cycle appears low, about 1/3 ML per cycle at all potentials. This is not the expected 1/2 ML behavior and indicates that the dependence of Te coverage on its deposition conditions needs more study. There is a short plateau between −1.1 and −1.0 V, similar to that seen in Fig. 16 for ZnSe formation, where again the Zn coverage is significantly higher than the

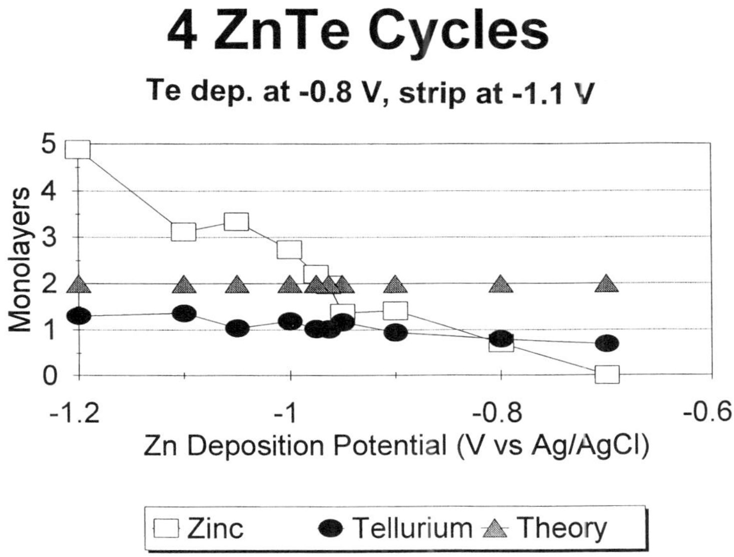

FIG. 17. Zinc and tellurium coverages, per cycle, after four ECALE cycles, as a function of the Zn-deposition potential. Te atomic layers were formed by initial deposition of a couple of ML of Te at −0.8 V, followed by reductive dissolution of the excess Te at −1.1 V.

corresponding chalcogenide. However, between −0.95 and −0.8 V the coverages per cycle for Zn and Te coincide at about 1/3 ML per cycle. Above −0.8 V, the coverage of Zn falls off as expected. It is not clear why the Te coverage remains high at −0.7 V when the Zn coverage is decreasing. It does indicate a need to better understand the Te atomic layer formation portion of the ECALE cycle. This problem with the Te atomic layers is probably responsible for the low CdTe coverages observed with the automated system (discussed in the next section). Given the data in Fig. 17, a Zn-deposition potential of near −0.9 V is probably a good starting point for subsequent studies.

Finally, given the starting conditions determined from Figs. 14 to 17, the dependence of coverages on the number of cycles performed was studied (Fig. 18). A linear dependence is expected for an ALE process; twice the number of cycles should result in twice the coverage. Linear behavior

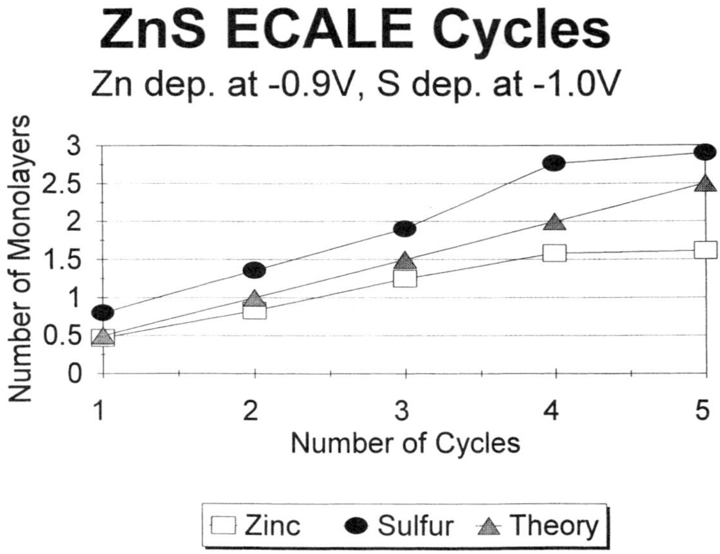
ZnS ECALE Cycles
Zn dep. at -0.9V, S dep. at -1.0V
Number of Monolayers
0
0.5
1
1.5
2
2.5
3
1
2
3
4
5
Number of Cycles
Zinc
Sulfur
Theory

(A)

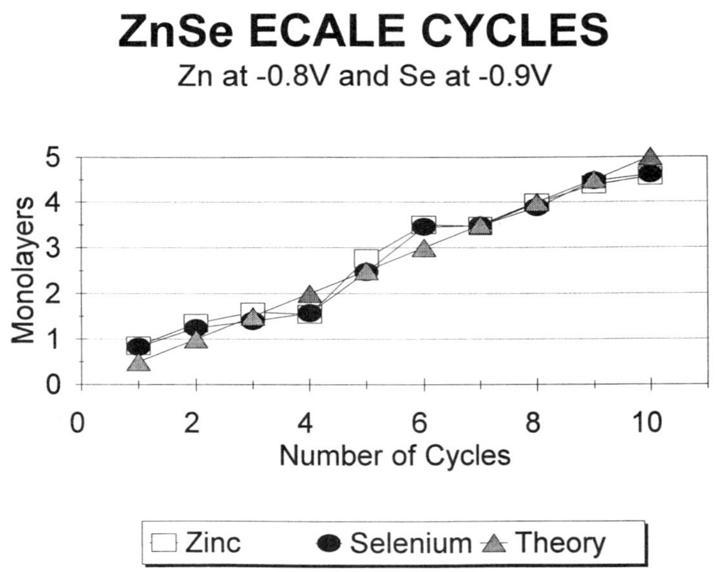
ZnSe ECALE CYCLES
Zn at -0.8V and Se at -0.9V
Monolayers
0
1
2
3
4
5
0
2
4
6
8
10
Number of Cycles
Zinc
Selenium
Theory

(B)

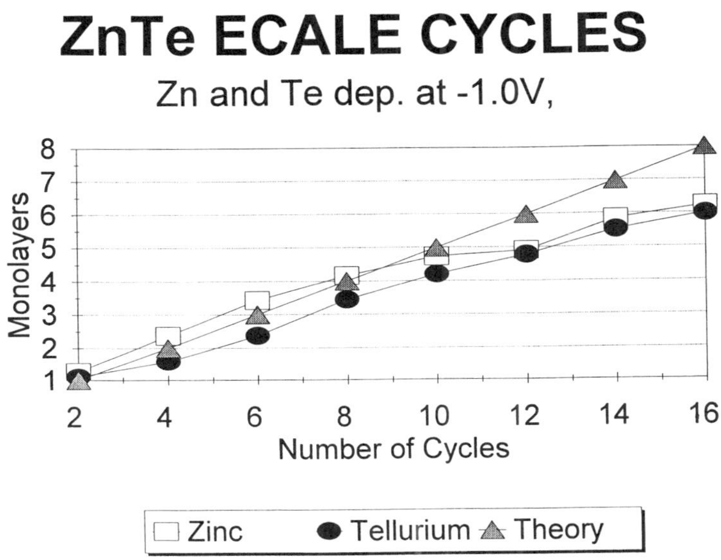

(C)

FIG. 18. Graphs of total coverage determined coulometrically as a function of the number of cycles: (A) ZnS, (B) ZnSe, and (C) ZnTe.

is evidenced in each of the graphs in Fig. 18, given the error inherent in the measurements. The slopes in the graphs of ZnSe and ZnS formation are very close to the expected 1/2 ML per cycle. In the case of ZnTe the slope is a little low, in line with the low Te atomic layer coverages discussed previously with regard to Fig. 17. In each graph there is some negative deviation from the line at the higher number of cycles. This may be due to problems in quantifying stripped amounts as the peaks start to overlap. It may also be a problem with operator error, as each cycle involves 12 steps (Fig. 12), and the more steps performed, the more chances for mistakes.

The deviations from the ideal evident in the studies above may be very important and real, or they may be a result of problems with the operator, with quantification, or with the experimental set-up. They do, however, provide a direction for an initial ECALE cycle to be used in the next stage: an automated deposition system to form thin films that can be analyzed with techniques other than coulometric stripping.

III. THIN FILM FORMATION USING ECALE

In considering the limits to the quality of deposits that may be formed electrochemically, a number of issues come up, including the quality of the reagents used: reactants, electrolytes, solvents, and substrates. Most of the reagents presently used in the formation of thin films of these same compounds by other commercial methods can be used in the formation of thin films electrochemically. Some of the same substrates used to form deposits via CVD or MBE can also be used for electrodeposition. Reactant precursors used in an electrochemical method are generally ionic and thus differ from the volatile species used in a metal organic chemical vapor deposition (MOCVD) reactor, but the absolute amounts used in the two cases can be very similar. In addition, if electronic grade metals are the basis for the CVD precursors, then the same grade metals can serve as the basis for the ionic precursors used in an electrodeposition method, generated by electrochemical dissolution.

Another frequently raised concern about purity involves the fact that electrodeposition takes place in a condensed phase, with a solvent in contact with the substrate and deposit. The solvent used for the present studies is water, and it is used copiously in the processing of compounds and devices. The electronics industries are well aware of how to obtain very high-purity water. The point is that the purity issues in an electrodeposition method are the same issues being addressed in presently used methodologies. There does not appear to be anything inherently dirty about electrodeposition.

A second factor controlling the absolute quality that can be achieved electrochemically is the mechanism. As discussed in the introduction, there may be fundamental flaws in the precipitation and codeposition methodologies (although, as mentioned, there have been some surprisingly good-quality films produced recently by codeposition [38,39]. In an ideal ECALE cycle, each element should be deposited under equilibrium conditions from a separately optimized solution. In addition, the deposition process can be thought of as decomposed into a series of steps, each step being a variable in the cycle and thus an adjustable parameter. This provides a large parameter space to work with, and the cycle can serve as a window into the mechanisms of compound electrodeposition in general. Most other electrodeposition methodologies are more limited (Table 1).

The logistics of forming thin films using ECALE revolve around alternating the solutions and potentials in a cycle. As stated earlier, manually forming deposits with much over 10 cycles proved tedious. Some work in this area has thus necessarily focused on development of an automated deposition system, where each cycle can be performed reproducibly.

Initially, a thin layer flow cell (Fig. 19) was used in this group to study the ECALE formation of compounds [158] and in studies of electrochemical digital etching [312,313]. Wei and Rajeshwar [130] used a flow cell system to deposit compound semiconductors as well, however, the major intent of that study was to form superlattices by modulating the deposition of CdSe and ZnSe. Their study appears to be the first example of the use of a flow electrodeposition system to form a compound semiconductor superlattice.

Besides the electrochemical flow cells just mentioned, other hardware for sequential solution-deposition scenarios has been developed to form

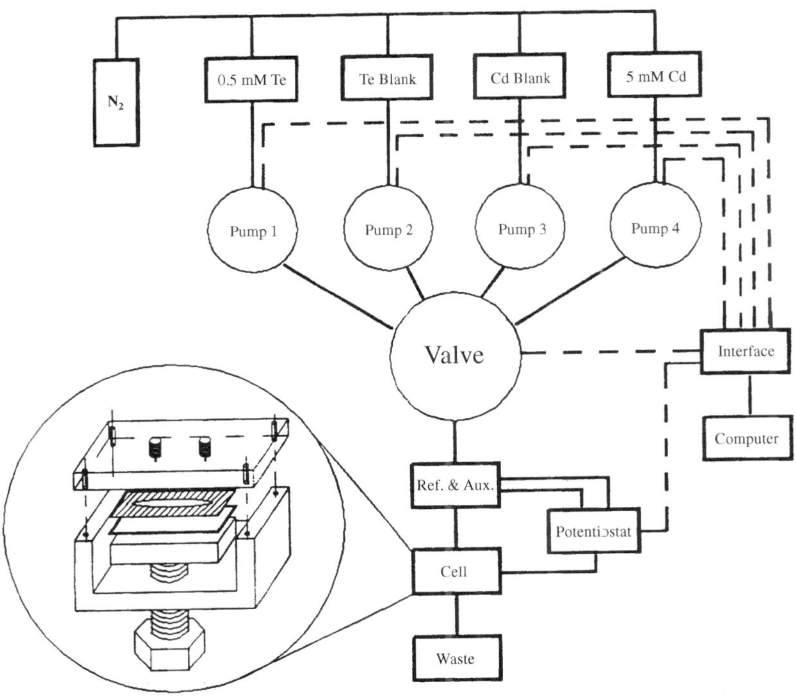

FIG. 19. Automated flow electrodeposition system, initial design. Thin layer cell design.

deposits by the SILAR method (Fig. 2) described in the introduction [28–32]. Related hardware has also been used by workers concerned with the construction of magnetic superlattices of Cu and Ni [314,315]. One method for forming Cu-Ni superlattices is referred to as the two-bath method, which involves using separate baths for depositing the individual layers and alternately exposing the substrate to the two baths. The two bath instruments were not designed to deposit atomic layers necessarily, but alternate 1-nm or thicker films of Cu and Ni. One particularly interesting two-bath instrument made use of a rotating deposition cell, where the substrate was rotated over two distinct deposition zones with rinsing stations in between. Thus, in a single cycle, a full period of the superlattice was produced [314]. An analogous system might be made to work for the formation of deposits using ECALE if sufficient control over deposition potentials and rinsing conditions can be achieved.

As mentioned above, initial work in automating ECALE involved use of a thin-layer flow cell-deposition system. A thin layer cell was chosen because it allowed very small amounts of solution to be used in each step, minimizing the total volume of solution used to form a deposit. Development of an automated thin layer flow deposition system began early on and proceeded by a long sequence of incremental improvements (Fig. 19) [158].

The hardware consisted of a set of low-volume pumps attached to Pyrex glass solution reservoir. Two solution reservoirs were used for each element—one for the reactant precursor and one for a corresponding blank solution—to rinse the system. The pumps fed a valve, which selected the solution to be sent to the cell. The cell consisted of two plates (Fig. 19) between which the substrate and a gasket were sandwiched. A small slot cut in the gasket material was used to define the deposition cavity. A variety of gasket materials was used, including Teflon, silicone rubber, and Viton rubber. Teflon is inert but does not seal as well as the other two materials. The cell plates were compressed in a steel vice, reproducibly tightened using a torque wrench. The plates were plumbed in a number of different patterns, but were generally tapped 1/4″−28 so that conventional low-pressure Peek fittings and tubing could be used. The basic design involved solution flowing in one side of the cavity and out the other. Additionally, a small compartment housing the reference and auxiliary electrodes was attached. Several positions for the compartment were investigated, however, it was generally placed downstream of the cell so that products from the auxiliary and leakage from the reference would not encounter the substrate. The major drawback to the downstream placement of the compartment appears to be IR drops associated with the thin layer. A second position for the compartment was in the center of the slot in the face plate. This position cut down on IR drop problems, but required a porous Vycor plug be inserted into the face plate. The plug may have resulted in some cross-contamination from ions trapped in the Vycor.

A typical program used in forming these deposits is shown in Fig. 20. The top of the figure shows the aliquots of solution being introduced into the flow cell and times involved. The bottom half of the figure shows the potentials applied to the cell and the times. In Fig. 20, the initial step was to introduce five aliquots of the Te deposition solution while the potential was at open circuit. Open circuit was used during the introduction so that the Te did not deposit differentially across the substrate surface, such as at the cell entrance. The Te-deposition solution was then held in the cell at −0.8 V for 60 seconds, while all the Te was deposited. Open circuit was then used again to introduce the Te blank solution so that excess Te solution in the tubing could be flushed through the cell without it depositing differentially across the surface. Next, the Te blank was held quiescent in

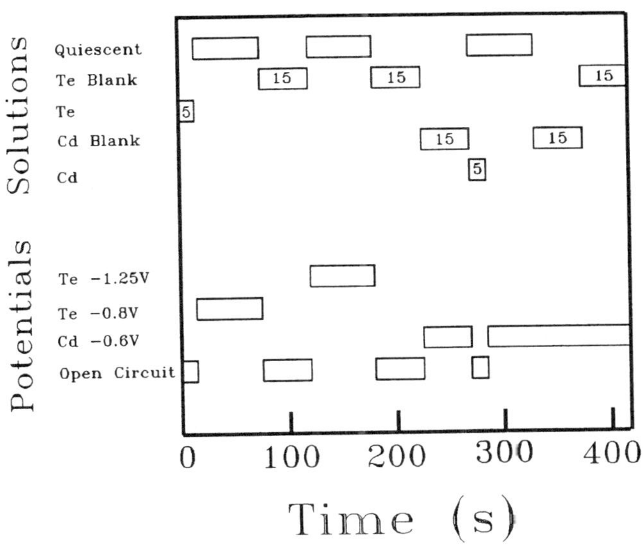

FIG. 20. Program of potentials and solutions amounts in one ECALE cycle for CdTe deposition. (From Ref. 44.)

the cell at -1.25 V, while excess Te was reduced to Te^{2-}. The resulting Te^{2-} was flushed from the cell with more Te blank, again at open circuit. Next, the Te blank was then replaced with a Cd blank at -0.6 V, and then five aliquots of the Cd^{2+} solution were introduced at open circuit. Finally, the Cd^{2+} solution was held quiescent in the cell for 60 seconds at -0.6 V and then flushed from the cell using the Cd blank and then the Te blank to end the cycle.

The principle of ALE is that deposits are formed one atomic layer at a time using surface-limited reactions. That appears to be the case with ECALE. The name electrochemical atomic layer epitaxy, however, suggests that the deposits should be epitaxial, and the first 100 cycle deposits appeared anything but epitaxial (Fig. 21a). These initial deposits looked like they were composed of a large number of particles, which fell out of solution.

There turned out to be several reasons for the particles, the first being conproportionation. As described in the introduction, conproportionation has been a problem with the codeposition of Se-containing compounds for a long time [Eq. (2)] [122,132], as the deposition potentials used were sufficiently negative to result in some Se^{2-} formation. As both Se^{2-} and $HSeO_3^-$ were present in solution at the same time, conproportionation resulted in the formation of traces of elemental Se in the deposits. Since Te is reduced to Te^{2-} at a significantly more negative potential than Se is to Se^{2-} (Fig. 10), conproportionation was not a significant problem during codeposition studies of CdTe (Table 1).

The initial program used to deposit the CdTe film shown in Fig. 21a involved a couple of cycle steps that were different than those shown in Fig. 20. Initially, the $HTeO_2^+$ aliquot was held at a potential where the atomic layer of Te was formed directly. That is, a more negative deposition potential was used so that the atomic layer would form and all the excess $HTeO_2^+$ would then be reduced to Te^{2-} in a single step [Eq. (8)]. That procedure, however, resulted in both $HTeO_2^+$ and Te^{2-} being present in the cell at the same time and, thus, conproportionation. It appears that a large number of the particles shown in Fig. 21a were the result of forming Te nuclei by conproportionation, which subsequently settled out on the surface and were coated with successive monolayers of CdTe.

The solution to the conproportionation problem was adoption of the program shown in Fig. 20, where the Te atomic layers were deposited using the two steps described in the last section: initial deposition of several atomic layers worth of Te at a potential sufficiently positive that no Te^{2-}

FIG. 21. SEM micrographs of two deposits, each formed with 50 cycles and the same deposition hardware, but with differing ECALE programs. The conditions are as follows: (a) deposit atomic layer of Te at -1.25 V directly; (b) deposit bulk Te at -0.8 V first, followed by rinsing out excess $HTeO_2^+$, and subsequently stripping off the excess Te in the corresponding blank. (From Ref. 158.)

is formed [Eq. (6)] and then reductive dissolution of excess Te to Te^{2-} at a second, more negative potential in the Te blank solution [Eq. (7)]. The result of this program change was formation of the deposit shown in Fig. 21b for 100 cycles.

In addition to conproportionation, there were indications that some of the particles shown in Fig. 21a resulted from cross-contamination. There were several problems with the pumps, valves, and cell design, which appear to have resulted in some cross-contamination. Homogeneous precipitation of CdTe should result any time Te^{2-} ions mix with Cd^{2+}. The most significant cross-contamination problem came from a solenoid-actuated Teflon distribution valve used to select the solution sent to the cell. The internal construction of the valve (Fig. 22a) resulted in a significant amount of intermixing. That valve was subsequently replaced by a rotating distribution valve (Fig. 22b), which eliminated the intermixing. After several changes in valves, pumps, and plumbing, cross-contamination was decreased significantly, as can be seen from the deposit shown in Fig. 23. Figure 23 shows a deposit made with 100 cycles in a system using the rotating distribution valve and a program very close to that shown in Fig. 20. There are still a number of obvious defects and particles in the deposit, a majority of which can be attributed to the commercial polycrystalline Au foil used in the early studies, which had a very high defect density to begin with.

The decision as to which substrates to use is ongoing. The surface chemistry of Au in solution is relatively well known. The Au foil was rugged and could be reused—the problem was the surface finish. Polishing, etching, and electropolishing were all investigated but resulted in too much variation from piece to piece. There are numerous reports in the literature that high-quality Au surfaces can be formed on mica [316–319], and studies have been performed. The problem is that the Au would frequently lift off during long runs due to the large number of rinses performed. The best solution, so far, has been the use of polished Si(100) wafers coated with Au with a thin layer of Ti to promote adhesion. Switching from foil to these wafers greatly enhanced reproducibility, because variations in the substrate were no longer a significant factor. The substrates are not perfect, but although they affect the quality of the deposit, under visual inspection the resulting deposits appear reproducible.

The images in Fig. 24 were all run on Au on Si wafers using the thin layer flow cell deposition system. Figure 24A is an optical micrograph of one of the best deposits formed with the thin layer flow cell system.

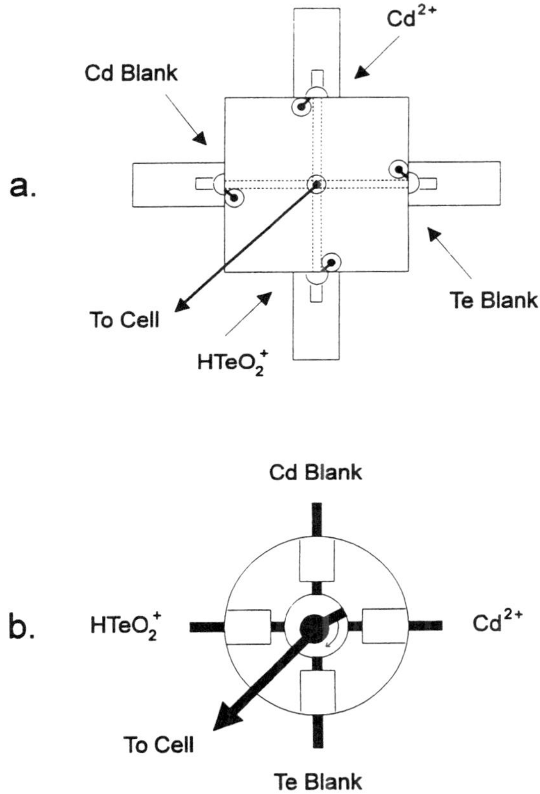

FIG. 22. Cross section of the distribution valves: (a) the sclenoid driven valve (b) the inert rotatable valve. The former has a design that can cause cross-contamination. (From Ref. 158.)

In addition to the problems discussed above, several others were associated specifically with using a thin layer flow cell–deposition system, including edge effects (Fig. 24B), effect of bubbles (Fig. 24C), and IR drop problems. The bubbles appeared to result from small amounts of gas that either leaked in through the fittings, became entrained in the solution, or were dissolved in the solutions to begin with. The bubbles appeared during the deposition and significantly interfered with the fluid flow (Figs. 24B and C). The edges of the gasket also cause problem with fluid flow, which

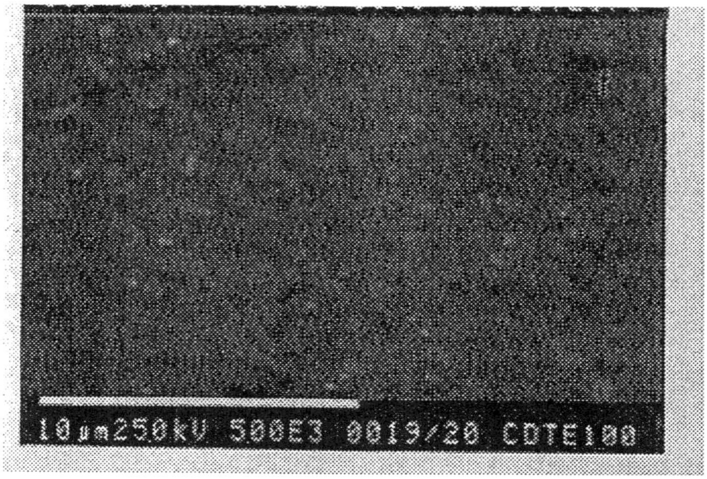

FIG. 23. An SEM micrograph of a deposit formed using 100 cycles and the new hardware.

showed up as decreased deposition near the deposit edges (Fig. 24C). IR drop problems in the cell were always suspected, due to the dimensions of the cavity. Measured values of the IR drop were around 1 kΩ, depending on the configuration and solution composition. However, to investigate the importance of the IR drop, deposits were formed with the auxiliary/reference electrode compartment positioned in the center of the cell faceplate and both at the outlet and the inlet with no obvious change in the deposits.

Overall, the main problem with the thin layer flow cell studies was reproducibility. It was nearly impossible to produce equivalent deposits 2 days in a row. The investigations became an endless series of tests to figure out how to improve the substrate, get rid of bubbles, get the right gasket material, and deal with possible IR drop problems. Although some problems were solved and some deposits were formed, very little progress was made toward a better understanding of the mechanism of compound electrodeposition, or ECALE [158].

The best solution to the bubble, edge, and IR drop problems was to switch to a larger cell (Fig. 25) to get rid of the thin layer configuration (Fig. 19). The two obvious drawbacks are that more solution is needed for every rinse and that potential control is lost between each rinse, using the

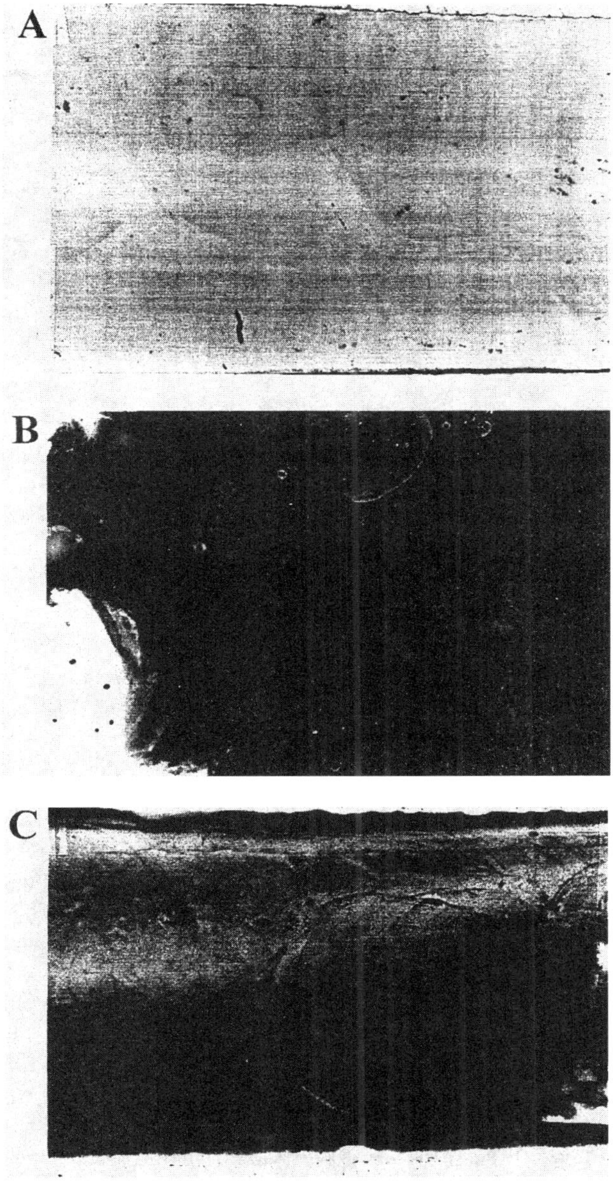

FIG. 24. Optical micrographs of CdTe deposits formed using the thin layer flow cell, with 200 cycles: (A) high-quality deposit; (B) effect of bubbles in the cavity; (C) edge effects and some bubble effects.

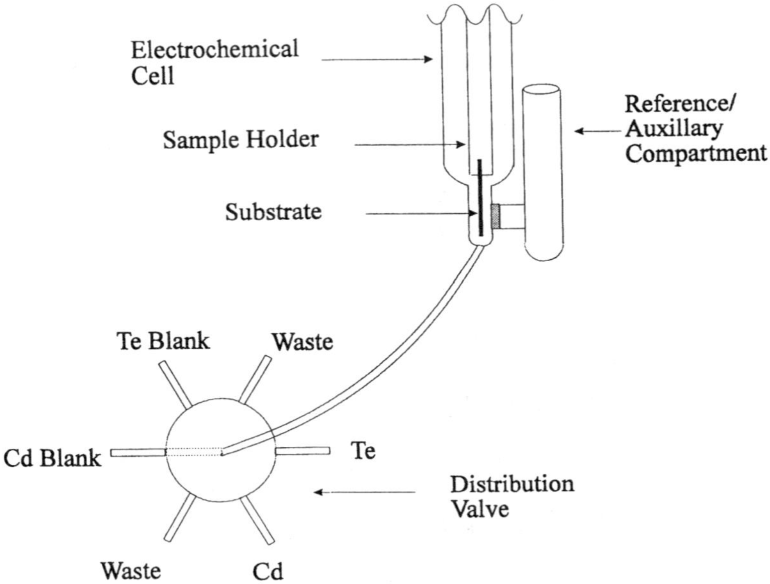

FIG. 25. Thick layer deposition cell for automated system.

cell design shown in Fig. 25. The solution volume problems have not been addressed as yet. So far, the problem with loss of potential control with each rinse does not appear to be a big one. However, there are several cell designs being considered that allow potential control during solution exchange. The design in Fig. 25 is simple and straightforward, consisting of an H-cell with a small rectangular portion at the bottom for the substrate. The compartment for the reference and auxiliary electrodes is attached across from the deposit via a fine glass frit. At the bottom of the H-cell the distribution valve is attached using capillary tubing. The valve is used to both fill and drain the cell. Having the frit in the cell is somewhat problematic, in that some ions get held up in the frit, resulting in cross-contamination. At present this does not appear to be a major problem. The rest of the hardware is very similar to that used for the thin layer flow cell system (Fig. 19). Each solution reservoir has a separate pump feeding into the distribution valve. The computer is used to select the solution to be delivered via the pumps and distribution valve. The control program is essentially the same as that shown in Fig. 20, except that the aliquot sizes are larger.

Immediate benefits of switching to the larger-volume cell (Fig. 25) include the formation of larger-area deposits, 2–3 cm^2, versus the 0.1–0.2 cm^2 deposits obtained using the thin layer flow cell. The thicker layer of solution eliminated problems with IR drops, bubbles, and edge effects except at the top of the deposit. Some nonuniformity at the top of the deposit resulted from small variations in the levels of different solutions pumped into the cell. Rinse solution levels, for instance, were always raised to a slightly higher level than the reactant solutions in order to make sure that the reactants were completely removed after each deposition.

Given reproducibility from run to run, the dependencies of deposit composition and structure on various ECALE cycle variables were investigated, including deposition potentials, deposition times, temperature, solution compositions, rinsing procedures, substrate dependence, annealing, and photoeffects. CdTe serves as a test compound in these studies, as it was the first compound studied by this group and it is the compound about which the most has been learned. Initial TLEC studies of CdTe formation, similar to those described in the last section for the formation of the Zn chalcogenides, involved manually performing 1–10 cycles and stripping the deposits to determine the total amounts of Cd and Te [160,161]. Some thin films were formed, as well, using the thin layer flow cell–deposition system described above [158]. In addition, UHV-EC studies of the formation of the first few monolayers of CdTe on Au electrodes were performed [159,162,320].

One of the most significant benefits of having these larger deposits is that an array of analytical techniques can be used to investigate the deposits. The first analysis, however, is always made visually. Visual inspection of deposits formed with 50–200 cycles can be used to get a clear idea of the homogeneity. A good deposit will have a homogeneous color, except at the very top (discussed above) and on some deposits at the bottom, where drops of solution sometimes hang up. These interference colors are quite striking, and a very good indicator of deposit thickness. The deposits with the H-cell are visually evident after only 15 cycles, brown after 50, mauve after 100, purple after 150, and dark blue after 200 cycles. Our standard deposits are formed with 200 cycles, and the dark blue deposits appeared to have the best quality formed under these conditions.

The potential used to deposit Cd in the formation of CdTe is an important factor in determining deposit morphology and composition. Figure

26A is a graph of the relative coverages of Cd and Te in deposits formed using different Cd-deposition potentials. The coverages were measured using electron probe microanalysis (EPMA). Absolute coverages by this technique are difficult to determine, as the thin CdTe films constitute only a small portion of the sampled volume. The 30 kV electron beam stimulates x-ray emission from up to several μm deep in the substrate. However, for the thin films, the Cd and Te signals are directly proportional to their coverages, having been corrected for the elemental sensitivity factors determined using a single crystal of CdTe. A sigmoid curve is apparent in Fig. 26, suggesting that reactivity over a large potential range is controlled by a surface-limited reaction. Figure 27 displays the stoichiometry of the deposits, the Cd/Te ratio, as a function of potential. At −0.5 V and above, the coverage of Cd and Te are down, as well as the Cd/Te, indicating the potential is too positive for the Cd to deposit. Between −0.55 and −0.65 V, there is a short plateau in the coverage (Fig. 26A), where the Cd/Te ratio is 1. The ratio remains 1 up to −0.75 V (Fig. 27), even though there is a significant increase in both the Cd and Te coverages between −0.65 and −0.75 V. Between −0.75 and −0.85 V, the coverages continue to rise, with the Cd rising faster than the Te and the Cd/Te ratio becoming 1.5. Below −0.85 V the coverage of Cd increases still more rapidly, as bulk Cd is deposited.

X-ray diffraction (XRD) has also been used to study the deposits used in Figs. 26 and 27. Figure 28 is a set of XRD patterns for which the Cd-deposition potential had been adjusted. A thin film attachment was used with a Sintage diffractometer so that glancing incident XRD patterns could be obtained. It was determined experimentally that Au and Si substrate features could be essentially eliminated using an incident angle near 0.25°. As expected, use of a Cd deposition potential of −0.4 V resulted in no CdTe deposit. However, a strong feature corresponding to zinc blende CdTe(111) planes was evident near 24° for deposits formed using a Cd deposition potential of −0.55 V or below. In addition, two significantly smaller peaks characteristic of the CdTe(220) and (311) planes were observed near 40°

FIG. 26. (A) Relative coverages of Cd and Te as a function of the Cd-deposition potential. Each deposit was formed using 200 cycles. Coverages were measured using EPMA. (B) Absolute coverages vs. the Cd UPD potential, measured by dissolving deposits and running ICP-MS on the resulting solutions.

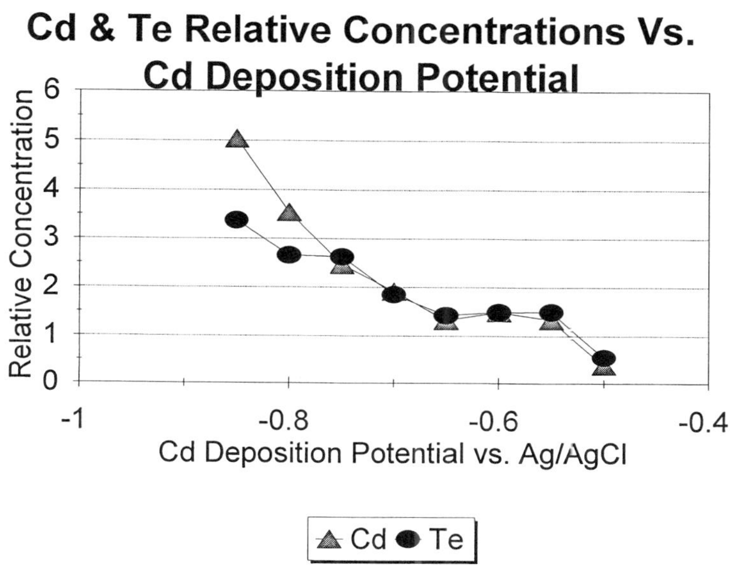
Cd & Te Relative Concentrations Vs. Cd Deposition Potential
Relative Concentration
0
1
2
3
4
5
6
-1
-0.8
-0.6
-0.4
Cd Deposition Potential vs. Ag/AgCl
Cd
Te

(A)

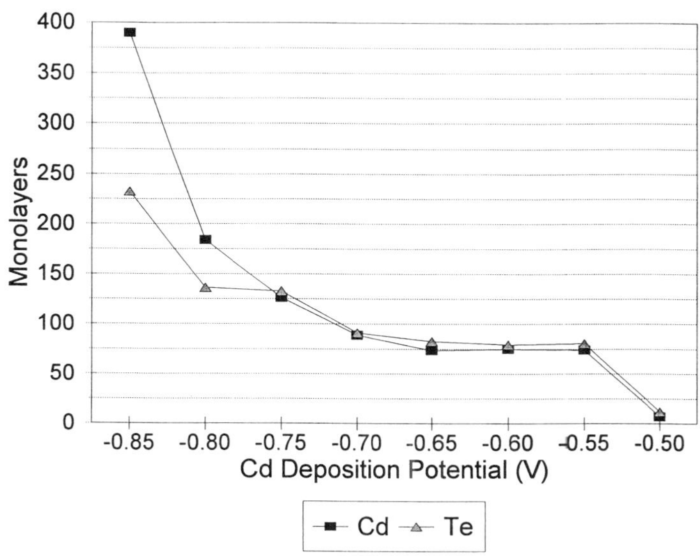
Monolayers
0
50
100
150
200
250
300
350
400
-0.85
-0.80
-0.75
-0.70
-0.65
-0.60
-0.55
-0.50
Cd Deposition Potential (V)
Cd
Te

(B)

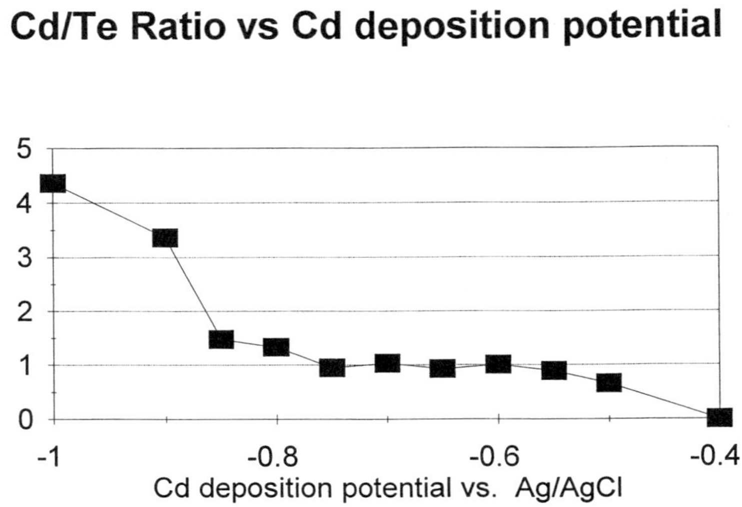

FIG. 27. Cd/Te ratio observed using EPMA for deposits formed using 200 cycles as a function of the Cd deposition potential.

and 47°, respectively. Clearly the deposits show a predominance of (111)-oriented crystallites. Hexagonally oriented deposits are expected, given that the Au on Si substrates used had a predominance of (111) surface facets. However, recent work using single-crystal Au substrates oriented to the (111), (100), and (110) planes all showed the same preference for forming CdTe(111) planes [311]. Also evident in Fig. 28 for deposits formed below −0.85 V is an array of peaks assignable to Cd, indicating that the potential for bulk Cd deposition was exceeded.

By simply adjusting the Cd-deposition potential, a broad range of deposit compositions and morphologies were formed. The fact that no deposits formed when the Cd potential was too high clearly indicates the dependence of Te deposition on the previously deposited Cd. Using slightly lower potentials, where Cd UPD does occur, there is a 0.2 V plateau (Fig. 27) where stoichiometric deposits were formed, but in Fig. 26A only a 0.1 V plateau in coverage is evident. At present, the best deposits made with this cell appear to be those formed using Cd-deposition potentials in that 0.1 V window—between −0.55 and −0.65 V.

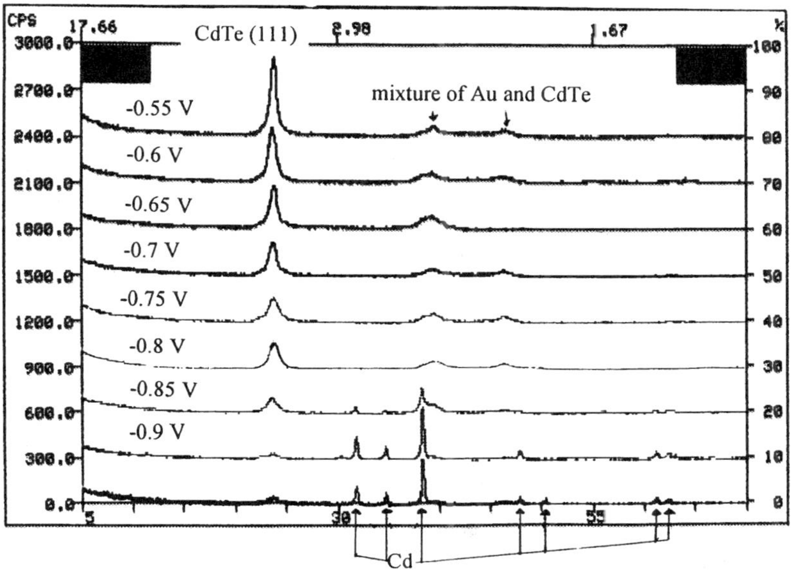

FIG. 28. XRD patterns for CdTe films grown with 200 cycles as a function of the Cd deposition potential. Omega had been optimized for increased surface sensitivity in each case.

Ideally, a coverage of one monolayer of CdTe should be formed with each cycle. Recently, studies were performed where the deposits used to perform the EPMA investigations (Figs. 26A and 27) were dissolved and used to run investigations of the total amounts of Cd and Te using ICP-MS. Those results for deposits formed at different Cd UPD potentials are shown in Fig. 26B. As discussed above, the plateau between -0.55 and -0.65 V results in what seem to be the best deposits. The ICP-MS results give absolute coverages, and from Fig. 26B it is evident that the optimal deposits are not 1 ML per cycle, but more like 0.4 ML per cycle [311,321].

That other than 1 ML per cycle appears to be the optimal deposition rate is not unusual for ALE, however [17–21]. The reasons for this low growth rate in the present studies are not yet clear. Similar studies in the formation of CdSe have been shown to result in growth rates of 0.8 ML per cycle. As described in the next section, the lattice match of CdSe with Au is much better than that of CdTe with Au, suggesting that lattice matching may be a controlling factor. There are, however, significant differences in the potentials that can be used to deposit the chalcogenides in these two cycles, and they may be the controlling factors [311,321]. The rinsing steps are presently under study as well.

CdTe deposits formed using Cd potentials between −0.65 and −0.75 V appear to have the best morphology after 200 cycles. SEM micrographs for some of these 200 cycle deposits (Fig. 29) show changes in surface morphology on the 10-μm scale as a function of the Cd-deposition potential. For the most part the deposits corresponding to the plateau between −0.55 to −0.65 V look very flat, with a few small white specks scattered across the surface. The larger feature, observed in Fig. 29B (−0.6 V), was used by the SEM operator to focus the microscope and is probably a dust particle. For deposits formed at −0.7 V, the number of small particles has increased significantly and suggests the beginning of three-dimensional nucleation and growth, clearly outside ideal ALE conditions. The flowery

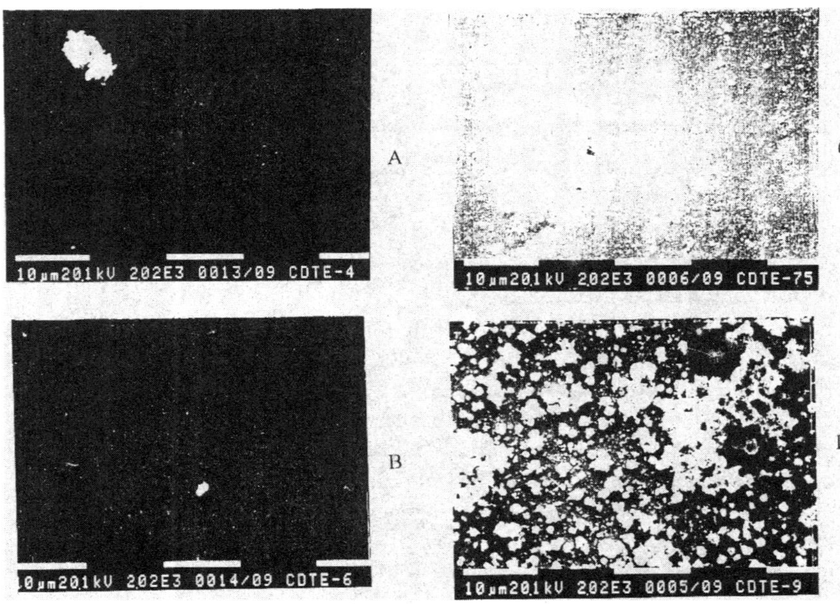

FIG. 29. SEM micrographs of CdTe deposits formed using 200 cycles, and adjusting the Cd deposition potential: (A) −0.4 V; (B) −0.6 V; (C) −0.75 V; (D) −0.90 V.

deposits obtained with Cd potentials below −0.8 V are consistent with the formation of some crystalline Cd. Figure 30 shows two images taken with a high-resolution field emission SEM. The first image shows the clean Au on Si substrate. Evident is a series of little bumps, and Fig. 31 is a scanning tunneling micrograph (STM) of the same surface. From Fig. 31 and others, it is apparent that the substrates being used are optically flat, yet are covered with 40–50 nm wide bumps that are about 20 nm in height. Higher-resolution STM images have shown these bumps to be crystallographically ordered but to have a very high density of short terraces. Atomically resolved images of the terraces show them to be well ordered and (111) oriented. The second field emission SEM image (Fig. 30B) shows an equivalent substrate on which 200 cycles of CdTe deposition have been performed using a Cd potential of −0.6 V. The similarity of the two images is encouraging because it is consistent with epitaxial deposition. The deposits have essentially the same morphology; on the scale of the micrograph, however, the image of the deposit (Fig. 30B) is a little less clear than that of the clean substrate (Fig. 30A). The difference might be a simple focus problem. Alternatively, it could be a charging problem or a difference in the electron-scattering characteristics of the two surfaces, due to the presence of the compound. However, given the morphology evident in Fig. 31, a high substrate-induced defect density is expected in the deposits. The difference between the images in Fig. 30 may result from the presence of a large number of small CdTe crystallites induced by the lattice mismatch with the Au and substrate step density. A problem with the CdTe deposits matching up with the highly stepped surface could help account for the low growth rate described above. Work is underway to switch substrates. In the case of Au substrates, it is well known that Au on mica can have very large terraces, as can Au on glass under certain sets of conditions.

As previously discussed, Te atomic layer formation in the present cycle is a two-step process. The first step is deposition of a few monolayers of bulk Te, while the second involves reduction of excess Te from the surface in a blank solution at fairly negative potentials. The dependence of the deposit composition on the potential used to form the first few monolayers of bulk Te is graphed in Fig. 32. At potentials above −0.7 V, the coverage per cycle for both Cd and Te drops off. The potentials are well within the range for bulk Te deposition, so the decrease is not a problem with depositing Te. The problem is more likely that Cd is being lost while the Te is being deposited, because the potentials are sufficiently positive that the Cd may not be stable. Deposits formed in the plateau region (be-

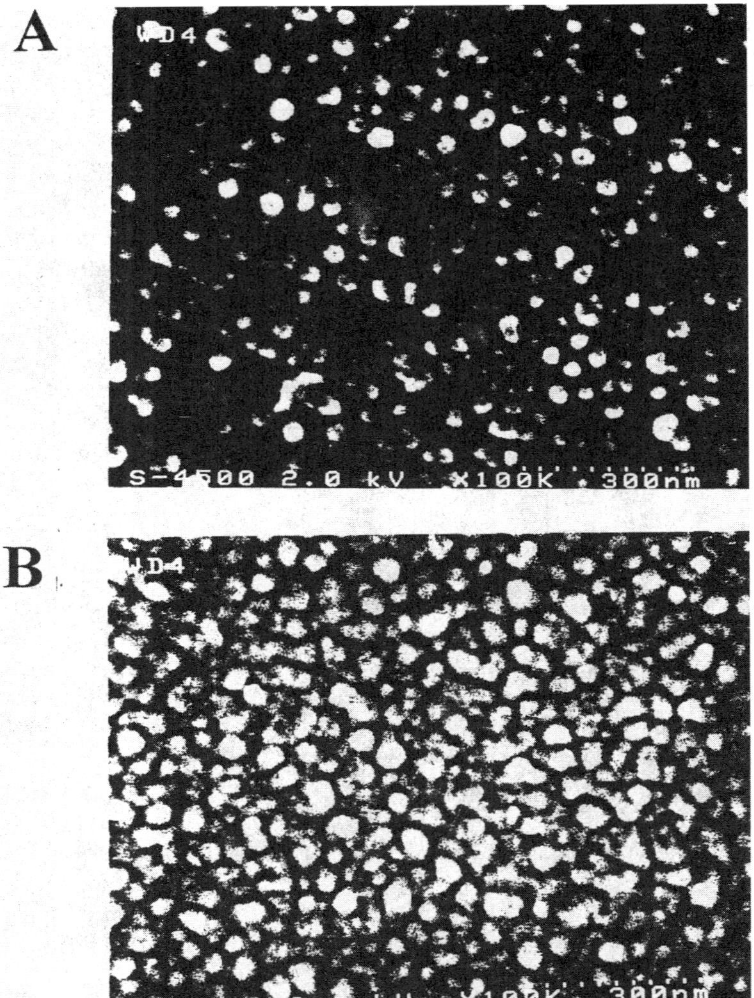

FIG. 30. SEM micrographs taken with a microscope with a field emission tip: (A) clean Au on Si substrate; (B) CdTe deposit formed using 200 cycles on a substrate equivalent to that shown in A.

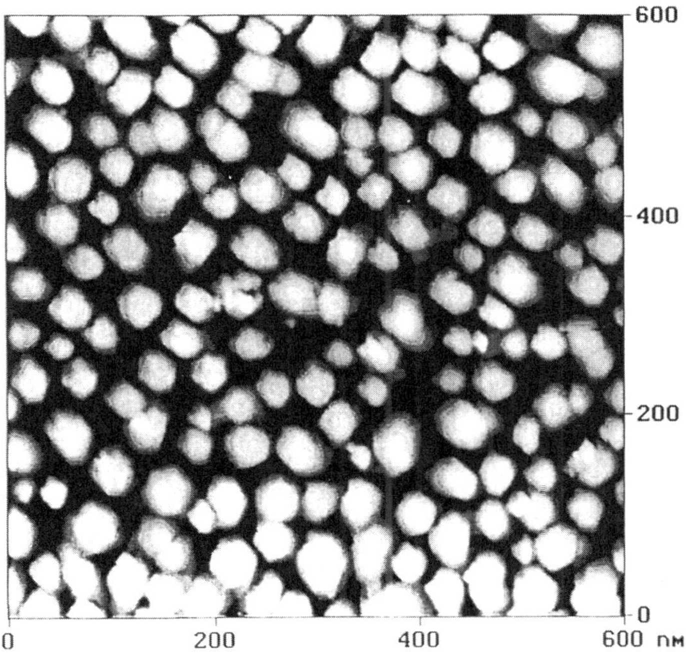

FIG. 31. STM micrograph of Au on Si substrate. The vertical range used in the image was 23 nm.

tween −0.8 V and −0.7 V) were equivalent to the optimal deposits previously formed. Using potentials more negative than −0.8 V introduced problems with conproportionation again [Eq. (2)], as evidenced by the decrease in the Cd/Te ratio (Fig. 33), because some Te^{2-} is formed at these low potentials, which then reacts with the $HTeO_2^+$ to form Te particles. Alternatively, the decreased Cd/Te ratio may simply result from too much Te deposition, so that the resulting Te layers are not completely reduced under the condition used in Figs. 32 and 33.

The second step in the formation of atomic layers of Te is the reduction of the excess Te. Figure 34 shows, at least for the first Te atomic layer on Au, that bulk Te is reduced from the surface at potentials below −0.8 V. The reduction is rapid, occurring within a few seconds, for a reduction potential of −1.4 V (Fig. 35). The graph in Fig. 36 shows the coverages

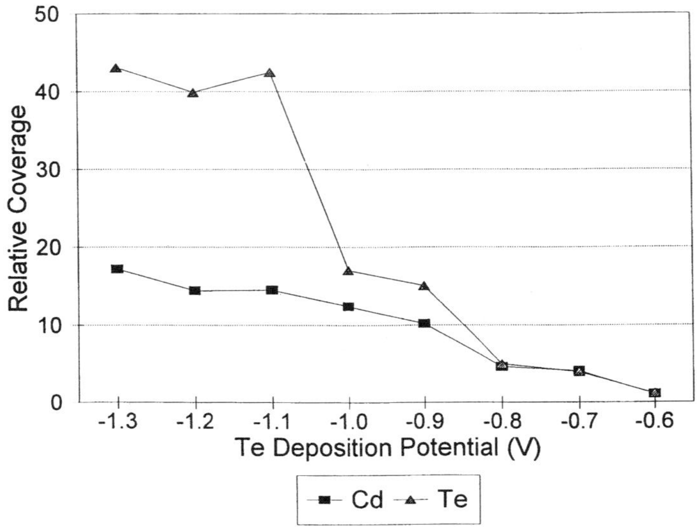

FIG. 32. Relative coverages of Cd and Te in deposits formed using 200 cycles each as a function of the potential used to form the first few monolayers of Te as a first step in the formation of Te atomic layers. Data obtained using EPMA.

of Cd and Te for a series of deposits made with 200 cycles each as a function of the potential used to reductively remove the excess Te. At potentials above −1.0 V the coverage per cycle goes up as expected, since not all the bulk Te deposited in the previous step is stripped. The corresponding Cd/Te ratio also drops (Fig. 37) because the Cd is not reacting quantitatively with the excess Te. Stripping the Te at potentials below −1.1 V results in optimal deposits. The plateau in Fig. 36 is large, stretching to below −1.4 V. It might be expected that even the Te bound to Cd could be stripped, given a sufficiently negative stripping potential. The fact that the coverage does not drop is an indication of the stability of the Te bound up as CdTe. In addition, the electrode is somewhat nonpolarizable due to solvent decomposition in the borate solution, preventing the substrate potential from adjusting sufficiently negative to reduce Te bound to Cd.

One of the attributes of an ALE growth mode is that the number of cycles determines the thickness of the deposit. Double the number of cycles and you double the thickness of the deposit. Figure 38A is a graph of the

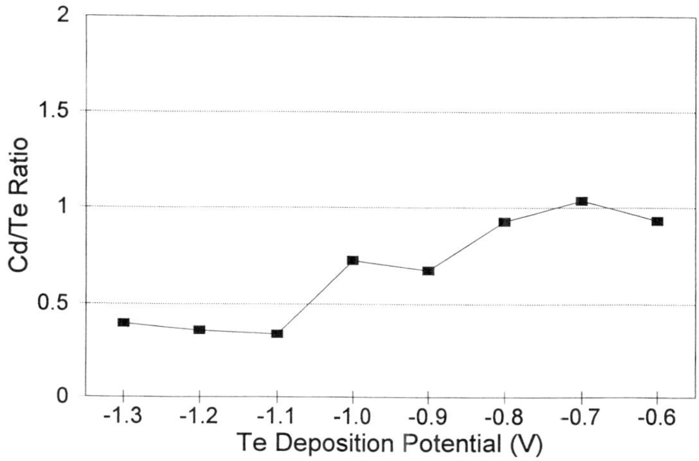

FIG. 33. Cd-to-Te ratio for deposits formed using 200 cycles each as a function of the potential used to form the first few monolayers of Te as a first step in the formation of Te atomic layers. Data obtained using EPMA.

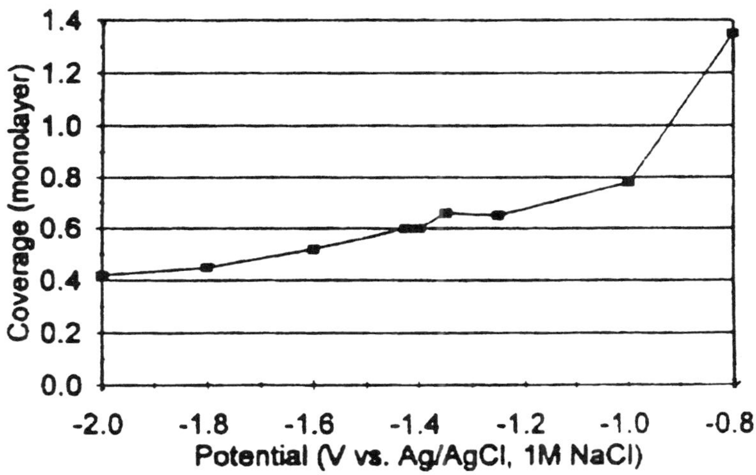

FIG. 34. Graph of the Te coverage remaining on the surface after reductions at various potentials. Each point results from the initial deposition of 1.3 ML at −0.8 V, followed by the indicated reduction in the Te blank solution.

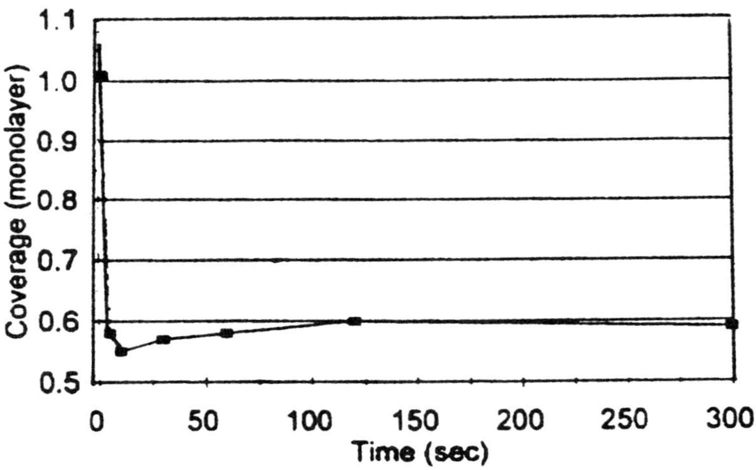

FIG. 35. The Te coverage remaining after reductive dissolution, as a function of time, at −1.4 V on the Au(100) surface.

relative coverages of Cd and Te as a function of the number of cycles. The graph is linear as expected, passing through the origin. The stoichiometry of the 1000-cycle deposit is a little Te-rich, however. XRD of this deposit showed that elemental Te was clearly present, suggesting that the conditions developed to deposit 200 cycles may not have been optimal for the deposition of 1000, and that the optimal conditions can change as the deposit grows thicker. Similar studies of CdSe deposition show linear growth as a function of the number of cycles (Fig. 38B). For CdSe (Fig. 38B), the 1000-cycle deposit shows no elemental Se with XRD but does show a slight excess of Se with EPMA. Very few 1000-cycle deposits have been formed so far, and further studies are needed to understand these changes in deposit composition.

These changes in composition could simply require fine-tuning of the deposition cycle for the thicker deposits. Alternatively, the conditions needed to grow optimal layers may change as the deposits get thicker. One of the nice things about using the automated deposition system is that the program could be easily modified so that a small change is made in a particular step in the deposition cycle, with each cycle as the deposit is formed. The Cd potential could be changed a mV each cycle as the semiconducting

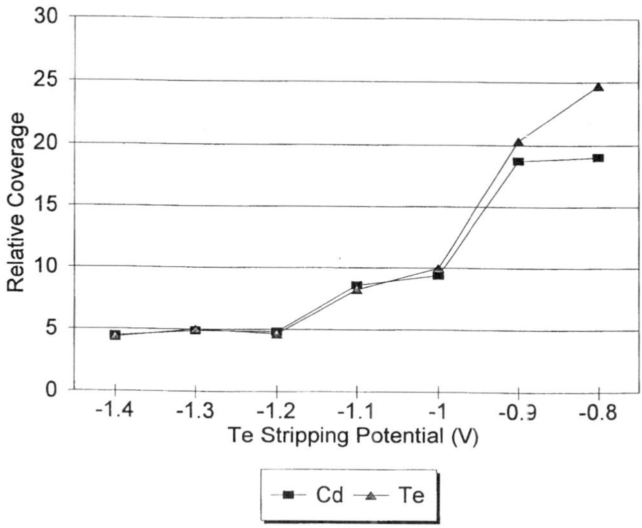

FIG. 36. Graph of the coverages of Cd and Te formed using 200 cycles as a function of the potential used to reductively remove excess Te as the second step in the formation of Te atomic layers.

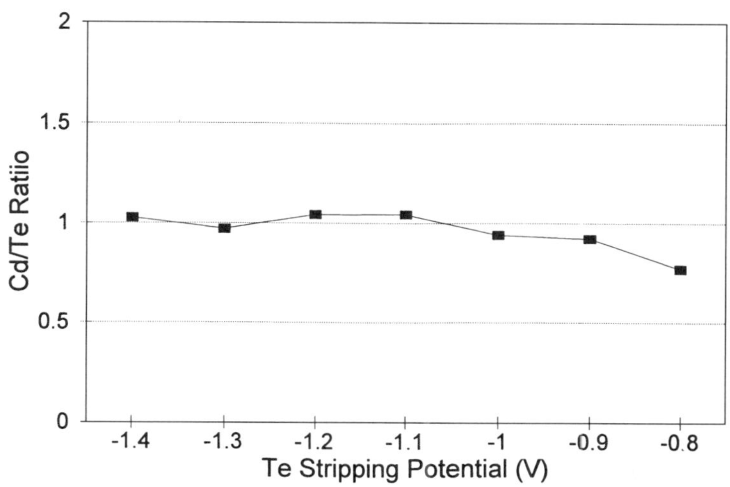

FIG. 37. Graph of the Cd/Te ratio for the same deposits shown in Fig. 36. Deposits formed using 200 cycles as a function of the potential used to reductively remove excess Te as the second step in the formation of Te atomic layers.

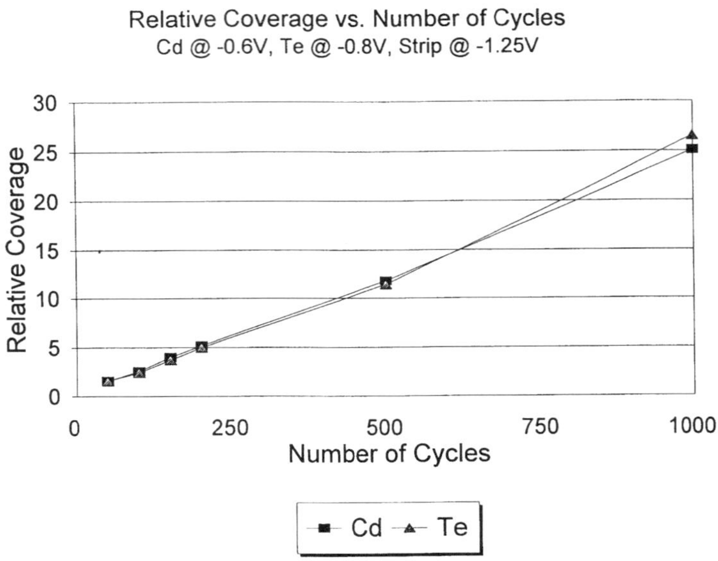

(A)

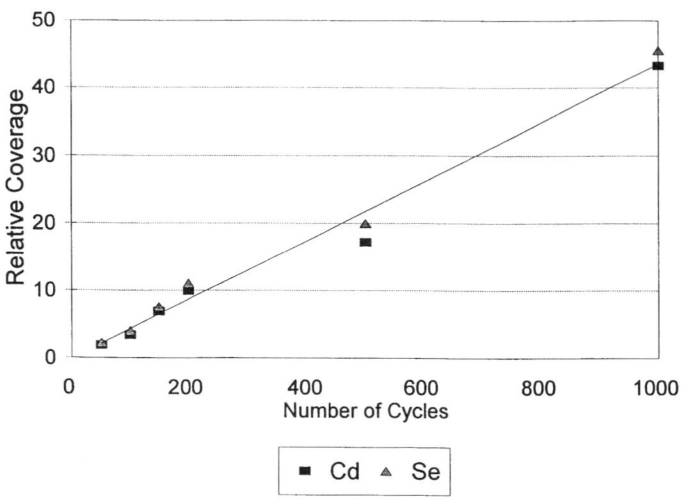

(B)

FIG. 38. (A) Graph of Cd and Te relative coverages as a function of the number of cycles used to form the deposits. (B) Graph of the Cd and Se relative coverages as a function of the number of cycles used to form the deposits.

film grows. There have been some interesting reports describing the codeposition of CdTe, where the open circuit potential was checked every few minutes during the codeposition of CdTe, and the deposition conditions were adjusted accordingly during the deposition [289]. An analogous feedback procedure might be incorporated into an ECALE deposition cycle if some measurement step can be devised.

Presently work is continuing on formation of CdTe thin films in an effort to understand the dependence of the deposit composition and structure on the cycle variables. Recent work involving changes in the cell design and deposition program has resulted in 0.95 monolayers of CdTe being deposited with each cycle, and publication of those results will be forthcoming. In addition, work is being performed on the deposition of ZnS, ZnSe, CdS, and CdSe.

Raman studies of CdS films formed using ECALE have been performed by Boone and Shannon [59]. Those studies concluded that good-quality bulk CdS was being formed and in a layer-by-layer manner.

IV. SURFACE CHEMISTRY IN THE ECALE CYCLE

The macroscopic approach to investigating ECALE, forming thin films, and analyzing them was discussed in the last section. In this section, studies designed to provide an atomic-level description of the growth of atomic layers in an ECALE cycle are described.

As mentioned, coverage is defined as the ratio of the number of deposited atoms relative to the number of substrate surface atoms. In the studies presented in the previous two sections, polycrystalline Au electrodes were used. Most of the work described in the present section has been performed using Au single-crystal electrodes. Each low index plane of Au has a different atomic surface density so that 1/2 ML coverage corresponds to a different number of atoms/cm^2 for each plane. Each compound studied has a number of possible orientations, which can be adopted at the deposit-substrate interface. Some compounds have multiple possible crystal structures; CdSe for instance can crystallize in the zinc blende or wurtzite structure. Deposition of a "monolayer" of a compound on Au is thus a relative term. In addition, there are questions concerning the dimensions and quality of the interface: Is there compression or expansion of the deposit structure to account for a lattice mismatch with the substrate? What are the densities of steps and defect sites? How clean is the substrate? How good is the assumption that each element is nucleating and growing two-dimensionally, that no three-dimensional growth is taking place? It should be clear that when 1/2 ML coverage on a polycrystalline substrate is discussed, it is an approximation. To say more requires detailed studies, such as described below, of the individual atomic layers on single-crystal substrates. The structure, composition, and morphology of the substrate and individual atomic layers of each of the elements making up the compound are being investigated with the intent of understanding their influence on deposit quality.

Studies carried out in this group concerning surface chemistry in an ECALE cycle have involved ultra-high-vacuum electrochemical (UHV-EC) techniques [322] and scanning tunneling microscopy [323–329]. The purpose of the UHV-EC techniques is to study the structure and composition of electrodes using surface-sensitive electron spectroscopes such as Auger electron spectroscopy (AES) and x-ray photoelectron spectroscopy (XPS or ESCA), as well as low-energy electron diffraction (LEED). LEED provides diffraction information for surface unit cells, while elemental coverages can be determined using AES. In addition, AES is very helpful in identifying any contamination problems, i.e., the presence of C or O signals. A carbon signal indicates that something was not clean, while an oxygen signal can indicate that some electrolyte has been ''emersed'' (withdrawn from solution) with the sample or that the surface has been oxidized. XPS provides information on the surface elemental composition, as does AES, but is generally slower. XPS has the advantage, however, that it can provide information on the oxidation states of surface species and is used mostly to answer question concerning the bonding environment of elements on the surface [330,331].

The idea of UHV-EC is to perform electrochemical experiments inside an antechamber attached to a UHV surface-analysis instrument, so that substrates can be examined using the above techniques both prior to and after electrolysis without their leaving the instrument or being exposed to air (Fig. 39). The challenge is to perform experiments where the surfaces studied closely resemble those that existed under potential control in solution, and thus gain information on the nature and mechanisms of interfacial process. There are significant limitations, however, and careful work and controls must be applied for the resulting data to be meaningful. For instance, if an adsorbed species is to be examined, it must be covalently, or ionically, bounded to the surface, because physadsorbed species (i.e., $\Delta H < 10$ kcal/mole) are easily pumped away when the substrates are transferred to UHV for analysis. UHV-EC is thus a method best suited for examining specifically adsorbed species or electrodeposits. In addition, there are

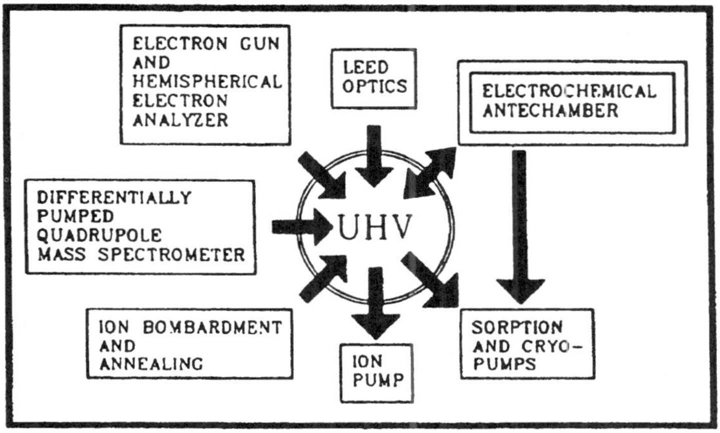

FIG. 39. Schematic diagram of UHV-EC instrument.

problems associated with studies in some potential regimes. Electrodes emersed from aqueous solutions at potentials below the formal potential for hydrogen stand a good chance of undergoing spontaneous oxidation upon loss of potential control. Indications of this are the appearance of a surface oxide where a reduced surface should have existed in solution or the dissolution of deposits that would otherwise be stable in solution under potential control [332]. Another general problem with UHV-EC studies is that the concentrations of reactants and electrolytes in the solutions must be kept low, generally below mM, because when the substrate is emersed from solution it brings with it a layer of solution, referred to as an "emersion layer." Upon evacuation, a layer of reactant or electrolyte can be left behind on the surface, depending on the viscosity of the solution, the geometry of the electrode emersion process, the hydrophobicity of the substrate surface, and the volatility of the species. The pump-down process generally helps to get rid of most of the emersed layer, because the solution layers explode when vacuum is applied, spraying small amounts of the solution around the inside of the antechamber and minimizing the amount left on the crystal. Experience has shown that most problems associated with emersed species go away if the total concentration is kept below mM, while problems are generally encountered when the total concentration exceeds 10 mM. Tricks like rinsing the electrode work in some situations, however, they can obscure the meaning of any results obtained.

STM studies give unparalleled atomic scale structural information in air, in situ (in solution and underpotential control), and in vacuum. Each environment has advantages. Imaging in air is the simplest. The drawbacks to imaging in air involve contamination, oxidation, humidity, and the fact that you can only look at ''snapshots'' of the deposition process. The deposition takes place in solution at some potential and must be emersed for imaging in air. If the deposits are stable in air, then good images may be possible [159,333,334]. Images taken in situ can be ''movies,'' where the conditions are adjusted on the fly, providing a sequence of images of the deposition taking place. Problems with in situ imaging are mostly associated with the tip, in that the tip must be insulated and compatible with the solution. The coating process is another step in the production of a tip, lowering the chances of producing one capable of atomic resolution. With respect to solution compatibility, both the insulating coating and the tips must be stable in solution. In in situ studies, the tips are electrodes and have their own potential-dependent reactivity with the solution. W tips are quite popular for achieving atomic resolution images, however, they can oxidize at potentials above 0 mV. Se deposition, for example, normally takes place at potentials above 0 mV, so that W does not work well, because the tip must be held at a still higher potential to prevent bulk Se from depositing on it. STM studies in UHV are advantageous because UHV is clean and oxygen-free, allowing less stable surfaces to be investigated as well as ready access to other surface-sensitive techniques such as LEED and AES. As with studies in air, however, only snapshots of the surface are obtained, making it difficult to follow the deposition sequence.

The first surface studies of ECALE cycle chemistry focused on CdTe deposition [159,162,320]. The structures of the first atomic layers of Te deposited on Au single-crystal surfaces were studied [320], as were the structures of the deposits after subsequent deposition of atomic layers of Cd to form CdTe monolayers [159,162]. UHV-EC studies of Cd atomic layer formation on Au were attempted but generally resulted in surfaces displaying diffuse LEED patterns and variable amounts of Cd and O. This is consistent with the problems, previously mentioned, of depositing reactive species at potentials negative of the formal potential for hydrogen. It was difficult to emerse the Cd-coated Au substrates without the Cd spontaneously oxidizing. The emersed crystal is at least momentarily encased in a layer of liquid at open circuit, prior to evacuation. Te, on the other hand, was quite stable and well behaved, displaying a sequence of ordered atomic layers on each of the Au low-index planes [320].

The influence of the initial Te atomic layer structure on the resulting CdTe monolayer was also investigated [159,162]. A series of experiments was performed where Te atomic layers of various coverages and structures were first deposited, and then Cd UPD was performed on top. Ordered CdTe deposits were formed on each of the low-index planes. What was interesting was that the initial Te structure or coverage had very little to do with the structure of the resulting CdTe monolayers, i.e., the LEED patterns observed for the CdTe deposits did not change appreciably as a function of coverage or structure of the initial Te atomic layer.

On Au(100), for instance, a $c(2 \times 2)$-CdTe structure was observed (Fig. 40). This structure ideally consists of 1/2 MLs of both Cd and Te [159,162]. The CdTe LEED pattern was not a function of the initial Te coverage or even the subsequent Cd coverage; only the clarity of the pattern changed with coverage. For instance, in the case where Cd was deposited on the 1/3 ML coverage Te structure, such as the Au(100) $(2 \times \sqrt{10})$-Te, there should only be enough Te for the resulting CdTe structure to cover two thirds of the surface (Fig. 41A). The remaining third of the surface should be covered by Cd UPD. The two thirds of the surface covered with CdTe produces a $c(2 \times 2)$ pattern, while the one third covered with Cd UPD contributes diffuse intensity, as it spontaneously oxidizes upon emersion, resulting in disordered domains. On the other hand, if the initial Te structure contained 2/3 ML of Te, only 1/2 ML is needed to form the first monolayer of CdTe. The excess Te may have contributed to the formation of a partial second layer and the formation of CdTe islands (Fig. 41C). Island formation

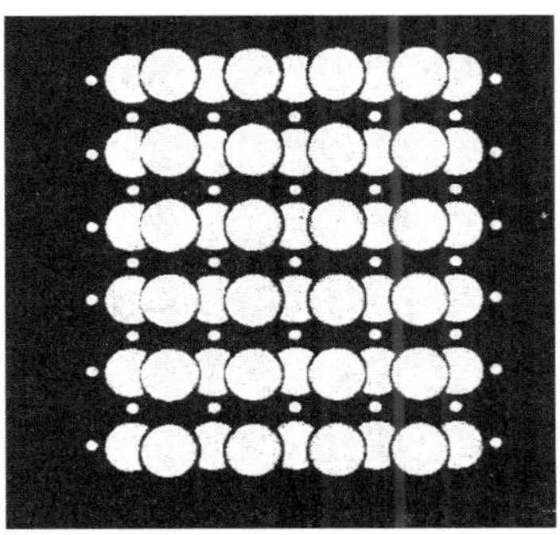

FIG. 40. Proposed structure for CdTe monolayer formed on Au(100). The structure corresponds to a $c(2 \times 2)$, with a coverage of 1/2 ML each of Cd and Te.

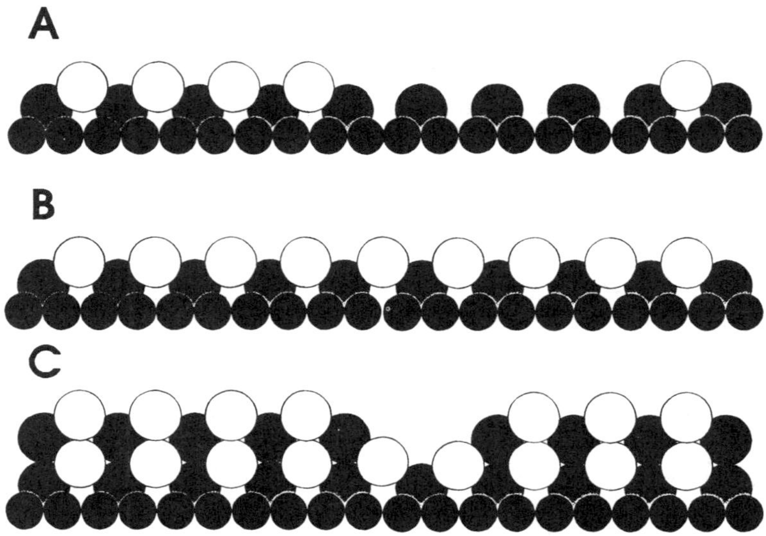

FIG. 41. Schematic drawing of possible scenarios for CdTe deposits formed on increasing initial coverages of Te: (A) 1/3; (B) 1/2; (C) 2/3.

is a form of surface roughing, which also contributes to diffuse scattering in LEED patterns, depending on the number and sizes of the islands. This helps explain why the LEED patterns become diffuse if too much Te is present in the initial layer. Studies have shown that the Cd can react completely with the Te, even if multiple monolayers of Te are present initially [291].

From the above results it is clear that the amount of Te deposited in the first atomic layer on the Au surface is critical but not structure controlling. Control over the coverage of Te in subsequent cycles should not be as critical, as it should be naturally controlled by the previously deposited Cd atomic layer. The first Te layer, however, involves the reaction of Te with the Au substrate to form a Au-Te surface compound, not CdTe. The coverages of the Te atomic layer structures formed on Au have nothing to do with the formation of stoichiometric CdTe. A homogeneous atomic layer of Te on Au at the correct coverage [1/2 ML on Au(100)] (Fig. 41B), with no pits or islands, should produce the lowest defect density in the subsequently formed CdTe film, according to the structure proposed in Fig. 40.

Similar studies of the formation of the first monolayer of GaAs have also been performed [157,252,253]. The chemistry of arsenic facilitates the ECALE cycle, as it is easily reduced to H_3As at negative potentials (Fig. 42):

$$As_{(bulk)} + As_{(UPD)} + 3e^- + 3H^+ \Leftrightarrow H_3As + As_{(UPD)} \quad (10)$$

UPD of As on Au single crystals resulted in well-ordered atomic layers, as observed with LEED. In addition, subsequent UPD of Ga results in formation of an ordered GaAs monolayer. The structure of the monolayer is similar to that for CdTe, in that it consists of a 1/2 ML each of Ga and As on Au(100). The observed LEED pattern, however, was not a c(2 × 2), but a p(2 × 2) structure (Fig. 43). This larger unit cell has been proposed to result from dimer formation in the top layer (Fig. 44), which is consistent with previous UHV studies of GaAs single crystal surfaces [335–339]. Those results concerned mostly the formation of a single monolayer of

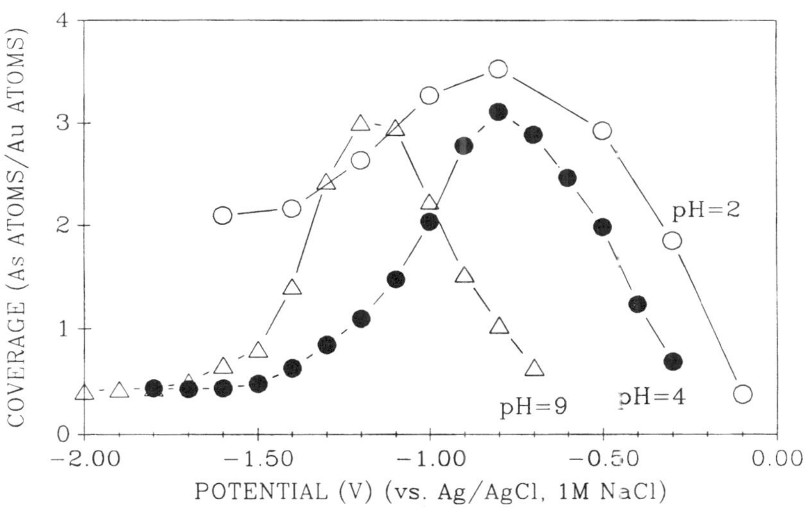

FIG. 42. As coverage vs. deposition potential and pH in a Au TLEC. (From Ref. 252.)

FIG. 43. Picture showing (2 × 2) LEED pattern from Au(100) surface supporting a monolayer of GaAs. Beam energy was 40 eV.

GaAs, as the deposition of multiple layers proved difficult since, at that time, the hardware resulted in deposits being left at open circuit, briefly, during each solution change. In the case of the CdTe deposits described above, Cd oxidized spontaneously when emersed in the absence of Te (Fig. 45B) but did not oxidize if Te was present as well (Figs. 45D and E). It was concluded that the Cd did not oxidize because it was stabilized by bonding to the Te. In the present case, the Ga did not appear to be sufficiently stabilized by bonding with the As to keep it from oxidizing under open-circuit conditions. Deposition hardware in which potential control is not lost with each rinse should facilitate the formation of thicker GaAs deposits, however, and those studies are planned.

The nature of the material that results after the first few ECALE cycles has been questioned by some—whether the surface is covered with separate domains or islands of Cd and of Te or with a monolayer of CdTe, for instance. The electrochemical results are fairly definitive on this point, in that if there were islands of Cd present on the surface, bulk Cd would be expected to strip from the surface at potentials below −0.7 V, which

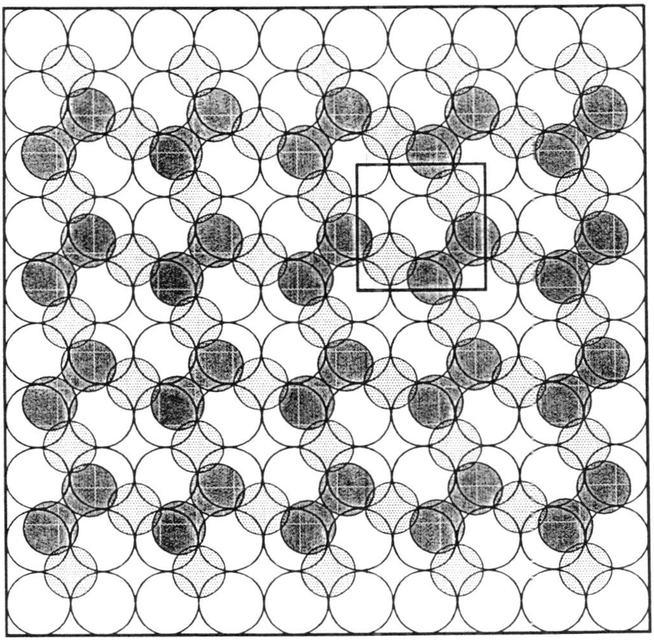

FIG. 44. Proposed structure for Au(100) (2 × 2)–GaAs structure.

is not observed. No Cd stripping was observed below −0.6 V from the CdTe deposits. Additionally, Fig. 46 is a set of XPS spectra of the 3d transitions of Te. Figure 46B is from a Te UPD layer on Au, while Fig. 46D is from bulk Te; they show essentially the same chemical shift. The spectrum in Fig. 46C is of a Te atomic layer after deposition of an atomic layer of Cd on top, and displays a shift of 0.4 eV to lower binding energy, indicating destabilization of the Te electrons consistent with the formation of CdTe [340].

The early surface studies described above indicated that compounds were probably being formed with the first cycle and that the deposits had some degree of order. LEED, however, is an averaging technique. A surface can have significant amounts of disorder, relatively small domains, and still give a reasonable LEED pattern. STM was also used in those studies of compound monolayers, however the images collected were generally of very small areas of the deposits—5–15 nm on a side. Those STM studies

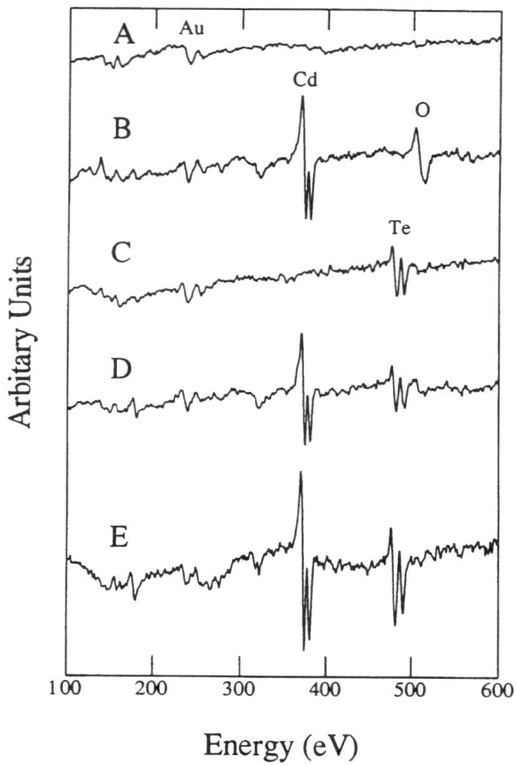

FIG. 45. Auger spectrum for Au(100): (A) after ion bombardment and annealing, (B) after emersion following Cd UPD, (C) after emersion following first Te UPD, (D) after emersion following Cd UPD on first Te UPD, and (E) after emersion following Cd UPD on second Te UPD. (From Ref. 162.)

did provide information on the atomic arrangements in the unit cells identified with LEED but did not address how much of the surface they covered or involve investigation of other features present on the surface, such as steps, pits, and islands. One of the last images taken in those studies is shown in Fig. 47. The image clearly shows a (2×2)-Te structure, but steps, pits, islands, and a series of phase boundaries are evident as well. Given that one monolayer of Te displays so many defect features, it is hard to envision epitaxial deposits after 50 cycles of CdTe formation.

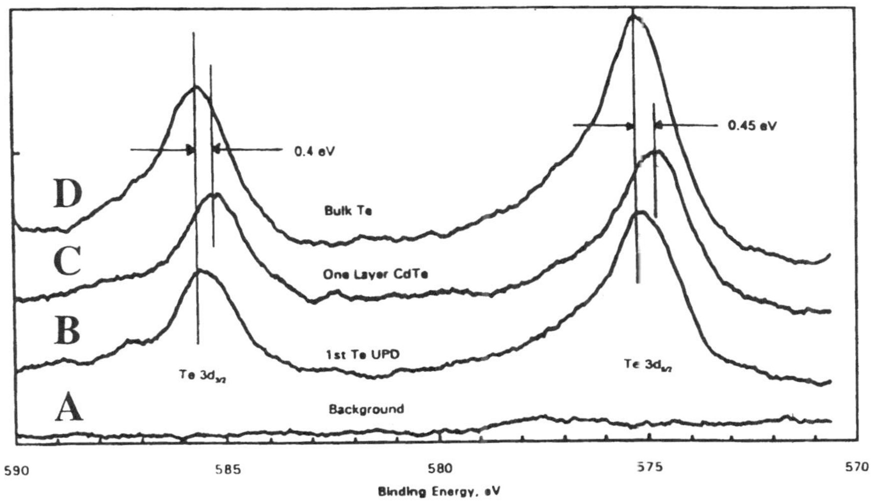

FIG. 46. XPS spectra of Au(100) after (A) ion bombardment and annealing, (B) first Te UPD, (C) Cd UPD on first Te UPD, and (D) deposition of bulk Te. (From Ref. 331.)

A study designed to better understand and remove many of the defects evident in Fig. 47 was performed [309]. With a single crystal, the step density is generally controlled by the misalignment. As delivered, single crystals are frequently misoriented by several degrees, so the first step to decreasing the substrate defect density was to reorient the crystal to within a fraction of a degree of the Au(100) plane. The crystal was then repolished mechanically with successively finer grades of polishing cloth, finishing with an electropolish [341]. The standard cleaning procedure used for the substrate in Fig. 47 involved an initial flame annealing and water quenching, followed by a number of oxidation-reduction cycles (ORCs) in acid. Figure 48 shows an STM image of the resulting clean Au surface and displays a large number of monoatomically high-Au islands. Islands such as these have been observed by a number of workers after ORC treatments. One explanation of the origin of these islands involves the reconstructions that take place when Au(100) is reduced. A corrugated hexagonal overlayer structure [307,308,342–344] is formed on Au(100) at low potentials, which contains 20–24% more surface atoms than the unreconstructed surface

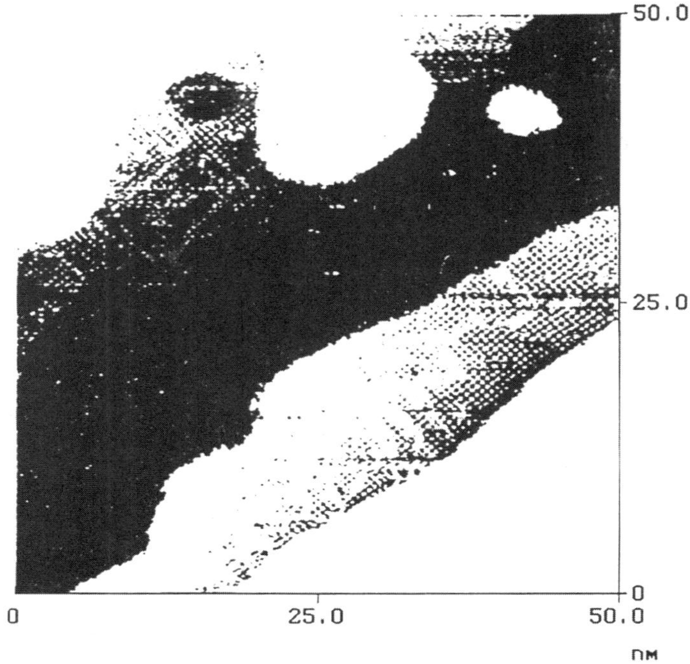

FIG. 47. STM image of a (2 × 2)-Te structure on Au(100), ¼ coverage. Image depicts the heterogeneity of the surface.

[345–347]. When the electrode potential is switched to positive, the reconstruction is relaxed, releasing the extra Au atoms to form Au islands. The more ORCs performed, the more islands are formed.

The need for ORCs to clean the surface was not necessary in the STM studies, as the electrodes were flame annealed prior to each experiment. Flame annealing does not, however, leave a perfect surface under most conditions, although it is generally much better than that resulting from ORCs [348,349]. There have been several studies showing that electrode surfaces can be atomically leveled, or annealed, using an electrochemical treatment [347,350,351]. In the case of Au, halide solutions appear to promote the leveling, either by increasing the mobility of the surface Au atoms [347,351] or by a small amount of etching [309]. Use of dilute KI solutions by this group have resulted in complete removal of the islands

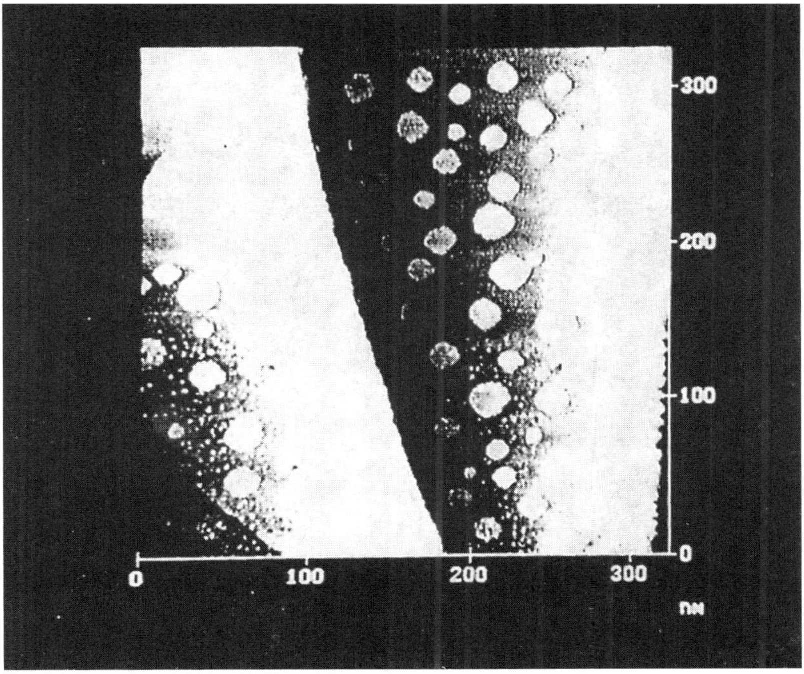

FIG. 48. STM micrograph showing monoatomically high Au islands on a clean Au(100) surface, V_b = 109.9 mV, i_t = 1.0 nA, Z range = 5.0 nm.

formed during ORCs and formation of 300 × 300 nm atomically flat terraces (Fig. 49). These surfaces, however, are coated with an atomic layer of I atoms [301,352–356].

There are ways to remove an atomic layer of I atoms. It can be reduced off:

$$I_{(ads)} + e^- \Leftrightarrow I^- \tag{11}$$

but the low potentials needed also result in reconstruction of the surface and then island formation upon lifting of the reconstruction. Alternatively, the I atom layers can be oxidatively removed and converted to iodate:

$$I_{(ads)} + 3H_2O \Leftrightarrow IO_3^- + 6H^+ + 5e^- \tag{12}$$

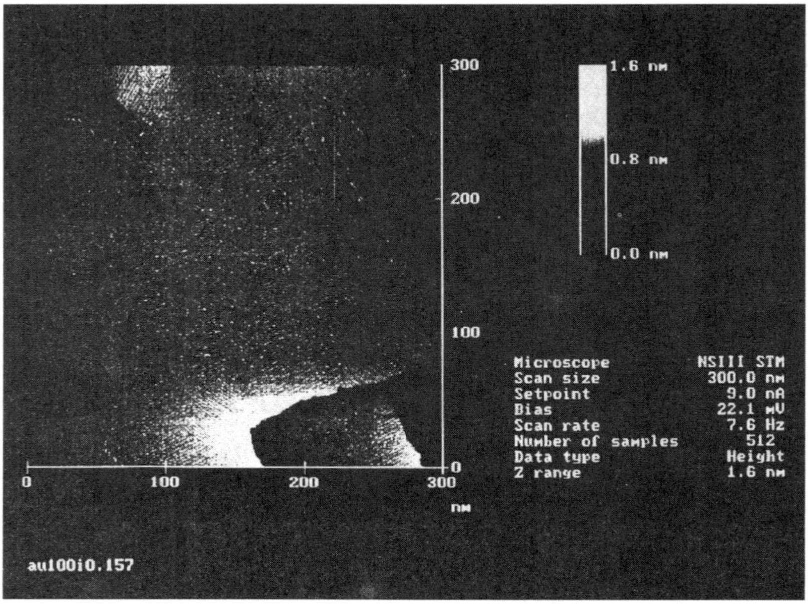

FIG. 49. STM image of I coated Au(100) surface showing large atomically flat terraces.

The oxidation, however, occurs commensurate with Au oxide formation, and oxidation can result in roughening of electrode surfaces, as occurs for Pt [357]. A third method that is more compatible with the formation of chalcogenide atomic layers and that avoids roughening the substrate involves deposition of a chalcogenide layer on top of the halide layer [309]. Figure 50 shows voltammetry for the deposition of Te from a $HTeO_2^+$ solution on clean Au(100) (Fig. 50a) and on I-coated Au(100) (Fig. 50b). Two Te-reductive UPD features are evident on the clean Au (0.325 and 0.0 V), as well as a bulk Te-deposition feature (below 0.0 V). On the I-coated Au(100) electrode, the first Te UPD feature is gone and the second has been noticeably suppressed. The subsequent oxidative Te stripping features in Fig. 50a and b are essentially the same, indicating that there is very little difference between the resulting Te layers. Studies following the surface composition with AES indicated that the I atom layer is completely displaced by the depositing Te atoms (Fig. 51). However, the resulting Te

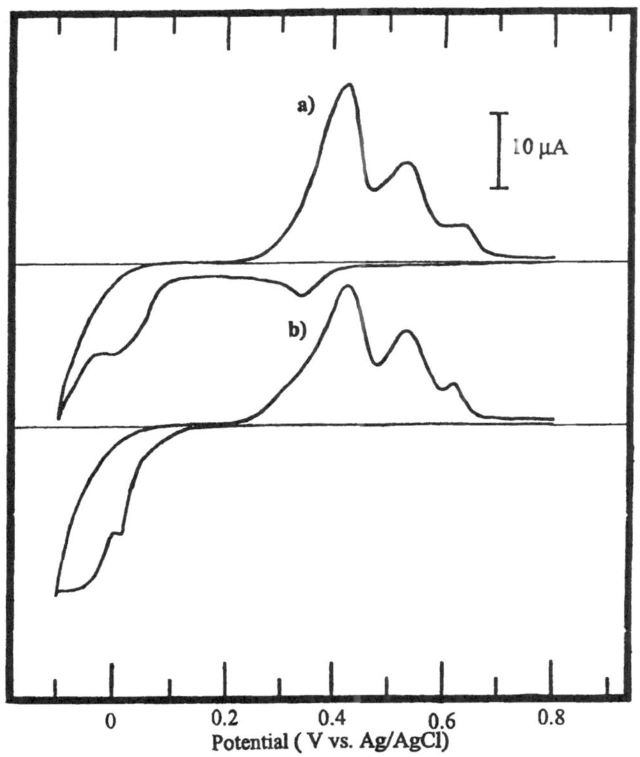

FIG. 50. Current-potential curves in 0.2 mM TeO_2^+, in 10 mM H_2SO_4: (a) Te deposition on clean Au(100); (b) Te deposition on I-treated Au(100).

layer generally contains more than a monolayer of Te, which can be removed by either oxidation at 0.400 V or reduction below −0.900 V, to produce a Te atomic layer. The layers appear well ordered and with significantly lower defect densities than those formed on substrates subjected to ORC cleaning.

More recent work involving surface chemistry in an ECALE cycle centers on the formation of the first few monolayers of CdSe. Again, the chalcogenide serves as the first atomic layer due to Cd instability upon emersion. Figure 52C is the Auger spectrum for a Au(100) crystal covered with Cd UPD, while Fig. 52D is the spectrum for Cd UPD on the crystal

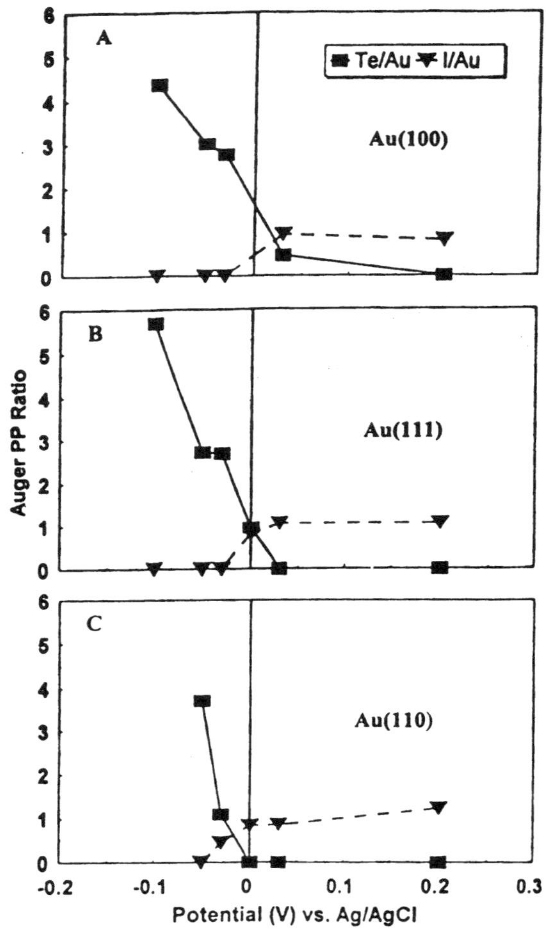

FIG. 51. AES intensities for Te and I as a function of potential for the three low-index planes: (A) Au(100); (B) Au(111); (C) Au(110).

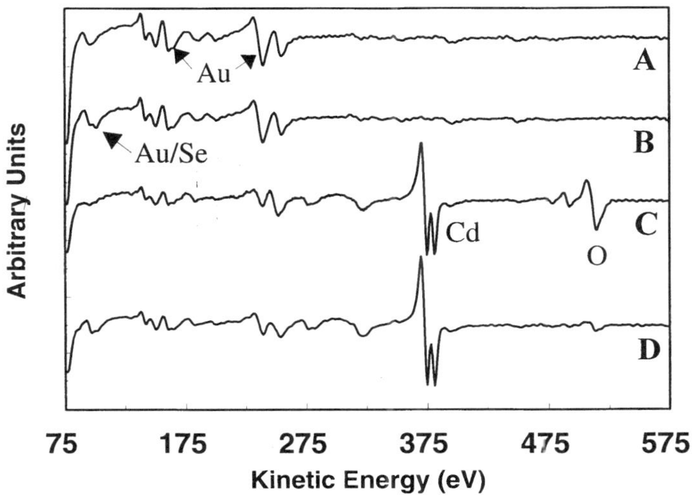

FIG. 52. Auger spectra for a Au(100) single crystal: (A) after cleaning; (B) covered with an atomic layer of Se; (C) covered with an atomic layer of Cd; (D) an atomic layer of Se covered with an atomic layer of Cd.

precoated with an atomic layer of Se. As in the case of CdTe formation (Fig. 45), the Cd Auger transition at 375 is accompanied by one at 500 eV for oxygen when no other chalcogenide is present (Fig. 52C). The oxygen signal is nearly absent from Fig. 52D, where the Cd has reacted with previously deposited Se.

Se forms a number of different atomic layer structures on each of the Au low-index planes (as does Te) as a function of time and potential. The following discussion of Se deposition will concentrate on just the (100) face [295,334], although similar results have been observed for Au(111) and, to a lesser extent, Au(110) [333]. As a function of coverage, the structures formed on Au(100) were a p(2 × 2)-Se at 1/4 coverage, a (2 × $\sqrt{10}$)-Se at 1/3 coverage, a c(2 × 2)-Se at 1/2 coverage, and a (3 × $\sqrt{10}$)-Se at 8/9 coverage [295,334]. LEED patterns, STM images, and proposed structures are displayed in Figs. 53, 54, and 55, respectively. The low-coverage p(2 × 2)-Se and (2 × $\sqrt{10}$)-Se structures are direct analogs of structures formed by Te on Au(100) [162,320], and a S structure similar to the higher coverage (3 × $\sqrt{10}$)-Se has been observed to form on Au(111) [358–360].

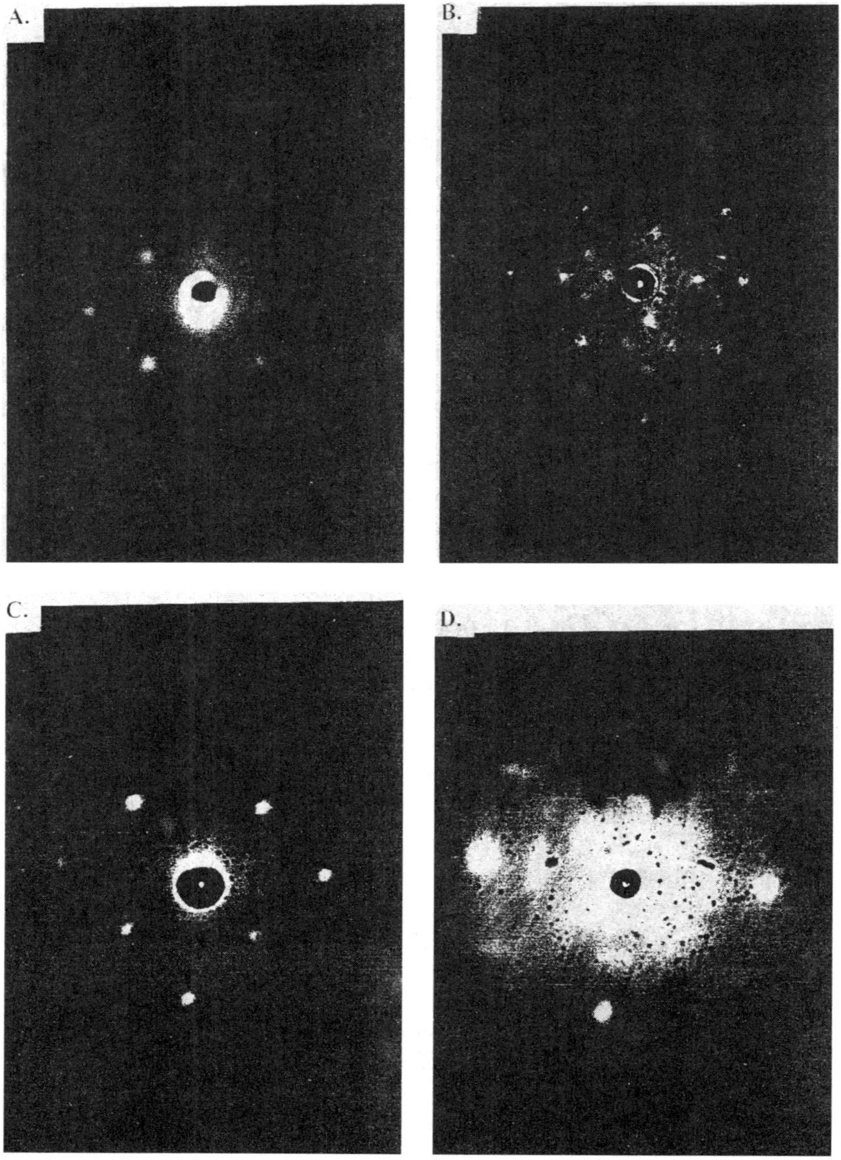

FIG. 53. LEED patterns for Se adsorbed on Au(100): (A) p(2 × 2)-Se, 1/4 ML coverage, 36.7 eV; (B) (2 × $\sqrt{10}$)-Se, 1/3 ML coverage, 33.5 eV; (C) c(2 × 2)-Se, 1/2 ML coverage, 32.9 eV; (D) (3 × $\sqrt{10}$)-Se, 8/9 ML coverage, 33.3 eV; (From Ref. 295.)

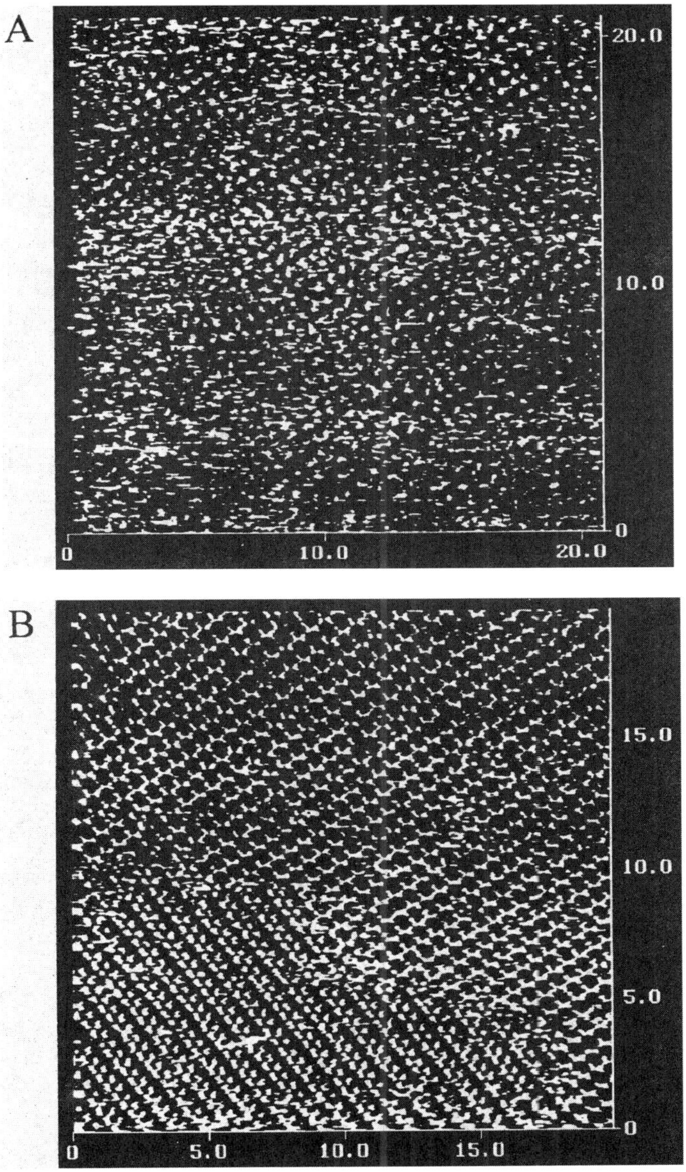

FIG. 54. STM micrographs of Se atomic layers on Au(100): (A) p(2 × 2)-Se, 1/4 coverage; (B) $(2 \times \sqrt{10})$-Se, 1/3 coverage; (C) c(2 × 2)-Se, 1/2 coverage; (D) increasing number of Se_8 rings; (E) close-packed Se_8 rings, $(3 \times \sqrt{10})$-Se; (F) pits formed at high Se coverages.

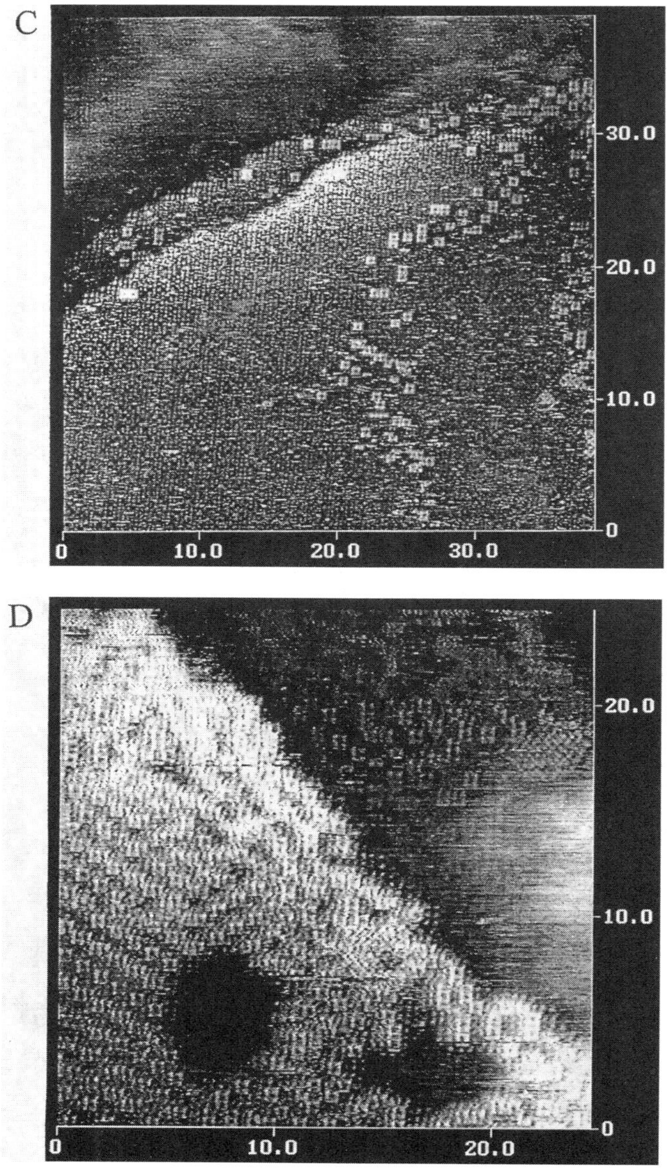

FIG. 54. Continued.

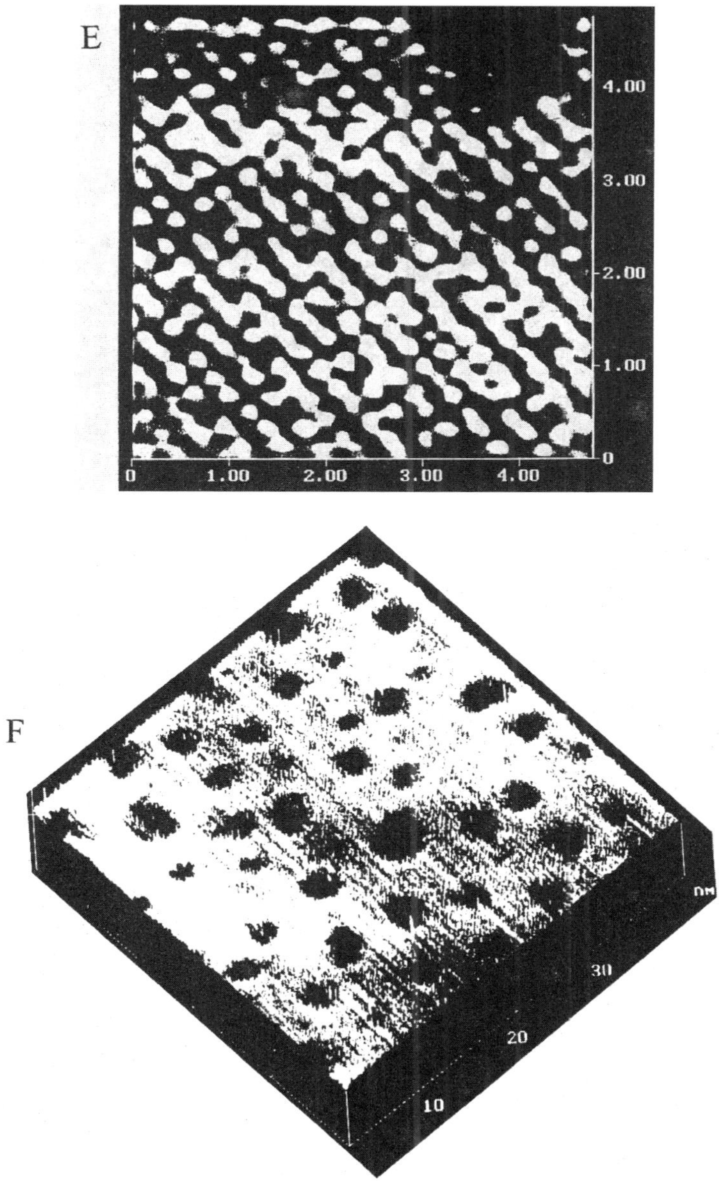

FIG. 54. Continued.

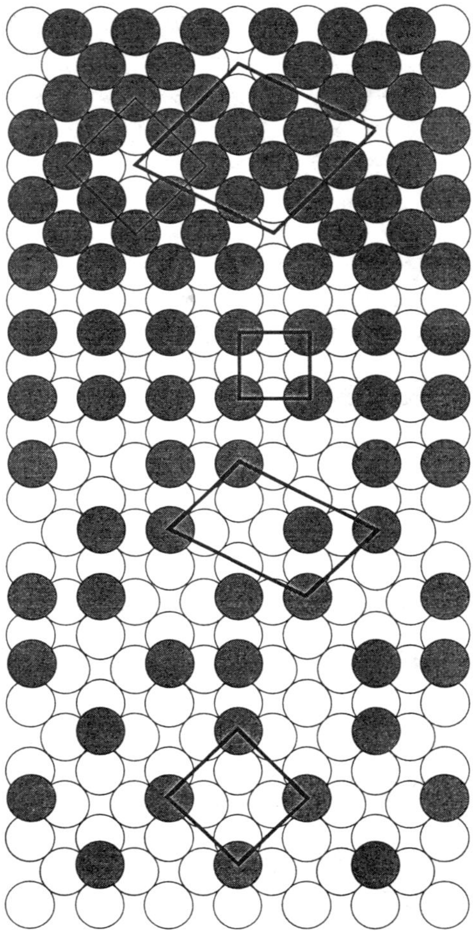

FIG. 55. Diagram of structures proposed for Se atomic layers formed on Au(100).

As previously mentioned, the first atomic layer, in contact with the substrate, appears to be the most critical. Electrochemically there are three different ways to form the first atomic layer of Se, and graphs depicting each are displayed in Fig. 56. The first method is straightforward—the direct reductive deposition of Se atomic layers from a $HSeO_3^-$ solution. The voltammetry for Se deposition is shown in Fig. 57 and displays two feature, C_1 and C_2. C_1 could be thought of as a "UPD" feature because it

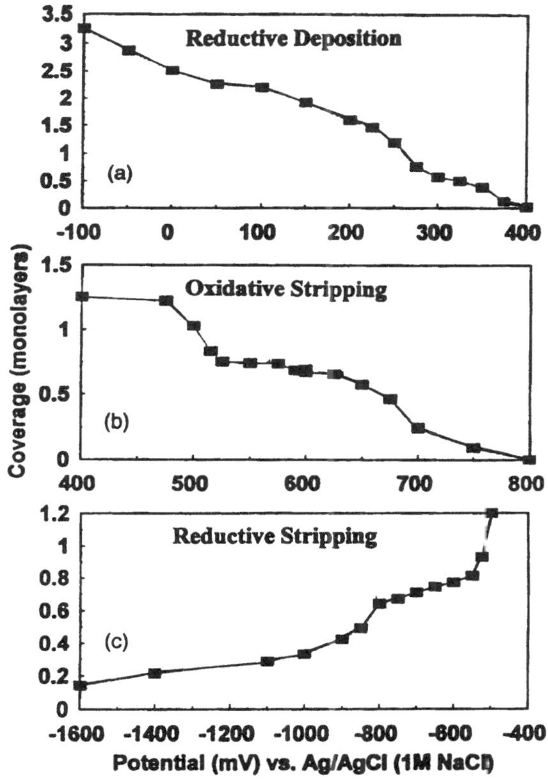

FIG. 56. Se coverage vs. potential curves on Au(100). Coverage values were obtained by coulometry, using anodic stripping voltammetry of deposit. (a) Reductive deposition of Se. (b) Anodic stripping of bulk Se (bulk = ~1.25 monolayers). (c) Cathodic stripping of bulk Se (bulk = ~1.25 monolayers). (From Ref. 295.)

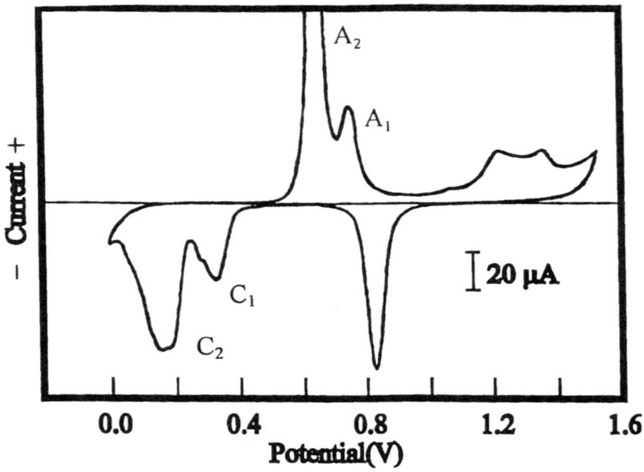

FIG. 57. Voltammetry of a Au "tri" crystal in 1 mM $HSeO_3^-$.

corresponds to the deposition of less than a monolayer of Se. C_2 appears to also result from a surface-limited reaction, but corresponds to more than 2 ML of Se. These features show up as small plateaus in Fig. 56a, which is a graph of the coverage observed after sitting for 2 minutes at the indicated deposition potential. Examination of the voltammetry alone might lead to the conclusion that UPD results in a single 1/2 ML structure. The second, larger feature (C_2) results in a coverage considerably higher than is normally associated with UPD. That is, UPD results from stabilization of a deposit via proximity to the substrate [22,23], and there are instances where a second monolayer does show some stabilization [26,361], however, in the present case (peak C_2) a coverage of over two Se monolayers is observed. It is interesting to note that the measured equilibrium potential for Se in this solution is about 0.4 V, positive of even C_1. Evidently, even the "UPD" feature has been deposited at an overpotential. Features in the reduction voltammetry (Fig. 57) appear to be very much a function of slow kinetics and will be discussed later.

There are two other methods of producing atomic layers of Se, both involving two steps. The first step in both methods is the deposition of about 1.25 ML of Se, while the second step involves stripping off the excess Se to leave the desired atomic layer. Figure 56b shows the resulting Se coverages left after oxidative stripping at various potentials. A broad plateau is visible at 3/4 ML coverage, and there is some indication of a halt near 1/4 ML. A similar graph is shown in Fig. 56c for the reductive stripping of excess Se in a borate solution, leaving an atomic layer. The same plateaus observed in Fig. 56B for oxidative stripping are observed in Fig. 56c for reductive stripping: one near 3/4 coverage and the other between 1/3 and 1/4 coverage [295].

A first atomic layer of Se can be formed by any one of the three methods. In construction of an ECALE cycle, however, the second and successive atomic layers of Se must be formed by the reductive stripping method (Fig. 56c), because use of a positive deposition or stripping potential would result in loss of the previously deposited group II element, as discussed previously.

The structures diagrammed in Fig. 55 can be formed by any one of the three methods described above; however, kinetics play a very important role in determining the homogeneity of the surface [333,334]. The fact that each of the structures shown in Fig. 55 can be formed at the same overpotentials (feature C_1 in Fig. 57) makes it difficult to isolate one to the exclusion of any others. The low-coverage structures p(2 × 2), (2 × $\sqrt{10}$), and c(2 × 2) are all difficult to form homogeneously over the surface. Deposition of less than 1/4 ML of Se generally resulted in no discernible structure, probably due to a high Se atom surface mobility at such a low coverage. Similar studies of Au(111) evidenced the same problem imaging the low-coverage Se unless higher resistance conditions were employed, whereupon some very interesting clustering of Se was observed at coverages below 1/4 ML [333]. As the Se coverage increased to a 1/4 ML on Au(100), the p(2 × 2) structure (Fig. 54A) was imaged but was frequently found to coexist with domains of the (2 × $\sqrt{10}$) (Fig. 54B). In Fig. 54C, at 1/3 coverage, domains of the (2 × $\sqrt{10}$) clearly coexist within domains of the c(2 × 2), and some Se_8 rings are present as well.

At coverages above 1/2 ML, the number of Se_8 rings increases (Fig. 54D). Present understanding of the rings is that they consist of a square arrangement of eight Se atoms and appear to be bound to the Au substrate in the first Se layer. STM height measurements show height differences of 0.05–0.1 nm, not the 0.25 nm expected if the rings represented a second layer of Se. As the number of rings increased still further, they began to coalesce into a close-packed structure, some domains of which are visible in Fig. 54D. It is interesting to note that the close-packed rings are rotated 45° from the isolated rings. Sequential STM images of low-density Se_8

rings showed that they are mobile, at least under the imaging conditions, that is, they act like molecules, with structural integrity. Evidently when they pack together, they rotate to achieve a better fit. Analogous Se_8 rings in crystalline Se adopt a chair conformation [362]; however, bonding with the Au substrate appears to facilitate the planar configuration.

Coincident with the formation of Se_8 rings was the formation of pits on the surface (Figs. 54D, E, and F). Cross sections of the pits indicated them to be a single atom deep (about 0.3 nm) and to have the same Se_8 ring structure on their floor. Pit formation appears to be reversible and connected to the appearance and disappearance of the ring structures on the surface. For well-prepared Au surfaces, no pits are visible until the rings begin to form. If the Se coverage is decreased by oxidation, the rings appear to go away as well. The pits account for about 10% of the total surface area at Se coverages near 1 ML. Similar pits are frequently observed on Au surfaces where long-chain alkane thiols have been adsorbed [363–369]. Those pits are usually a single atom deep as well and have a similar distribution over the surface. Several explanations have been proposed, including contamination or defect sites [363,364], small amounts of etching [366–369], and the idea of a surface reconstruction where the lateral extent of the surface Au atoms actually shrinks due to bonding with the thiols [365,370]. At present, the last two explanations fit most closely with the observations described here. A small amount of dissolution could easily account for the observed pits. Alternatively, for 10% of the surface to be covered with pits, a change of only 3% in the Au-Au distance would be required, which is difficult to rule out with STM. Elemental Se_8 rings naturally have a Se-Se bond distance of 0.233 nm [362], while the Se-Se distances in the square Se_8 ring structures on Au (Fig. 54) appear to coincide more closely with the 1×1 arrangement on the Au surface or a Se-Se distance of close to 0.29 nm. The increased Se-Se distance in the adsorbed Se_8 rings probably involves significant strain associated with forming a layer commensurate with the Au surface. This strain could in turn account for the compression of the surface Au atoms and the observed pits.

Ideally, the starting point for the formation of CdSe deposits using an ECALE cycle would be formation of a homogeneous, stable, 1/2 coverage Se structure, as described previously in the case of CdTe. Isolation of the pure Au(100)c(2×2)-Se with a Se coverage of 1/2, as described above, was difficult to achieve under all conditions investigated. Multiple structures were nearly always present simultaneously on the resulting surfaces due to the sluggish kinetics of Se electrodeposition. However, as previously described in the case of CdTe formation, the structure of the chalcogenide atomic layer may not be as important as the overall chalcogenide coverage. The mobility of the chalcogenide during the subsequent deposition of the Cd then becomes very important. The Se atoms must move from areas of high coverage, such as where the Se_8 rings are formed, to areas with a lower Se coverage, such as domains of the 1/3 coverage $(2 \times \sqrt{10})$-Se structure. If this mobility were not observed, domains of the Se_8 rings would probably result in islands of CdSe more than a monolayer thick.

Voltammetry for Cd deposition on the Au ‘‘tri’’ crystal (a single crystal that has been oriented and polished on three faces, each to a different low index plane) is shown in Fig. 58A. A very broad UPD feature, peaked at 0.0 V is evident. The Cd coverage of the UPD feature is a function of the low potential limit and the background corrections chosen for the integration, but is nominally about 1/3 ML or less [371]. The formal potential for Cd appears to be near −0.75 V, as similar scans to −0.8 V and below show a sharp stripping spike in the oxidative scan near −0.75 V. There is a gradual increase in the Cd-deposition current beginning at −0.4 V (Fig. 58A) as the potential is scanned negatively. Examination of the literature concerning Cd UPD indicates that alloying of Cd with the Au electrode [372] frequently takes place, which could account for this current. This alloy formation, also occurring at an underpotential, is not strictly a surface-limited reaction as is the UPD. From the voltammetry (Fig. 58A) it appears that most alloy formation can be avoided if the potential is kept above −0.5 V.

For the initial studies of Cd deposition on Se atomic layers, the 1/3 coverage Au(111)$(\sqrt{3} \times \sqrt{3})$R 30°, the 1/2 coverage Au(100)c(2×2), and the 2/3 coverage Au(110)(3×2) (Figs. 59A, B, and C, respectively) were selected. For those structures the ideal coverages are 4.5×10^{14}, 5.8×10^{14}, and 5.5×10^{14} atoms/cm^2, respectively. Given the constraints of the Se-deposition kinetics previously discussed, these Se adlattices were formed by scanning to just past the Se UPD peak (C_1 in Fig. 57), where the potential was then held for 2 minutes. The Cd-deposition voltammetry shown in Fig. 58B was performed on the Au(100)c(2×2)-Se surface. The UPD feature is significantly increased relative to Cd UPD on clean Au (Fig. 58A) and has shifted negatively by 200 mV. The peak shift indicates that Cd at low coverages is more stable on clean Au than on the Se-coated Au. At −0.3 V, the Cd coverage for deposition on clean Au vs. Se-coated Au

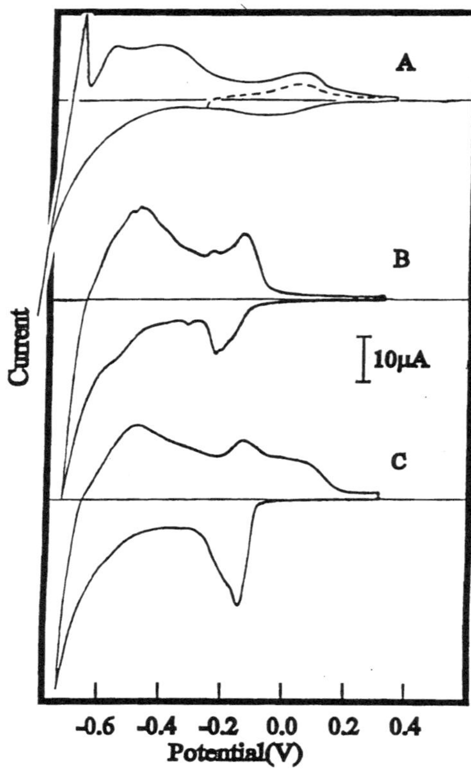

FIG. 58. Cd-deposition voltammetry from a mM $CdSO_4$ solution on (A) clean Au, (B) 1/2 ML of Se, (C) full ML of Se.

is 1/3 vs. 1/2 ML. As the potential is scanned to still more negative potentials, the current characteristic of alloy formation is again evident, regardless of the presence of the Se layer.

Ordered deposits were formed at −0.3 V on all three low-index planes of Au precoated with the Se adlattices listed above. The clearest LEED patterns were observed for deposits formed on Au(111), which happens to be the plane for which ordered Se structures were seen most infrequently with LEED. The Au(100) surface, which evidenced the sequence of Se atomic layers diagrammed in Fig. 55, showed the least tendency to

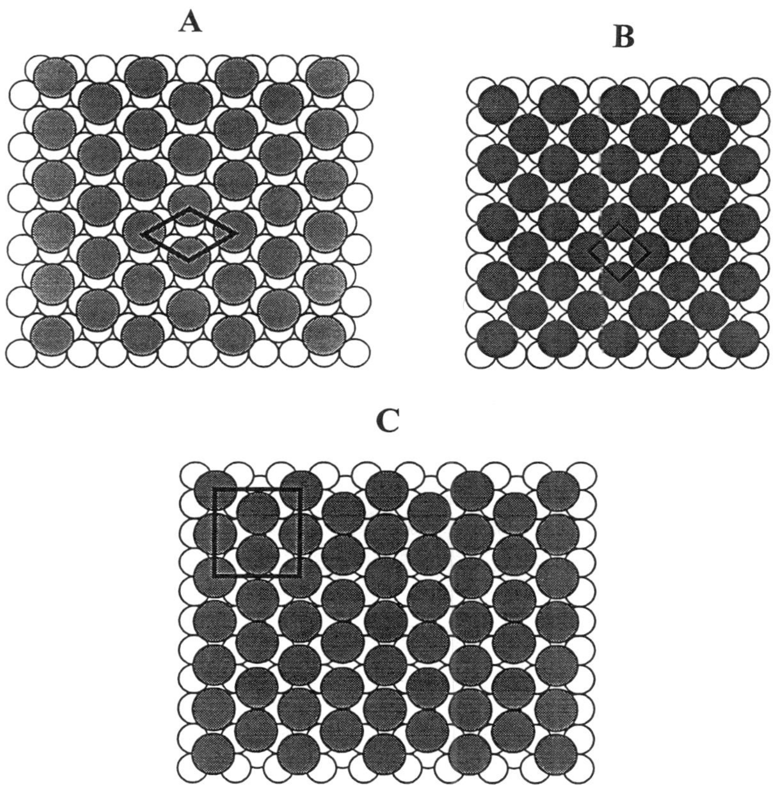

FIG. 59. Se structures on the Au low-index planes used as substrate for Cd deposition: (A) Au(111) $(\sqrt{3} \times \sqrt{3})R\ 30°$-Se, 1/3 coverage; (B) Au(100) $c(2 \times 2)$-Se, 1/2 coverage; (C) Au(110) (3×2)-Se, 2/3 coverage.

form ordered CdSe deposits. This is further evidence that the structures of the II-VI compound monolayers are independent of the structures of the initial chalcogenide atomic layers. Because the CdSe monolayers formed on Au(111) appear to be the most well ordered and understood, they will be described below.

Under the various Se-deposition conditions studied, a $(\sqrt{3} \times \sqrt{3})R\ 30°$-Se structure was the only ordered structure observed on Au(111) (Fig. 60). The $(\sqrt{3} \times \sqrt{3})R\ 30°$ unit cell should correspond to a Se coverage of

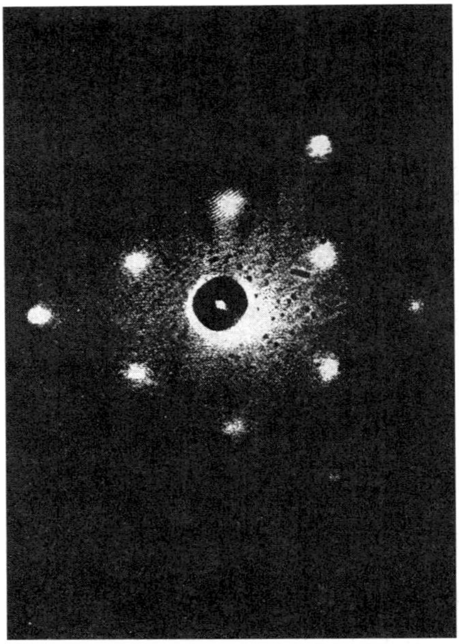

FIG. 60. LEED pattern of a Au(111)($\sqrt{3} \times \sqrt{3}$)R 30°-Se.

1/3 or possibly 2/3 ML. The LEED pattern was observed for deposits emersed at the outset of the UPD feature (Fig. 57) and was still evident for deposits emersed after conclusion of the UPD feature. Coulometry and Auger spectroscopy indicated a Se coverage of 1/3 (333), as did STM (Fig. 61A). Figure 62A is the LEED pattern observed when a potential of -0.25 V (Fig. 58B) was used to deposit Cd on the Au(111)($\sqrt{3} \times \sqrt{3}$)R 30°-Se structure. The unit cell responsible for Fig. 62A is a ($\sqrt{7} \times \sqrt{7}$)R 19.1°-CdSe. A proposed structure is shown in Fig. 63A and indicates a coverage of 3/7 for both Cd and Se. A similar structure was seen in the case of CdTe formation on Au(111) [159,162]. By depositing Cd at a more negative potential, below -0.45 V, a structure displaying a (3×3) unit cell was observed using LEED (Fig. 62B). STM studies, such as those of Se structures in air, were unworkable with the CdSe layers, as the Cd did not show sufficient stability in air. Some in situ STM results have been obtained where Se layers were first formed in a standard electrochemical H-cell and then transferred to a custom electrochemical cell designed for use with a

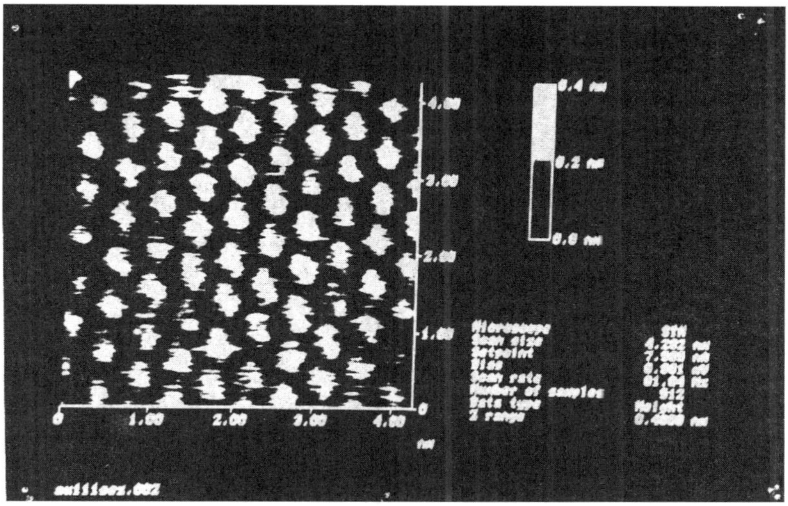

(A)

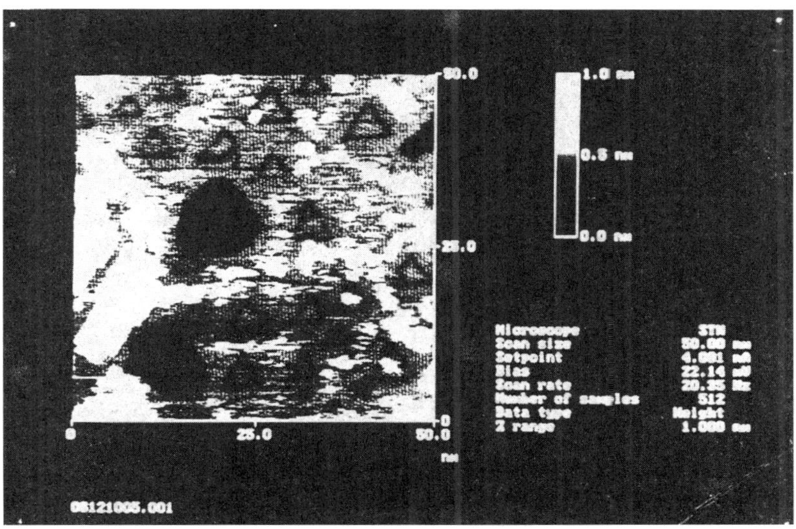

(B)

FIG. 61. STM micrographs of Au(111) supporting Se atomic layers: (A) just the Au(111)($\sqrt{3} \times \sqrt{3}$)R 30°-Se; (B) Au(111)($\sqrt{3} \times \sqrt{3}$)R 30°-Se and some domains of boxes.

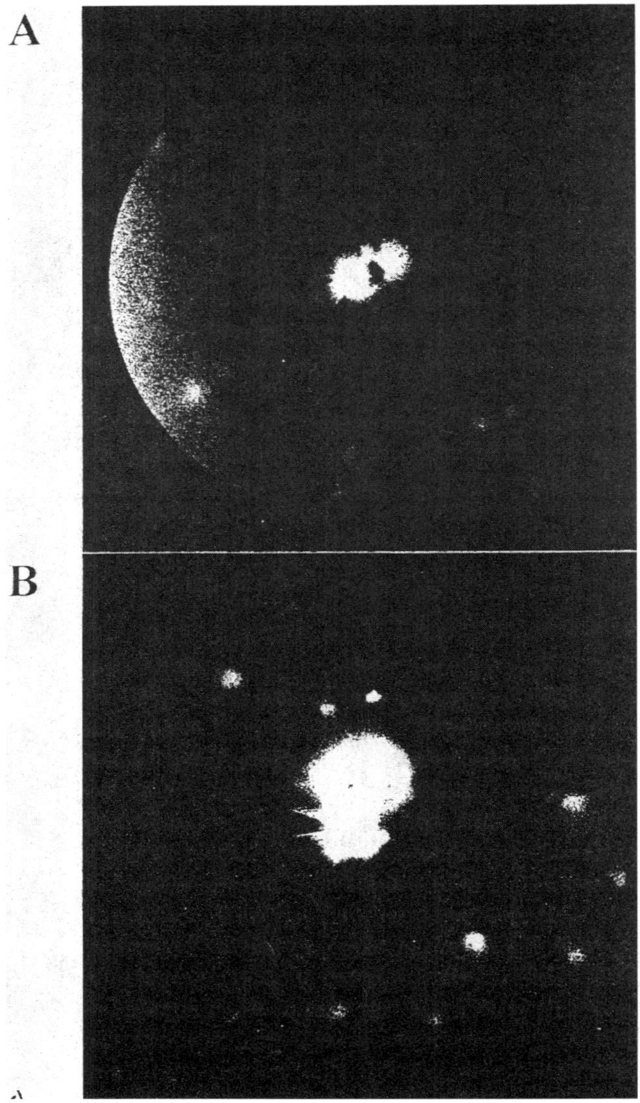

FIG. 62. LEED patterns for CdSe monolayers on Au(111): (A) Au(111)($\sqrt{7} \times \sqrt{7}$)R 19.1°-CdSe; (B) Au(111)(3×3)-CdSe.

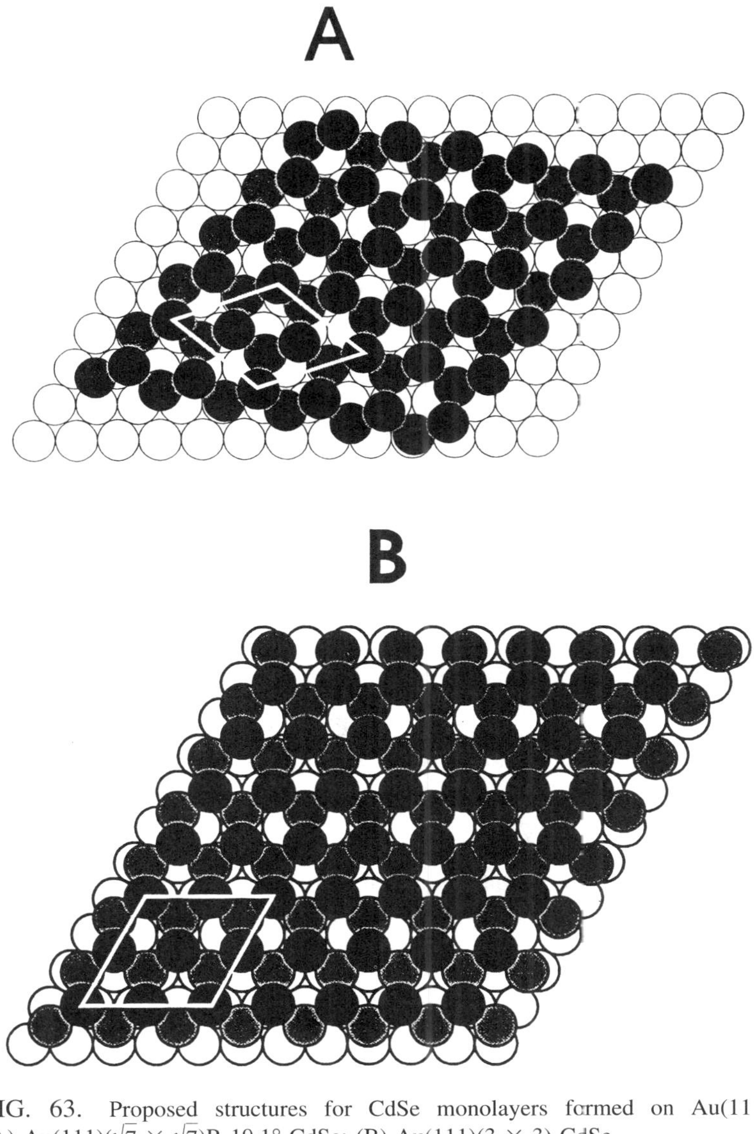

FIG. 63. Proposed structures for CdSe monolayers formed on Au(111): (A) Au(111)($\sqrt{7} \times \sqrt{7}$)R 19.1°-CdSe; (B) Au(111)(3 × 3)-CdSe.

Nanoscope III (373) (Fig. 64A). The Se structure shown in Fig. 64A has domains of both the $(\sqrt{3} \times \sqrt{3})R\ 30°$-Se structure at 1/3 coverage and Se_8 domains at 8/9ths coverage. The overall Se coverage on the surface was probably close to 0.6 ML. The image was obtained at -0.1 V in the Cd^{2+} solution, where no Cd had yet deposited. The image shown in Fig. 64B was taken on the same surface after shifting the potential to -0.4 V, where Cd had deposited. The image displays a surface that has an orientation and dimensions consistent with the (3×3) LEED pattern (Fig. 62B). A proposed structure is shown in Fig. 63B that suggests slightly higher Cd and Se coverages, 4/9 (0.444), vs. the 3/7 (0.429) in the $(\sqrt{7} \times \sqrt{7})R\ 19.1°$-CdSe structure (Fig. 63A). An equivalent structure has been reported by other workers using TEM in a study of electrodeposited CdSe nanoclusters on Au(111) substrates [115,125]. This increase in the Cd coverage to 0.444 has resulted in a structure closely resembling the hexagonal basal plane of CdSe (Wurtzite) and involves a lattice mismatch of only 0.6% between the CdSe and Au.

Also evident in Fig. 64B are small light-colored clusters close to 1/2 nm in size. At present it appears that these clusters are extra CdSe formed because too much Se was present on the surface initially (Fig. 64A). As mentioned above, the Se coverage prior to Cd deposition appeared close to 0.6 ML—significantly greater than the 0.44 needed for the structure diagrammed in Fig. 63B. Studies have shown that Cd will react nearly stoichiometrically with up to several monolayers of Se [107]. These results are consistent with the discussions of CdTe formation in Fig. 41C, where a second layer begins to form and results in island on the surface.

In the work described above concerning the formation of CdSe monolayers, the Se was the first atomic layer formed. Cd can be used as the first atomic layer as well, although the structures of Cd UPD layers on Au are not well characterized. Subsequent Se deposition resulted in a structure with a (3×3) unit cell on Cd-coated Au(111), very similar to that formed when the Se was deposited first (Fig. 62B). There is some evidence that when the Se is deposited first, it remains under as the Cd is deposited, and conversely, that when the Cd is deposited first it remains under subsequently deposited Se [333].

Progress on understanding the surface chemistry relevant to the formation of compound semiconductors is being made. One major issue is the genesis of defects that appear in deposits formed with the flow deposition system. Probable defect sources include the substrate quality, lattice mismatch problem, and problems associated with deposition of a compound

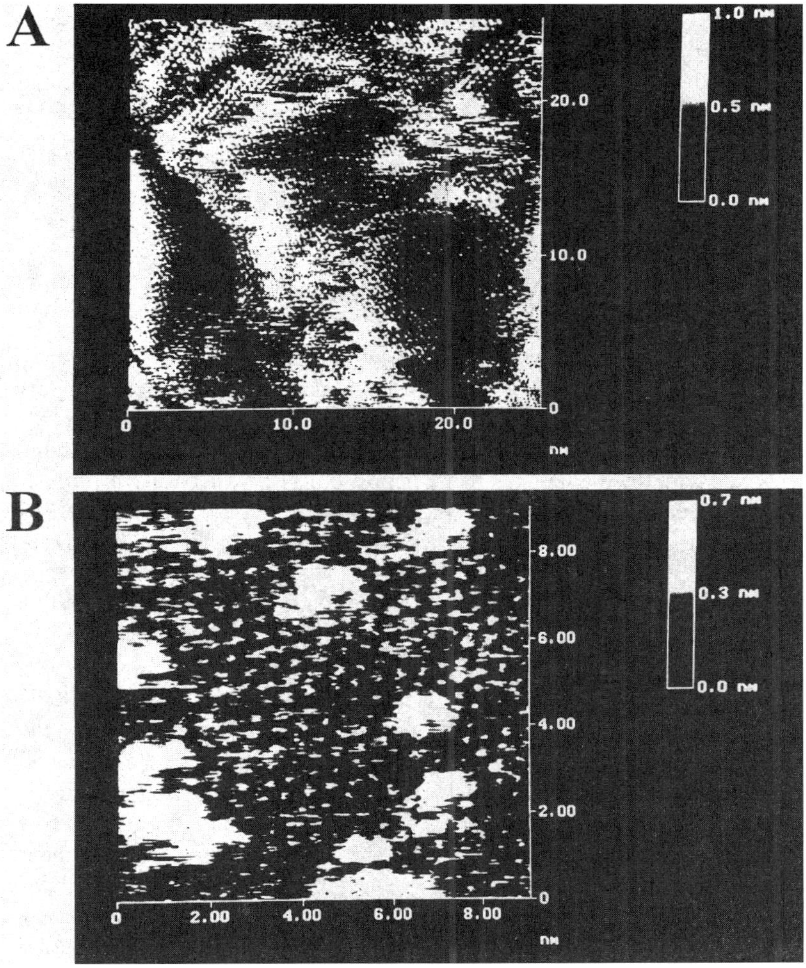

FIG. 64. (A) In situ STM image of Se layer formed on Au(111). Some domains of ($\sqrt{3} \times \sqrt{3}$)R 30°-Se are present as well as many domains of the Se_8 structure, in mM $CdSO_4$, at −0.1 V. (B) The same surface, only at −0.4 V. The hexagonal (3 × 3)-CdSe structure (Fig. 63B) is evident, as well as small puffs, probably the beginning of a second layer of CdSe.

on an elemental substrate [374–383]. Many of these problems are related to the choice of Au as a substrate. The next section describes attempts to better understand the electrochemistry of compound surfaces. Given sufficient control of compound surface chemistry in aqueous solutions, it may be possible to use them as lattice-matched substrates in subsequent deposition studies. Other efforts in this group concern the formation of other compounds and an increased effort to investigate the deposition process one atomic layer at a time using in situ STM.

V. DIGITAL ELECTROCHEMICAL ETCHING

Au is an excellent electrode material. It is inert in most electrochemical environments, and its surface chemistry is moderately well understood. It is not, however, the substrate of choice for the epitaxial formation of most compounds. One major problem with Au is that it is not well lattice matched with the compounds being deposited. There are cases where fortuitous lattice matches are found, such as with CdSe on Au(111), where the $\sqrt{3}$ times the lattice constant of CdSe match up with three times the Au (Fig. 63B) [115,125]. However, there is still a 0.6% mismatch. A second problem has to do with formation of a compound on an elemental substrate (Fig. 65) [384–387]. Two types of problems are depicted in Fig. 65. In Fig. 65A the first element incompletely covers the surface, so that when an atomic layer of the second element is deposited, antiphase boundaries result on the surface between the domains. These boundaries may then propagate as the deposit grows. In Fig. 65b the presence of an atomically high step in the substrate is seen to also promote the formation of antiphase boundaries. The first atomic layer is seen to be complete in this case, but when an atomic layer of the second element is deposited on top, a boundary forms at the step edge. Both of the scenarios in Fig. 65 are avoided by use of a compound substrate.

For deposition of II-VI compounds, three scenarios for lattice-matched compound substrates are readily available: (1) deposition on the compound itself, homoepitaxy; (2) deposition on a second II-VI compound that happens to be lattice matched, such as HgTe on CdTe; or (3) deposition on a corresponding III-V compound. There are two problems with homoepitaxy—one is why? Most of the time there is no need to grow a layer of a compound on a substrate of the same compound. There are cases where the substrate is grown by a method that produces an inexpensive but inferior material, and a high-quality ''epi'' layer of the same compound is grown

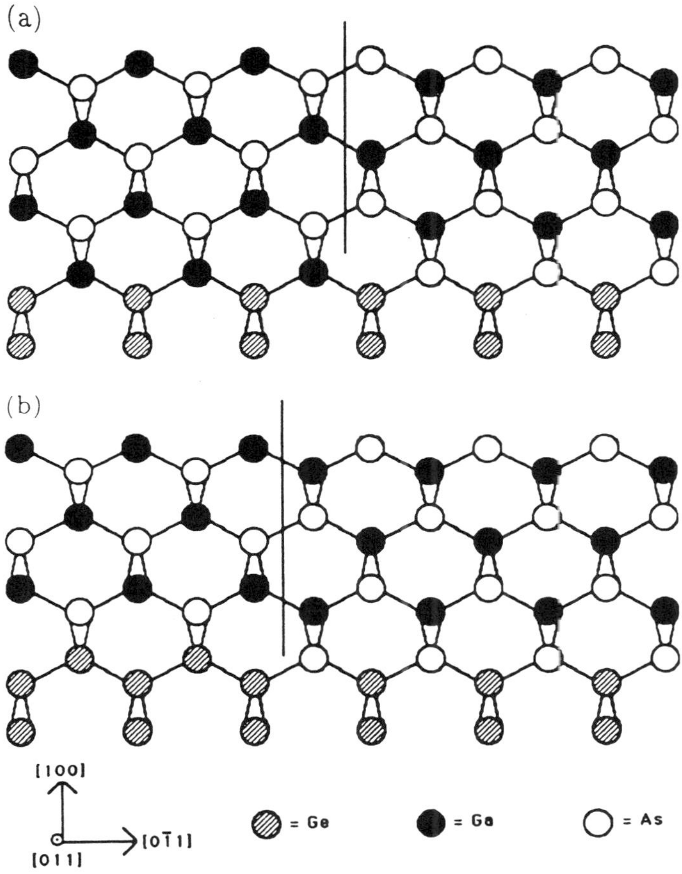

FIG. 65. Antiphase domain formation in polar on nonpolar epitaxy: (a) incomplete prelayer coverage, (b) odd step height. (From Ref. 387.)

on top. These epi layers can also be used to provide an independently doped layer, depending on the device structure. The second problem with homoepitaxy is that the ability to analyze the quality of the resulting material is decreased, because most techniques have trouble differentiating between the substrate composition or structure and that of the deposit.

The use of a lattice-matched II-VI compound as a substrate works if lattice-matched substrates are available. However, even in the case of HgTe on CdTe there is a slight mismatch. Frequently the solution has been to alloy in a third element to adjust the lattice constant. For example, substrates used to make MCT detectors ($Hg_xCd_{1-x}Te$) frequently have at least a very thin buffer layer of CdTe alloyed with Zn to provide a better lattice match [388,389].

A third lattice-matching scenario is to use a corresponding III-V compound for a substrate. For instance, the lattice constants for CdTe and InSb are both listed as 0.648 nm [390], and similar lattice matches are found between the other II-VI and III-V compounds. As with the II-VI compounds, there are a number of situations where the lattice constants of the substrate can be incrementally adjusted by alloying with a third element. In addition, some high-quality wafers of III-V compounds are available commercially at reasonable prices.

The problem with all three of the above scenarios is that they require an understanding of the surface chemistry of compound semiconductor in aqueous solutions. Much more is known about the surface chemistry and reactivity of Au in aqueous solutions. A prerequisite, then, to the use of a compound semiconductor as a substrate for compound electrodeposition is to gain a better understanding of the substrate's reactivity under electrochemically relevant conditions. Our initial studies of compound reactivity in electrochemical environments involved CdTe single crystals [391]. The electrochemistry of CdTe is reasonably well understood from electrodeposition studies (Table 1), and single crystals are commercially available.

For a compound semiconductor to be useful as a substrate in studies of electrodeposition, it is desirable that clean, unreconstructed, stoichiometric surfaces be formed in solution prior to electrodeposition. For CdTe, the logical starting point is the standard wet chemical etch used in industry, a 1–5% Br_2 methanol solution. A CdTe(111) crystal prepared in this way was transferred directly into the UHV-EC instrument (Fig. 39) and examined [391]. Figure 66B is an Auger spectrum of the CdTe surface after a 3-minute etch in a 1% Br_2 methanol solution. Transitions for Cd and Te are clearly visible at 380 and 480 eV, respectively, as well as a small feature due to Br at 100 eV. No LEED pattern was visible, however. As described previously, a layer of solution is generally withdrawn with the crystal as it is dragged (emersed) from solution (the emersion layer). After all the solvent has evaporated, the surface is left with a coating composed of the

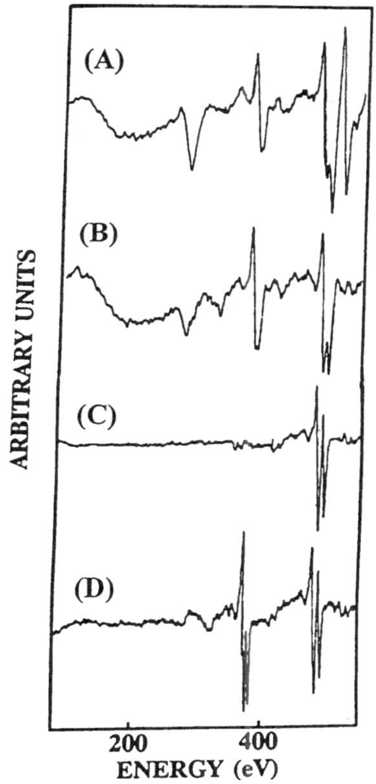

FIG. 66. Auger spectra: (A) freshly polished, oxidized, and contaminated (111)Cd surface of a CdTe crystal; (B) (111)Cd surface after a 3-min etch in 1% Br_2/CH_3OH solution; (C) Br_2/CH_3OH-etched surface after rinse in 1 mM H_2SO_4 at -0.1 V; (D) (111)Cd surface after electrochemical reduction at -2 V in mM H_2SO_4 for 10 min. (From Ref. 391.)

electrolyte initially contained in the emersion layer. In many cases it can be rinsed away prior to analysis using a dilute solution. The spectrum shown in Fig. 66C is for a Br_2/methanol etched crystal after a rinse in dilute H_2SO_4. Both the Br and Cd signals have disappeared. Evidently, the Cd and Br were present as $CdBr_2$ salt, which was easily rinsed from the surface. The low Auger signal for Br in Fig. 66B can be accounted for by its low Auger sensitivity factor [392]. Again, no LEED pattern was observed for the resulting Te-rich surface. That a Te layer was present is not surprising, as Cd is the less noble element and should be preferentially etched in the Br_2 solution. That the layer is disordered is consistent with having removed Cd atoms from the CdTe zinc blende lattice, leaving behind Te.

Figure 66D was obtained for the same CdTe(111) crystal after electrochemical reduction at -2.0 V for 2 minutes. Transitions for both Cd and Te are evident, and the Cd/Te peak height ratio is similar to that observed by other workers for stoichiometric CdTe [393,394]. In addition, well-ordered (1×1) LEED patterns (Fig. 67) were observed on both the CdTe(111)-Cd and CdTe(111)-Te faces. This is in contrast to CdTe surfaces prepared by ion bombardment, where postbombardment annealing was required to produce a LEED pattern, and the annealing appeared to result in formation of a reconstructed surface. In summary, well-ordered, clean, and unreconstructed CdTe surfaces have been produced using a wet etching/electrochemical treatment.

The results described above suggested that an electrochemical analog of digital etching might be achievable. Digital etching is a term used to describe methodologies where a material is decomposed an atomic layer at a time in a cycle [395–399]. It fits under the heading of atomic layer processing, along with ALE. The benefits are increased control over the etching process and etched depths. One obvious use for digital etching is where it is desirable to carefully and homogeneously remove some small number of monolayers, say 1–500. Historically, digital etching cycles have involved the alternated exposure of a substrate first to a gas source and second to an energetic beam of electron, ions, photons, or neutrals. Examples of gases include the halides, hydrogen halides, or halide sources such as SF_6. One scenario would be to expose a surface to a reactive gas, forming an adsorbed layer in a surface-limited reaction. After pumping away the excess, the surface is exposed to an energetic beam, promoting the formation of a volatile species composed of stoichiometric amounts of the adsorbed gas and the material being etched. The extent of etching is then limited by the amount of the previously adsorbed gas.

An electrochemical analog of digital etching can easily be envisioned, where first some reactant is adsorbed in a surface-limited reaction, and then the potential is switched to one where a product species is produced, stoichiometric in the adsorbate and the substrate. The results of the CdTe etching study described above suggested still another scenario, where no adsorbate is involved, just two electrochemical potentials. Figure 68 is a

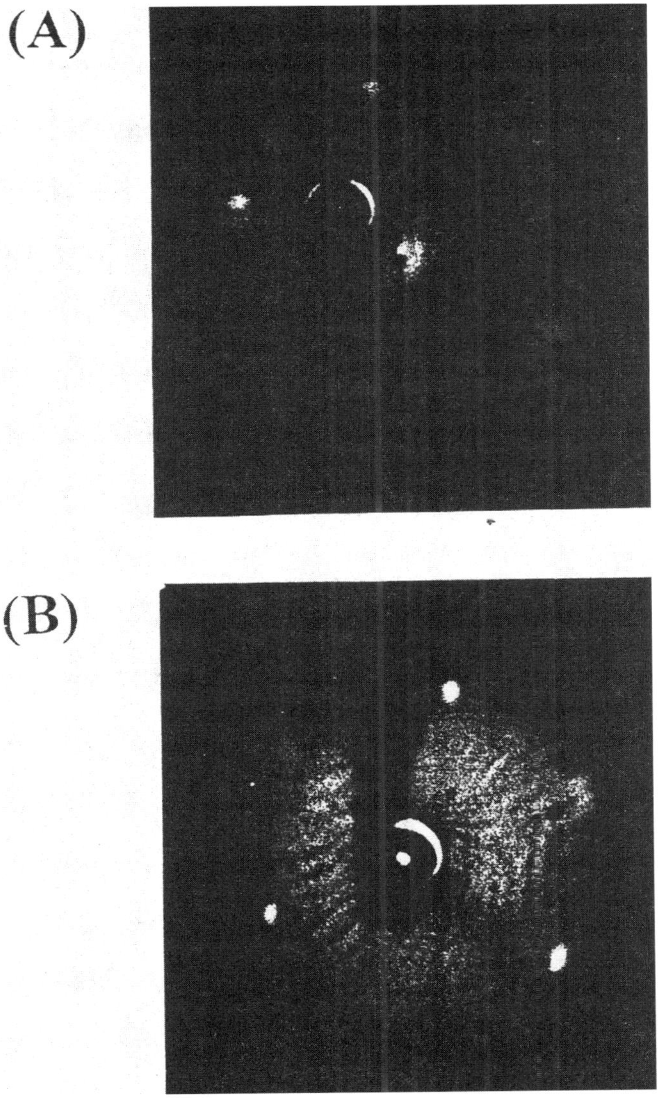

FIG. 67. LEED patterns: (A) (111)Cd surface after etching, rinsing, and electrochemical reduction (beam energy: 59 eV); (B) (111)Te surface after etching, rinsing, and electrochemical reduction (beam energy: 64 eV). (From Ref. 391.)

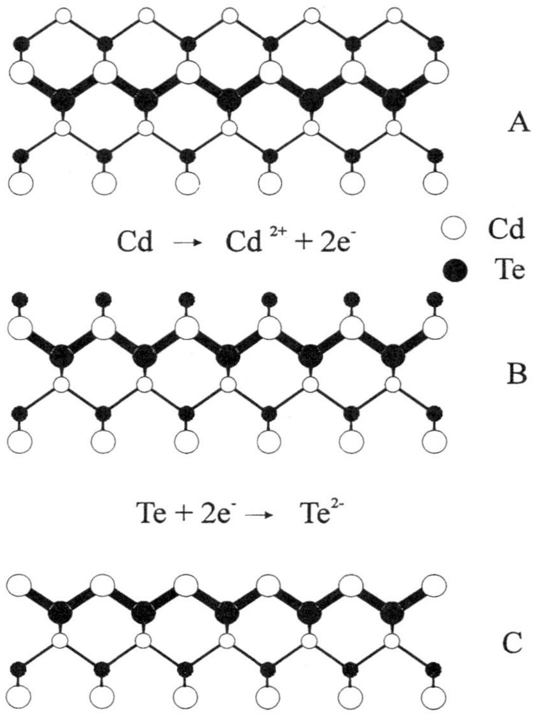

FIG. 68. Schematic illustrating the electrochemical digital etching process on CdTe(100). A) initial surface, B) after oxidative stripping of a Cd atomic layer, C) after reductive stripping of the Te atomic layer.

side view of a CdTe crystal. The structure is drawn such that the (100) plane of CdTe is the top, and we are looking down the (110) direction. The figure has been drawn to suggest that the top layer of atoms contains all Cd atoms. The premise of this electrochemical digital etching cycle is that there is a potential at which only the top layer of Cd atoms will be removed, because they are less stable than interior Cd atoms, less coordinated. Removal of the Cd atoms will expose a layer of Te atoms. From the above studies it appears that these undercoordinated Te atoms can be removed by electrochemical reduction to reform the Cd-terminated surface, completing removal of a single compound monolayer [44,313,400].

Studies of electrochemical digital etching in this group [44,313,400] have taken two tracks, the first involving UHV-EC studies of CdTe single-crystal surfaces such as those described above [391]. The second track involved the use of atomic force microscopy (AFM) to follow etch depths on CdTe single crystals subjected to 150 cycles. Patterned photoresist coated crystals were used so that the etch depth could be accurately determined. Given 0.648 nm as the lattice constant for CdTe (zinc blende) and definition of a monolayer of CdTe as being one atomic layer of Cd and one of Te (Fig. 68), a monolayer should be half the thickness of the unit cell in the (100) direction, or 0.328 nm thick. One hundred and fifty cycles of etching should then remove 52.2 nm of CdTe.

Ideal substrates for digital etching studies using the AFM, would start off atomically flat. However, preparation of an atomically flat CdTe surface is not a simple matter. Commercially polished CdTe has striations the size of the last grit used unless a wet chemical etch is used as a last polishing step. The homogeneity and extent of leveling produced by the wet etch, especially at an atomic level, is quite variable. Figure 69A is an AFM image of a CdTe crystal surface prior to electrochemical digital etching, after a Br_2/methanol etch. The roughness is considerable because the experiment is only expected to remove 50 nm of CdTe. This surface was then coated with a commercial photo-resist and developed so that it consisted of 2-μm-wide lines of photoresist 2 μm apart. This photo-resist–covered surface was then subjected to 150 cycles of electrochemical digital etching in a thin layer flow cell [44]. The photo-resist was then removed and imaged again with AFM (Fig. 69B). The measured height was close to that predicted above but on the high side, indicating that under the conditions used a little more than a monolayer of the compound was being removed each cycle.

In addition to studies using photo-resist–covered substrates and AFM, atomic level studies were performed to help identify the nature of the surface-limited reactions used to form the electrochemical digital etching cycle [313]. In those studies the dependence of etched amounts on the potential used for Cd oxidation was investigated using a UHV-EC instrument (Fig. 39).

Initial treatments of a zinc blende CdTe(100) crystal involved argon ion bombardment followed by a brief anneal. The resulting LEED pattern was a (1×1), indicating an unreconstructed surface. The crystal was then immersed into 50 mM K_2SO_4, pH 5.6. A voltammogram showing the initial reduction of the crystal is depicted in Fig. 70A, exhibiting no significant

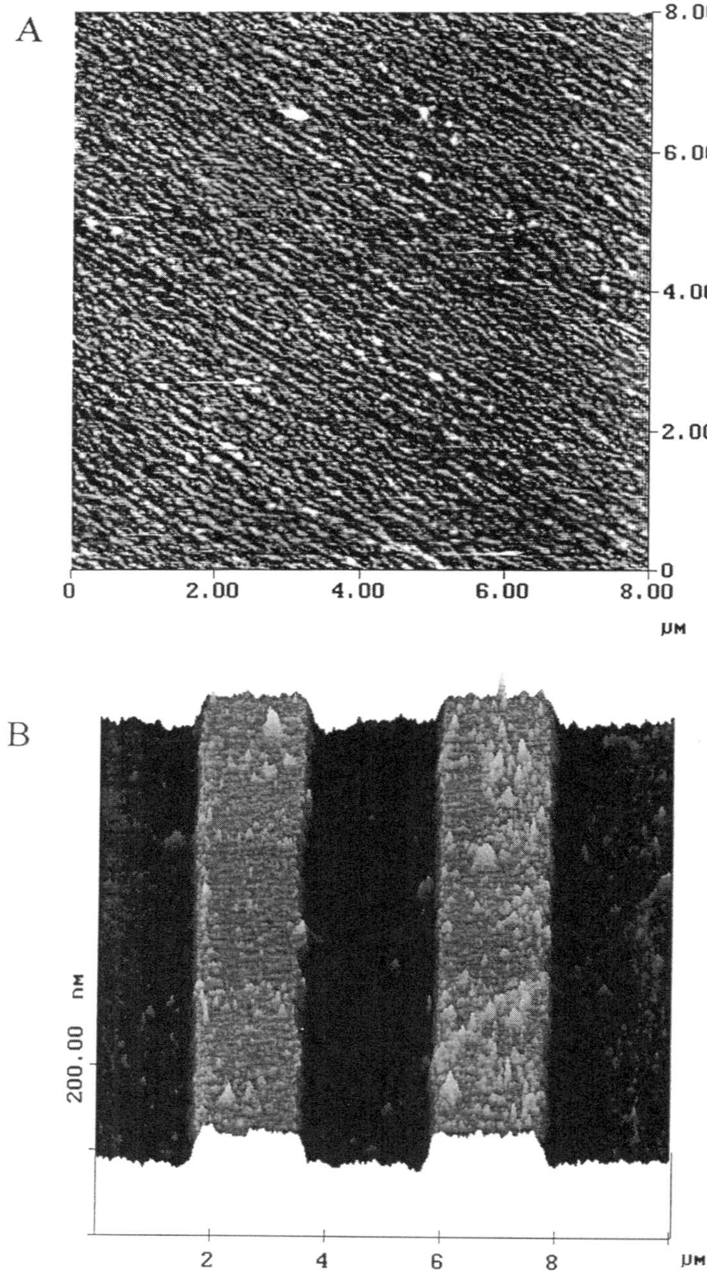

FIG. 69. AFM images of CdTe(100) surfaces (A) before etching and (B) after 150 cycles of electrochemical digital etching.

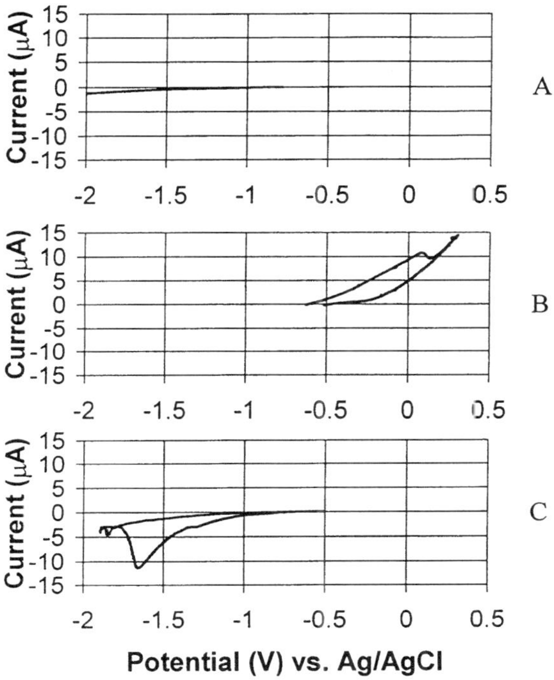

FIG. 70. Voltammograms on an argon ion–bombarded, annealed CdTe(100) surface in 50 mM K_2SO_4; pH = 5.6: (A) reduction from the open circuit potential to −2.0 V; (B) oxidation from the open circuit potential to +0.30 V and reversing to −0.55 V, under illuminated conditions; (C) reduction following (B) from open circuit potential to −1.8 V and reversing to −0.50 V.

reduction features. Subsequent examination of the surface with Auger spectroscopy evidenced no change in the surface composition. No change was observed in the LEED pattern either. The conclusion drawn was that the surface resulting from ion bombardment and annealing contained no excess (reducible) Te and that the resulting surface was Cd terminated.

When an equivalently prepared surface was first scanned in the positive direction, Fig. 70B was obtained. There appear to be two components to the oxidation feature: a peak at about 0.15 V superimposed on a slowly increasing background. Auger spectra of the resulting surface showed a marked decrease in the signal for Cd, indicating that some had been removed. In addition, the (1 × 1) LEED pattern was no longer present—

only diffuse intensity, indicating that some roughening of the surface resulted from the oxidation step.

Subsequent reduction of the surface after the oxidation step is shown in Fig. 70C. A new reduction feature is clearly visible. After scanning through this feature, the Auger spectrum looks as it did after ion bombardment and annealing. In addition, a LEED pattern was again visible. It appears that the initial oxidation step results in a Te-terminated surface, indicated by the presence of reducible Te, and that after its reduction the Cd-terminated surface is regenerated, suggesting that the digital etching cycle can be performed.

In order to gain a better understanding of the potential dependence of this process, a series of chronoamperograms were run (Fig. 71). In each case, the crystal was pretreated the same way, by ion bombardment and annealing. It was then stepped to the indicated potentials in the 50 mM K_2SO_4 solution and held. Two types of current appear to flow in the system. In the first 60 seconds, a transitory current flows, while in addition a constant background current flows the whole time. The background current is a function of potential, increasing significantly the more positive the oxida-

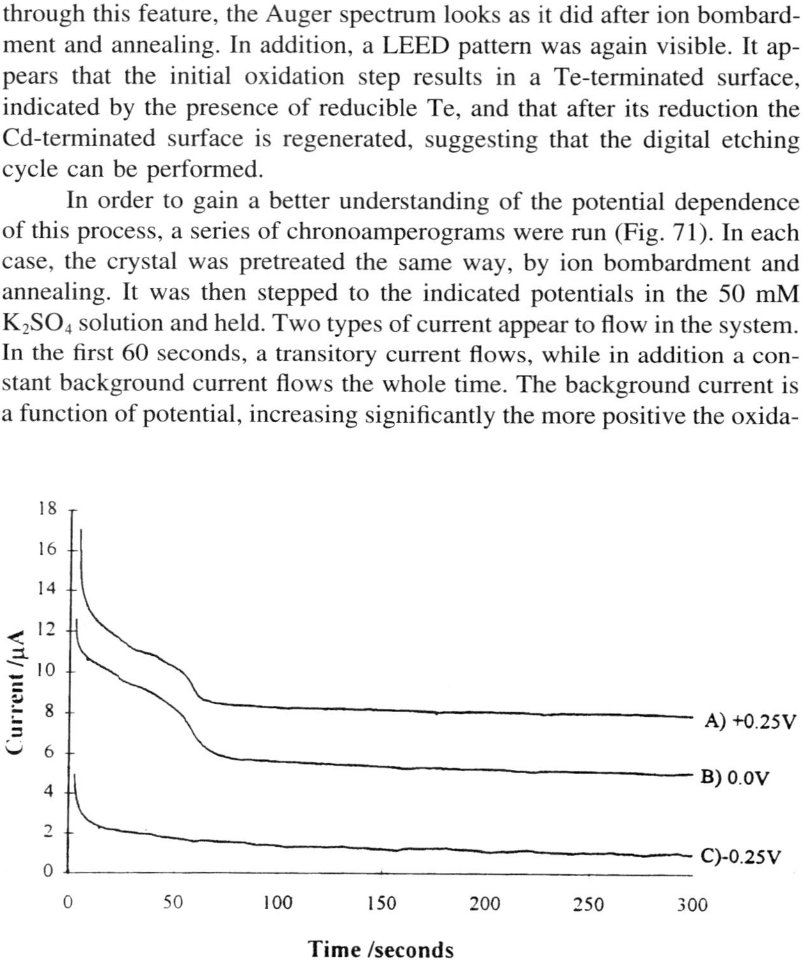

FIG. 71. Chronoamperograms of ion-bombarded, annealed CdTe(100) in 50 mM K_2SO_4 at different oxidative potentials for 5 minutes: (A) +0.25 V; (B) 0.0 V; (C) −0.25 V.

tion potential used. On the other hand, the charge associated with the transitory current is about the same in all experiments run using potentials above −0.25 V. After each potential step, a corresponding potential step was carried out to reduce off any Te made available by removal of the Cd. These data are summarized in Fig. 72.

In Fig. 72, there appears to be a break at about −0.2 V. Below −0.2 V, the oxidation charge matches the subsequent Te-reduction charge, as would be expected for a digital etching cycle. On the other hand, at potentials above −0.2 V, the total charge for oxidation goes up steeply and corresponds to the oxidation of multiple monolayers, not the single atomic layer anticipated. This oxidation charge has been further broken down into its two components: the transient oxidation charge and the background oxidation charge. If just the transient charge is considered, it follows the subsequent Te-reduction charge, as did the total charge at potentials below −0.2 V.

Assuming that the CdTe crystal starts with a Cd-terminated surface, the data in Figs. 70 to 72 suggest that at potentials below −0.2 V the charge corresponds to the oxidation of just the top Cd atomic layer, as does the

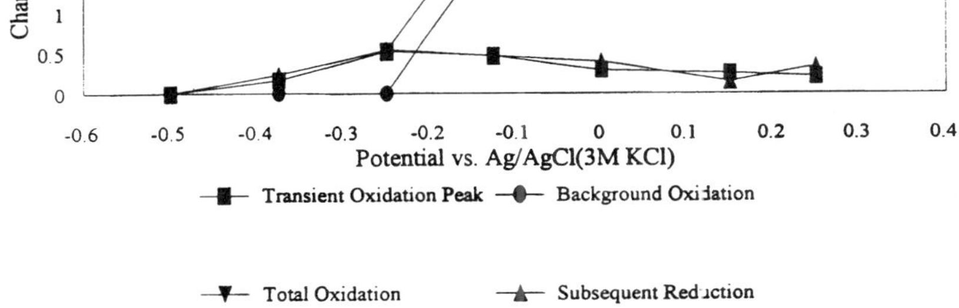

FIG. 72. Graph showing the charge passed, converted to monolayers, as a function of potential used for oxidation. Total oxidative charge has been separated into two components: transient oxidation and background oxidation. In addition, the subsequent reduction charge for Te is listed as well.

peak at 0.15 V in Fig. 70. However, as the potential is increased to above −0.2 V, a constant background current develops and the total charge increases dramatically to the point that it cannot be accounted for simply by removal of a Cd atomic layer. In addition, it does not appear reasonable that all the charge is due to Cd oxidation. If the increase in oxidation charge was all Cd, then significantly more Te would be available for reduction, which is not the case. The amount of Te subsequently reduced actually goes down slowly at potentials above −0.2 V. The increasing charge for potentials above −0.2 V is best understood if it is considered to arise from both Cd and Te oxidation. Elemental Te, however, begins to oxidize at about 0.2 V in this solution. To account for this 0.4 V shift in the oxidation potential for Te, it is suggested that the Te present on the surface after Cd has been electrochemically removed is destabilized, that is, if the Cd atoms in Fig. 68 were simply removed from the model, a network of doubly coordinated Te atoms would be left. These atoms might reconstruct to form elemental Te, etc., however, at room temperature it may be difficult due to limited mobility. If they do not reconstruct, then they should show significantly less stability than elemental Te and might oxidize at a significantly lower potential. At potentials above −0.2 V, where more than the top layer of Cd atoms is removed, the resulting destabilized Te atoms are oxidized as well in what might be referred to as underpotential etching (UPE). UPE would then be oxidation of an element at a potential under that required to oxidize it in its elemental form.

The idea of electrochemical digital etching described in this section is similar to the methodology developed by Kohl and co-workers [401–406]. Their studies involved the electrochemical photoetching of compound semiconductors such as InAs, InP, and GaAs, using a modulated potential. The main difference between the work described above and that of Kohl et al. is the intent. In the work described above, the removal of a single monolayer of the compound each cycle was the express goal and was felt to be achievable via the application of surface-limited reactions. The work of Kohl et al. was intended to perform macroscopic etching and is in fact a much more generally applicable methodology. Their work examined an extensive set of conditions, inclusive of those described above, and their studies have been useful in understanding some aspects of electrochemical digital etching. It is anticipated that these atomic-level studies [313] will benefit the development of a digital etching cycle and will provide some insight into the photo-electrochemical etching performed by Kohl et al.

Future directions for this work will involve other compounds, such as some of the III-Vs: InAs and InP. In addition, etching studies of the different low-index planes of a compound and the dependence on cycle conditions are planned. It is anticipated that individual planes will etch differentially as a function of the cycle conditions. Understanding this dependence may facilitate controlled anisotropic etching. It is also clear that the conditions for etching compounds differ and that device structures composed of more than one compound could be selectively etched.

VI. DIRECTIONS

There are a number of methods available for the formation of compound semiconductors by electrodeposition (Table 1). Electrochemical ALE promises better control over the deposition process by breaking it into a series of individual steps, with each step becoming a point of control. The questions raised include: What kind of deposit quality can be achieved? What compounds can be formed? What device structures can be formed? What niches will electrodeposition fill in the field of compound semiconductor device formation?

Possible advantages to an electrochemical deposition method include (1) that deposits can be formed near equilibrium, at room temperature; (2) that the hardware required is cost-efficient; (3) that the waste produced consists of an aqueous solution with a low concentration of the reactant ions, reasonably easy to treat, and relatively harmless unless consumed; (4) that there are no poisonous gases; and (5) that the hardware is easy to clean.

From the previous sections of this chapter, it should be clear that the substrate is very important and will continue to be a major area of study. Improvements in the quality of Au substrates can be made by switching to Au that is vapor-deposited on mica or glass. However, a very important direction of study will be towards using lattice-matched compound semiconductor substrates, and that work is closely tied with studies of electrochemical digital etching [44,312,313,400].

Cycles have been worked on for a number of different compounds, including CdTe [44,159–162,311,321], CdSe [106–108,321], CdS [60–62,321], ZnTe [44,279], ZnSe [279], ZnS [279,407], PbSe, CuSe [306], InSe [306], and $CuInSe_2$ [306]. However, thin films of reasonable quality have, so far, only been formed of CdTe [158,311,321], CdSe [321], and CdS [59,321]. The main problem is that there is a significant effort needed to get a flow deposition system up and running and producing deposits.

Then there is a large variable space to be examined, so that it takes a significant amount of time to investigate and optimize a single compound. At present, then, the main limitation to the formation of more compounds is time.

Other compounds to be investigated include the III-V compounds, such as InAs, InP, and GaAs [155,157,252,253]. The stability of these compounds in aqueous solutions is questionable, but with relatively simple modifications in the hardware a system where potential control is maintained during the whole deposition can be constructed. Work is underway to deposit the Zn-based II-VI compounds as well [279], and initial work has been encouraging.

As mentioned in Sec. III, initially a thin layer flow cell was used in the automated deposition systems, which proved problematic [158]. Presently, a simple H-cell configuration is being used, with much better results [311,321]. There are a number of drawbacks to the present configuration, however, including the volumes of solution needed to fill the cell and the fact that potential control is lost between each deposition step when the solutions are drained from the cell. There appears to be significant room for improvement in the deposition hardware. One direction presently being investigated is the use of a flow cell. The new flow cell would not be considered a thin layer cell, however, a significantly thicker gap would be used between the electrode and the far side of the cell. It is anticipated that a cell of this geometry will have fewer problems with fluid flow, similar to the present H-cell configuration, yet potential control will not be lost with each rinse, allowing the formation of less stable compounds.

Studies of the dependence of deposit structure on cycle conditions are continuing. Most of the work along this line has been performed with CdTe [311]. Optimizations of CdSe and CdS are continuing as well and look promising [321]. That is, it appears that significant improvements in deposit structure will be possible. Studies of this type with most of the compounds mentioned above represent the bulk of the studies needed to answer the questions raised at the beginning of this section. These studies involve the formation of thin film deposits under systematically varying conditions followed by characterization.

Given that high-quality deposits of a number of compounds can be formed, device structure formation should be pursued. Obvious structures include diodes, such as p-CdTe on CdS, which is an important photovoltaic structure [15,39,63,64,84,85,88,89]. Questions include: Can high-quality CdTe be formed on CdS? What will the structure of the CdS be if grown on ITO? Can p-CdTe be formed on the CdS directly, or will it be n-type, requiring an annealing step to convert it to p-type? Type conversion for electrodeposited CdTe has historically been achieved by annealing [164]. One of the main benefits, however, of an electrochemical method is the low temperature needed to form films. It is desirable, then, to achieve p-type CdTe during the electrodeposition process, either by optimizing the deposition conditions or by an electrochemical doping scheme, and to avoid annealing.

One natural direction for this work to go is toward the formation of superlattices. As very thin films can, in principle, be formed reproducibly and at low temperatures, superlattices would be a logical application. There are significant problems in placing one compound on another in most deposition schemes because they generally involve elevated substrate temperatures, promoting interdiffusion. Since electrodeposition is a low-temperature technique, very little interdiffusion should occur. Rajeshwar et al. [130] have performed some studies in this area using a flow deposition system to modulate between ZnSe and CdSe, forming a superlattice. These studies showed that it could be done. The modulation period of those deposits where relatively large, however, compared with what should be possible using an ALE method for electrodeposition.

As mentioned, studies of substrates are closely connected with studies of digital electrochemical etching. At present, most work has been performed on CdTe substrates. A number of other suitable compounds should be looked at for various reasons. Kohl et al. [401–406] have shown that InP and GaSb can be photoelectrochemically etched using a modulated potential program. These compounds would be good candidates for digital etching studies.

Substrate orientation should be examined to determine if some planes are preferentially etched. If there is preferential etching taking place, what is its dependence on the etching cycle conditions? The hardware being used for these studies should also be investigated. Very little has been done to optimize the flow cell. It is anticipated that a hydrodynamic electrode system such as a rotating disk or wall jet should work as well.

One of the most intriguing directions for electrochemical digital etching involves the selective etching of device structures composed of multiple layers of different compounds. Some compounds contain less noble elements, which will be the first to oxidize, possibly using conditions that leave behind the compound containing the more noble element, while the other compound might contain an element that is more easily reduced then the corresponding element in the first compound, so that that it could be selectively removed by controlling the reductive step. It would then be possible to remove either compound in the presence of the other by controlling the cycle potentials used.

REFERENCES

1. W. A. Bryant, J. Mater. Sci. *12*:1285 (1977).
2. K. K. Schuegraf, *Handbook of Thin-Film Deposition Processes and Techniques*, Noyes, Park Ridge, NJ, 1988.
3. W. Kern, in *Microelectronic Materials and Processes* (R. A. Levy, ed.), Kluwer Academic, Dordrecht, 1989.
4. F. S. Galasso, *Chemical Vapor Deposited Materials*, CRC Press, Boca Raton, FL, 1991.
5. A. Y. Cho and J. R. Arthur, Prog. Solid State Chem. *10*:157 (1975).
6. K. Ploog, Ann. Rev. Mater. Sci. *11*:171 (1981).
7. A. Y. Cho, Thin Solid Films *100*:291 (1983).
8. E. H. C. Parker, *The Technology and Physics of Molecular Beam Epitaxy*, Plenum Press, New York, 1985.
9. M. A. Herman and H. Sitter, *Molecular Beam Epitaxy: Fundamentals and Current Status*, Springer-Verlag, Berlin, 1989.
10. M. B. Panish and H. Temkin, Ann. Rev. Mater. Sci. *19*:209 (1989).
11. J. Y. Tsao, *Materials Fundamentals of Molecular Beam Epitaxy*, Academic Press, Inc., Boston, 1993.
12. G. Hodes and D. Cahen, Solar Cells *16*:245 (1986).
13. L. E. Lyons, G. C. Morris, D. H. Horton, and J. G. Keyes, J. Electroanal. Chem. *168*:101 (1984).
14. R. C. D. Mattei, D. Elwell, and R. S. Feigelson, J. Cryst. Growth *43*:643 (1978).
15. S. K. Das and G. C. Morris, Solar Energy Mater. *30*:107 (1993).
16. T. Suntula and J. Anson, U.S. patent # 4,058,430, ''Method for producing compound thin films.'' 1977.
17. C. H. L. Goodman and M. V. Pessa, J. Appl. Phys. *60*:R65 (1986).
18. S. P. DenBaars and P. D. Dapkus, J. Cryst. Growth *98*:195 (1989).
19. A. Usui and H. Watanabe, Annu. Rev. Mater. Sci. *21*:185 (1991).
20. T. F. Kuech, et al., eds., *Atomic Layer Growth and Processing*, Vol. 222, Materials Research Society, Pittsburgh, 1991.
21. S. Bedair, ed., *Atomic Layer Epitaxy*, Elsevier, Amsterdam, 1993.
22. K. Juttner and W. J. Lorenz, Z. Phys. Chem. N. F. *122*:163 (1980).
23. D. M. Kolb, In *Advances in Electrochemistry and Electrochemical Engineering,* Vol. 11, p. 125, H. Gerischer, and C. W. Tobias, eds., Wiley, New York, 1978.

24. G. V. Hevesy, Physik. Z. *13*:715 (1912).
25. L. B. Rogers, D. P. Krause, J. C. Griess, Jr., and D. B. Ehrlinger, J. Electrochem. Soc. *95*:33 (1949).
26. J. L. Stickney, S. D. Rosasco, D. Song, M. P. Soriaga, and A. T. Hubbard, Surf. Sci. *130*:326 (1983).
27. Y. F. Nicolau and J. C. Menard, J. Cryst. Growth *92*:128 (1988).
28. Y. F. Nicolau and J. C. Menard, J. Cryst. Growth *92*:128 (1988).
29. V. V. Klechkovskaya, V. N. Maslov, M. B. Muradov, and S. A. Semiletov, Sov. Phys. Crystallogr. *34*:105 (1989).
30. Y. F. Nicolau and J. C. Menard, J. Appl. Electrochem. *20*:1063 (1990).
31. Y. F. Nicolau, M. Dupuy, and M. Brunel, J. Electrochem. Soc. *137*:2915 (1990).
32. Y. F. Nicolau and J. C. Menard, J. Col. Int. Sci. *148*:551 (1992).
33. P. K. Nair, M. T. S. Nair, H. M. K. K. Pathirana, R. A. Zingaro, and E. A. Meyers, J. Electrochem. Soc. *140*:754 (1993).
34. J. M. Dona and J. Herrero, J. Electrochem. Soc. *141*:205 (1994).
35. M. Froment, M. C. Bernard, R. Cortes, B. Mokili, and D. Lincot, J. Electrochem. Soc. *142*:2642 (1995).
36. H. Cachet, H. Essaaidi, M. Froment, and G. Maurin, J. Electroanal. Chem. *396*:175 (1995).
37. H. Cachet, M. Froment, and G. Maurin, J. Electroanal. Chem. *406*:239 (1996).
38. A. Kampmann, P. Cowache, J. Vedel, and D. Lincot, J. Electroanal. Chem. *387*:53 (1995).
39. D. Lincot, et al., Appl. Phys. Lett. *67*:2355 (1995).
40. G. F. Fulop and R. M. Taylor, Ann. Rev. Mater. Sci. *15*:197 (1985).
41. G. Hodes, in *Physical Electrochemistry* (I. Rubinstein, ed.), 1995. Marcel Dekker, Inc., New York, p. 515.
42. T. Gruszecki and B. Holmstrom, Solar Energy Mater. Solar Cells *31*:227 (1993).
43. G. Hodes, Isr. J. Chem. *3*:95 (1993).
44. C. K. Rhee, B. M. Huang, E. M. Huang, E. M. Wilmer, S. Thomas, and J. L. Stickney, Mater. Manufact. Proc. *10*:283 (1995).
45. H. Gobrecht, H. D. Liess, and A. Tausend, Ber. Bunsenges. Phys. Chem. *67*:930 (1963).
46. J. J. Cuomo and R. J. Gambino, J. Electrochem. Soc. *115*:755 (1968).
47. A. Yamamoto and M. Yamaguchi, Jpn. J. Appl. *14*:561 (1975).
48. F. A. Kroger, J. Electrochem. Soc. *125*:2028 (1978).
49. R. D. Engelken, J. Electrochem. Soc. *135*:834 (1988).
50. R. C. DeMattei and R. S. Feigelson, in *Electrochemistry of Semiconductors and Electronics, Processes and Devices* (J. McHardy and F. Ludwig, eds.), Noyes Publications, Park Ridge, NJ, 1992, p. 1.

51. K. Rajeshwar, Adv. Mater. *4*:23 (1992).
52. R. K. Pandey, S. N. Sahu, and S. Chandra, *Handbook of Semiconductor Electrodeposition* (A. M. Hermann, ed.), Marcel Dekker, Inc., New York, 1996.
53. M. W. Verbrugge and C. W. Tobias, J. Electrochem. Soc. *132*:1298 (1985).
54. A. S. Baranski and W. R. Fawcett, J. Electrochem. Soc. *127*:766 (1980).
55. B. Miller, S. Menezes, and A. Heller, J. Electroanal. Chem. *94*:85 (1978).
56. B. Miller and A. Heller, Nature *262*:680 (1976).
57. C. D. Lokhande, V. S. Yermune, and S. H. Pawar, J. Electrochem. Soc. *138*: 624 (1991).
58. K. Shirai, Y. Moriguchi, M. Ichimura, A. Usami, and M. Saji, Jpn. J. Appl. Phys. *35*:2057 (1996).
59. B. E. Boone and C. Shannon, J. Phys. Chem. *100*:9480 (1996).
60. U. Demir and C. Shannon, Langmuir *10*:2794 (1994).
61. L. P. Colletti, D. Teklay, and J. L. Stickney, J. Electroanal. Chem. *369*:145 (1994).
62. U. Demir and C. Shannon, Langmuir *12*:594 (1996).
63. A. K. Turner, et al., Solar Energy Mater. *23*:388 (1991).
64. B. M. Basol, E. S. Tseng, and R. L. Rod., Proceedings 16th IEEE Photovolt. Special. Conf., San Diego, 1982, p. 805.
65. D. Ham, K. K. Mishra, and K. Rajeshwar, J. Electrochem. Soc. *138*:100 (1991).
66. A. S. Baranski and W. R. Fawcett, J. Electrochem. Soc. *131*:2509 (1984).
67. A. S. Baranski, M. S. Bennett, and W. R. Fawcett, J. Appl. Phys. *54*:6390 (1983).
68. F. Mondon, J. Electrochem. Soc. *132*:319 (1985).
69. A. S. Baranski, W. R. Fawcett, and A. C. McDonald, J. Electroanal. Chem. *160*:271 (1984).
70. D. K. Roe, L. Wenzhao, and H. Gerisher, J. Electroanal. Chem. *136*:323 (1982).
71. G. Hodes, I. D. J. Howell, and L. M. Peter, J. Electrochem. Soc. *139*:3136 (1992).
72. A. S. Baranski, W. R. Fawcett, A. C. McDonald, and R. M. D. Nobriga, J. Electrochem. Soc. *128*:963 (1981).
73. T. Edamura and J. Muto, Thin Solid Films *235*:198 (1993).
74. E. Fatas, P. Herrasti, F. Arjona, E. G. Camarero, and J. A. Medina, Electrochim. Acta *32*:139 (1987).
75. A. Darkowski and M. Cocivera, J. Electrochem. Soc. *134*:226 (1987).
76. H. Minoura, Y. Takeichi, H. Furuta, T. Sugiura, and Y. Ueno, J. Mater. Sci. *25*:472 (1990).
77. I. Markov, E. Valova, M. Ilieva, and I. Kristev, J. Crystal Growth *65*:611 (1985).
78. J. F. McCann and M. S. Kazacos, J. Electroanal. Chem. *119*:409 (1981).

79. E. Fatas and P. Herrasti, Electrochimica Acta *33*:959 (1988).
80. S. Preusser and M. Cocivera, J. Electroanal. Chem. *252*:139 (1988).
81. J. V. Windheim, A. Darkowski, and M. Cocivera, Can. J. Phys. *65*:1053 (1987).
82. S. S. Ou, O. M. Stafsudd, and B. M. Basol, Thin Solid Films *112*:301 (1984).
83. B. M. Basol and E. S. Tseng, Appl. Phys. Lett. *48*:946 (1986).
84. G. C. Morris, S. K. Das, and P. G. Tanner, J. Crystal Growth *117*:929 (1992).
85. S. K. Das, Thin Solid Films *226*:259 (1993).
86. E. Fatas, R. Duo, P. Herrasti, F. Arjona, and E. Garcia-Camarero, J. Electrochem. Soc. *131*:2243 (1984).
87. E. Fatas, P. Herrasti, F. Arjona, E. Garcia-Camarero, and M. Leon, J. Mater. Sci. Lett. *5*:583 (1986).
88. S. K. Das and G. C. Morris, J. Appl. Phys. *73*:782 (1993).
89. S. K. Das and G. C. Morris, J. Appl. Phys. *72*:4940 (1992).
90. G. P. Power, D. R. Peggs, and A. J. Parker, Electrochim. Acta *26*:681 (1981).
91. S. Dennison, Electrochim. Acta *38*:2395 (1993).
92. J. A. V. Windheim, H. Wynands, and M. Cocivera, J. Electrochem. Soc. *138*:3435 (1991).
93. G. C. Morris and R. Vanderveen, Solar Energy Mater. Solar Cells *27*:305 (1992).
94. P. V. Elfick, L. E. A. Berlouis, S. M. M. Donald, S. Affrossman, and P. Rocabolis, J. Phys. Chem. *99*:15129 (1995).
95. N. R. D. Tacconi and K. Rajeshwar, J. Phys. Chem. *97*:6504 (1993).
96. M. I. D. S. Pereira and L. M. Peter, J. Electroanal. Chem. *140*:103 (1982).
97. L. M. Peter, Electrochem. Acta *23*:165 (1978).
98. L. M. Peter, Electrochem. Acta *23*:1073 (1978).
99. D. Ham, Y. Son, K. K. Mishra, and K. Rajeshwar, J. Electroanal. Chem. *310*:417 (1991).
100. V. I. Briss and L. E. Kee, J. Electrochem. Soc. *133*:2097 (1986).
101. M. Krebs, M. I. Sosa, and K. E. Heusler, J. Electroanal. Chem. *301*:101 (1991).
102. J. Vedel, M. Soubeyrand, and E. Castel, J. Electrochem. Soc. *124*:177 (1977).
103. M. Krebs and K. E. Heusler, Electrochem. Acta *37*:1371 (1992).
104. T. Edamura and J. Muto, Thin Solid Films *266*:135 (1993).
105. A. S. Baranski, et al., J. Electrochem. Soc. *130*:579 (1983).
106. T. E. Lister and J. L. Stickney, Appl. Surf. Sci. *107*:153 (1996).
107. T. E. Lister and J. L. Stickney, Isr. J. Chem. *37*:287 (1997).
108. J. Jenkins and J. L. Stickney, unpublished results.
109. A. M. Kressin, V. V. Doan, J. D. Klein, and M. J. Sailor, Chem. Mater. *3*: 1015 (1991).

110. J. D. Klein, et al., Chem. Mater. *5*:902 (1993).
111. D. Liu and P. V. Kamat, J. Electroanal. Chem. *347*:451 (1993).
112. A. C. Rastogi, K. S. Balakrishnan, and A. Garg, J. Electrochem. Soc. *140*: 2373 (1993).
113. H. Wynands and M. Cocivera, Solar Energy Mater. *32*:187 (1994).
114. M. T. Litz, et al., J. Cryst. Growth *159*:54 (1996).
115. Y. Golan, L. Vargulis, I. Rubinstein, and G. Hodes, Langmuir *8*:749 (1992).
116. A. Albu-Yaron, D. Cahen, and G. Hodes, Thin Solid Films *112*:349 (1984).
117. K. K. Mishra and K. Rajeshwar, J. Electroanal. Chem. *273*:169 (1989).
118. A. T. Vasko, E. M. Tsikovdin, and S. K. Kovach, Ukr. Khim. Zhurnal *49*: 1070 (1983).
119. M. Tomkiewicz, I. Ling, and W. S. Parsons, J. Electrochem. Soc. *129*:2016 (1982).
120. G. J. Houston, J. F. McCann, and D. Haneman, J. Electroanal. Chem. *134*: 37 (1982).
121. R. P. Silberstein and M. Tomkiewicz, J. Appl. Phys. *54*:542 (1983).
122. M. Skyllas-Kazacos and B. Miller, J. Electrochem. Soc. *127*:869 (1980).
123. K. Colbow, D. J. Harrison, and B. L. Funt, J. Electrochem. Soc. *128*:547 (1981).
124. B. W. Sanders and M. Cocivera, J. Electrochem. Soc. *134*:1075 (1987).
125. Y. Golan, L. Margulis, G. Hodes, I. Rubinstein, and J. L. Hutchison, Surf. Sci. *311*:L633 (1994).
126. N. Muller, M. Abramovich, F. Decker, F. Iikawa, and P. Motisuke, J. Electrochem. Soc. *131*:2204 (1984).
127. M. Skyllas-Kazacos, J. Electroanal. Chem. *148*:233 (1983).
128. G. Hodes, J. Manassen, and S. Neagu, Thin Solid Films *90*:433 (1982).
129. G. Hodes, J. Manassen, and D. Cahen, Nature *261*:403 (1976).
130. C. Wei and K. Rajeshwar, J. Electrochem. Soc. *139*:L40 (1992).
131. R. A. Boudreau and R. D. Rauh, Solar Energy Mater. *7*:385 (1982).
132. M. Skyllas-Kazacos and B. Miller, J. Electrochem. Soc. *127*:2378 (1980).
133. B. M. Basol, J. Appl. Phys. *55*:601 (1984).
134. V. Krishnan, D. Ham, K. K. Mishra, and K. Rajeshwar, J. Electrochem. Soc. *139*:23 (1992).
135. Z. Loizos, A. Mitsis, N. Spyrellis, M. Froment, and G. Maurin, Thin Solid Films *235*:51 (1993).
136. M. Bouroushian, Z. Loizos, N. Spyrellis, and G. Maurin, Thin Solid Films *229*:101 (1993).
137. Z. Loizos, N. Spyrellis, G. Maurin, and D. Pottier, J. Electroanal. Chem. *269*:399 (1989).
138. R. K. Pandey and A. J. N. Rooz, J. Phys. D: Appl. Phys. *19*:917 (1986).
139. C. Wei, C. S. C. Bose, and K. Rajeshwar, J. Electroanal. Chem. *327*:331 (1992).
140. K. R. Murali, V. Subramanian, N. Rangarajan, A. S. Lakshmanan, and S. K. Rangarajan, J. Electroanal. Chem. *368*:95 (1994).

141. H. Wynands and M. Cocivera, J. Electrochem. Soc. *139*:2052 (1992).
142. J. P. Szabo and M. Cocivera, J. Electroanal. Chem. *239*:307 (1988).
143. J. P. Szabo and M. Cocivera, J. Electrochem. Soc. *133*:1247 (1986).
144. N. Ardoin and J. Winnick, J. Electrochem. Soc. *135*:1719 (1988).
145. M. Cocivera, A. Darkowski, and B. Love, J. Electrochem. Soc. *131*:2514 (1984).
146. F. Decker, N. G. Ferreira, and M. Fracastoro-Decker, J. Electrochem. Soc. *134*:1499 (1987).
147. F. Decker, J. R. Moro, and J. L. S. Ferreira, Ber. Bunsenges. Phys. Chem. *91*:408 (1987).
148. S. M. Babu, T. Rajalakshmi, R. Dhanasekaran, and P. Ramasamy, J. Crystal Growth *110*:423 (1991).
149. W. A. Gerrard and J. R. Owen, Mat. Res. Bull. *12*:677 (1977).
150. B. Miller, et al., J. Electrochem. Soc. *124*:1019 (1977).
151. M. Abramovich, et al., J. Solid State Chem. *59*:1 (1985).
152. S. Licht and J. Manassen, J. Electrochem. Soc. *132*:1076 (1985).
153. R. N. Bhattacharya, J. Appl. Electrochem. *16*:168 (1986).
154. J. M. Rosamilia and B. Miller, J. Electroanal. Chem. *215*:249 (1986).
155. D. W. Suggs, I. Villegas, B. W. Gregory, and J. L. Stickney, J. Vac. Sci. Technol. A *10*:886 (1992).
156. J. L. Stickney, B. W. Gregory, and I. Villegas, in *U.S. Patent*, #5,320,736, 1996.
157. D. W. Suggs, I. Villegas, B. W. Gregory, and J. L. Stickney, Mat. Res. Soc. Symp. Proc. *222*:283 (1991).
158. B. M. Huang, L. P. Colletti, B. W. Gregory, J. L. Anderson, and J. L. Stickney, J. Electrochem. Soc. *142*:3007 (1995).
159. D. W. Suggs and J. L. Stickney, Surf. Sci. *290*:375 (1993).
160. B. W. Gregory and J. L. Stickney, J. Electroanal. Chem. *300*:543 (1991).
161. B. W. Gregory, D. W. Suggs, and J. L. Stickney, J. Electrochem. Soc. *138*: 1279 (1991).
162. D. W. Suggs and J. L. Stickney, Surf. Sci. *290*:362 (1993).
163. I. Shih and C. X. Qiu, Mater. Lett. *3*:446 (1985).
164. B. M. Basol, J. Appl. Phys. *58*:3809 (1985).
165. J. V. Windheim, I. Renaud, and M. Cocivera, J. Appl. Phys. *67*:4167 (1990).
166. Y. Sugimoto and L. M. Peter, J. Electroanal. Chem. *386*:183 (1995).
167. R. D. Engelken and T. P. V. Doren, J. Electrochem. Soc. *132*:2910 (1985).
168. P. Carbonnelle, C. Labar, and L. Lamberts, Analysis *15*:286 (1987).
169. C. Wei, N. Myung, and K. Rajeshwar, J. Electroanal. Chem. *347*:223 (1993).
170. E. Mori and K. Rajeshwar, J. Electroanal. Chem. *258*:415 (1989).
171. W. Lin, K. K. Mishra, E. Mori, and K. Rajeshwar, Anal. Chem. *62*:821 (1990).
172. W. J. Danaher and L. E. Lyons, Aust. J. Chem. *37*:689 (1984).

173. W. J. Danaher and L. E. Lyons, Aust. J. Chem. *36*:1011 (1983).
174. D. Ham, K. K. Mishra, A. Weiss, and K. Rajeshwar, Chem. Mater. *1*:619 (1989).
175. A. C. Rastogi and K. S. Balakrishnan, J. Electrochem. Soc. *136*:1502 (1989).
176. K. S. Balakrishnan and A. C. Rastogi, Solar Energy Mater. *23*:61 (1991).
177. R. K. Pandey, G. Razzini, and L. P. Bicelli, Solar Energy Mater. *26*:285 (1992).
178. R. B. Gore, R. K. Pandey, and S. K. Kulkarni, Solar Energy Mater. *18*:159 (1989).
179. R. B. Gore, R. K. Pandey, and S. K. Kulkarni, J. Appl. Phys. *65*:2693 (1989).
180. J. Llabres and V. Delmas, J. Electrochem. Soc. *133*:2580 (1986).
181. D. A. Fiedler and H. P. Fritz, Electrochim. Acta *40*:1595 (1995).
182. J. K. You, et al., J. Electroanal. Chem. *405*:233 (1996).
183. S. S. Ou, A. Bindal, O. M. Stafsudd, K. L. Wang, and B. M. Basol, J. Appl. Phys. *55*:1020 (1984).
184. J. H. Chen and C. C. Wan, J. Electroanal. Chem. *365*:87 (1994).
185. H. Minoura, M. Kitakata, T. Sugiura, and M. Murayama, Bull. Chem. Soc. Jpn. *60*:2373 (1987).
186. J. A. V. Windheim and M. Cocivera, J. Phys. D: Phys. *23*:581 (1990).
187. J. A. V. Windheim and M. Cocivera, J. Electrochem. Soc. *138*:250 (1991).
188. A. Darkowski and M. Cocivera, J. Electrochem. Soc. *132*:2768 (1985).
189. J. V. Windheim and M. Cocivera, J. Electrochem. Soc. *134*:441 (1987).
190. L. Gheorghita, M. Cocivera, A. J. Nelson, and A. B. Swartzlander, J. Electrochem. Soc. *141*:529 (1994).
191. J. A. V. Windheim and M. Cocivera, J. Phys. Chem. Solids *53*:31 (1992).
192. M. Takahashi, K. Uosaki, H. Kita, and S. Yamaguchi, J. Appl. Phys. *60*:2049 (1986).
193. S. Dennison and S. Webster, J. Electroanal. Chem. *333*:287 (1992).
194. W. J. Danaher and L. E. Lyons, Nature *271*:139 (1978).
195. Y. Sugimoto and L. M. Peter, J. Electroanal. Chem. *381*:251 (1995).
196. M. Takahashi, K. Uosaki, and H. Kita, J. Appl. Phys. *58*:4292 (1985).
197. M. Takahashi, K. Uosaki, and H. Kita, J. Electrochem. Soc. *131*:2304 (1984).
198. K. Uosaki, M. Takahashi, and H. Kita, Electrochim. Acta *29*:279 (1984).
199. M. Takahashi, K. Uosaki, and H. Kita, J. Appl. Phys. *55*:3879 (1984).
200. C. Sella, P. Boncorps, and J. Vedel, J. Electrochem. Soc. *133*:2043 (1986).
201. A. Saraby-Reintjes, L. M. Peter, M. E. Ozsan, S. Dennison, and S. Webster, J. Electrochem. Soc. *140*:2880 (1993).
202. B. M. Basol, Solar Cells *23*:69 (1988).
203. A. Pal, J. Dutta, D. Bhattacharyya, S. Chaudhuri, and A. K. Pai, Vacuum *46*:147 (1995).
204. J. Dutta, D. Bhattacharyya, A. B. Maitil, S. Chaudhuri, and A. K. Pal, Vacuum *46*:17 (1995).
205. P. Sircar, Appl. Phys. Lett. *53*:1184 (1988).

206. J. M. Fisher, et al., J. Crystal Growth *138*:86 (1994).
207. J. M. Fisher, L. E. A. Berlouis, B. N. Rospendowski, P. J. Hall, and M. G. Astles, Semicond. Sci. Technol. *8*:1459 (1993).
208. B. Qi, D. Kim, D. L. Williamson, and J. U. Trefny, J. Electrochem. Soc. *143*:517 (1996).
209. V. Valvoda, J. Touskova, and D. Kindl, Cryst. Res. Technol. *21*:975 (1986).
210. J. Llabres, J. Electrochem. Soc. *131*:464 (1984).
211. G. Maurin, O. Solorza, and H. Takenouti, J. Electroanal. Chem. *202*:323 (1986).
212. P. Cowache, D. Lincot, and J. Vedel, J. Electrochem. Soc. *136*:1646 (1989).
213. G. Fulop, M. Doty, P. Meyers, J. Betz, and C. H. Liu, Appl. Phys. Lett. *40*: 327 (1982).
214. L. Ndiaye, P. Cowache, M. Cadene, D. Lincot, and J. Vedel, Thin Solid Films *224*:227 (1993).
215. G. Maurin and D. Pottier, J. Mater. Sci. Lett. *6*:817 (1987).
216. R. N. Bhattacharya, K. Rajeshwar, and R. N. Noufi, J. Electrochem. Soc. *132*:732 (1985).
217. R. D. Engelken and T. P. V. Doren, J. Electrochem. Soc. *132*:2904 (1985).
218. M. P. R. Panicker, M. Knaster, and F. A. Kroger, J. Electrochem. Soc. *125*: 566 (1978).
219. R. N. Bhattacharya and K. Rajeshwar, J. Electrochem. Soc. *131*:2032 (1984).
220. R. N. Bhattacharya and K. Rajeshwar, J. Appl. Phys. *58*:3590 (1985).
221. M. W. Verbrugge and C. W. Tobias, Am. Inst. Chem. Eng. J. *33*:628 (1987).
222. M. W. Verbrugge and C. W. Tobias, J. Electrochem. Soc. *134*:3104 (1987).
223. H. J. Gerritsen, J. Electrochem. Soc. *131*:136 (1984).
224. M. H. Miles and W. S. McEwan, J. Electrochem. Soc. *119*:1188 (1972).
225. G. C. Morris and R. Vanderveen, Solar Energy Mater. *26*:217 (1992).
226. K. Singh and A. K. Shukla, Solar Energy Mater. *30*:169 (1993).
227. B. M. Basol, V. K. Kapur, and M. L. Ferris, J. Appl. Phys. *66*:1816 (1989).
228. D. Pottier and G. Maurin, J. Appl. Electrochem. *19*:361 (1989).
229. S. Menezes, Appl. Phys. Lett. *61*:1564 (1992).
230. G. Hodes, T. Engelhard, C. R. Herrington, L. L. Kazmerski, and D. Cahen, Progr. Cryst. Growth Charact. *10*:345 (1985).
231. B. M. Basol, J. Vac. Sci. Technol. A *10*:2006 (1992).
232. C. Guillen and J. Herrero, J. Electrochem. Soc. *143*:493 (1996).
233. C. Guillen and J. Herrero, J. Electrochem. Soc. *142*:1834 (1995).
234. C. X. Qiu and I. Shih, J. Appl. Phys. *64*:758 (1988).
235. I. Shih, C. X. Qiu, S. N. Qiu, and J. F. Huang, J. Appl. Phys. *63*:439 (1988).
236. C. X. Qiu and I. Shih, Can. J. Phys. *67*:444 (1989).
237. R. P. Singh and S. L. Singh, J. Phys. D: Appl. Phys. *19*:1759 (1986).
238. C. X. Qiu and I. Shih, Can. J. Phys. *65*:1011 (1987).
239. C. Guillen and J. Herrero, J. Electrochem. Soc. *141*:225 (1994).
240. K. K. Mishra and K. Rajeshwar, J. Electroanal. Chem. *271*:279 (1989).

241. R. P. Singh, S. L. Singh, and S. Chandra, J. Phys. D: Appl. Phys. *19*:1299 (1986).
242. R. N. Bhattacharya, J. Electrochem. Soc. *130*:2040 (1983).
243. W. Jiang-Shan, T. Zheng, M. Tian-Ying, and S. Gong-Quan, Sci. China (series B) *35*:281 (1992).
244. C. D. Lokhande, J. Electrochem. Soc. *134*:1728 (1987).
245. S. Endo, Y. Nagahori, and S. Nomura, Jpn. J. Appl. Phys. *35*:L1101 (1996).
246. T. L. Chu, S. S. Chu, S. C. Lin, and J. Yue, J. Electrochem. Soc. *131*:2182 (1984).
247. C. D. Lokhande and S. H. Pawar, J. Phys. D *20*:1213 (1987).
248. A. M. Al-Dhafiri, P. C. Pande, G. J. Russell, and J. Woods, J. Crystal Growth *86*:900 (1988).
249. A. C. Rastogi and S. Salkalachen, J. Appl. Phys. *58*:4442 (1985).
250. R. I. Kazhlaeva, A. I. Alekperov, Z. A. Aliyarova, and K. V. Kirilyuk, Elektrokhimiya *21*:1197 (1985).
251. N. R. D. Tacconi, O. Medvedko, and K. Rajeshwar, J. Electroanal. Chem. *379*:545 (1994).
252. I. Villegas and J. L. Stickney, J. Electrochem. Soc. *139*:686 (1992).
253. I. Villegas and J. L. Stickney, J. Vac. Sci. Technol. A *10*:3032 (1992).
254. S. Chandra and N. Khare, Semicond. Sci. Technol. *2*:220 (1987).
255. I. G. Dioum, J. Vedel, and B. Tremillon, J. Electroanal. Chem. *139*:329 (1982).
256. M. C. Yang, U. Landau, and J. C. Angus, J. Electrochem. Soc. *139*:3480 (1992).
257. P. Andreoli, S. Cattarin, M. Musiani, and F. Paolucci, J. Electroanal. Chem. *385*:265 (1995).
258. R. C. D. Mattei, D. Elwell, and R. S. Feigelson, J. Crystal Growth *44*:545 (1978).
259. F. Paolucci, G. Mengoli, and M. M. Musiani, J. Appl. Electrochem. *20*:868 (1990).
260. C. L. Colyer and M. Cocivera, J. Electrochem. Soc. *139*:406 (1992).
261. M. Neumann-Spallart, G. Tamizhmani, and C. Levy-Clement, J. Electrochem. Soc. *137*:3434 (1990).
262. L. Jun, T. Zheng, and M. Tian-Ying, Sci. China (series B) *33*:1153 (1990).
263. J. Ramiro, E. Garcia, and E. G. Camarero, J. Mater. Sci. *31*:2047 (1996).
264. L. M. Peter, J. D. Reid, and B. R. Scharifker, J. Electroanal. Chem. *119*:73 (1981).
265. G. Mattsson, L. Nyhom, and L. M. Peter, J. Electroanal. Chem. *347*:303 (1993).
266. G. Mengoli, M. M. Musiani, and F. Paolucci, J. Electroanal. Chem. *332*:199 (1992).
267. J. Ortega and J. Herrero, J. Electrochem. Soc. *136*:3388 (1989).

268. G. Mengoli, M. M. Musiani, and F. Paolucci, J. Appl. Electrochem. *21*:863 (1991).
269. S. N. Sahu, J. Mater. Sci. *3*:102 (1992).
270. D. Elwell, R. S. Feigelson, and M. M. Simkins, J. Cryst. Growth *51*:171 (1981).
271. S. N. Sahu, J. Mater. Sci. Lett. *8*:533 (1989).
272. M. C. Hobson, Jr., and H. Leidheiser, Jr., Trans. Metallurg. Soc. AIME *233*: 482 (1965).
273. S. Massaccesi, S. Sanchez, and J. Vedel, J. Electroanal. Chem. *412*:95 (1996).
274. Y. Igasaki and T. Fujiwara, J. Crystal Growth *158*:268 (1996).
275. K. M. Totland and D. A. Harrington, J. Electroanal. Chem. *274*:61 (1989).
276. B. Scharifker, Z. Ferreira, and J. Mozota, Electrochem. Acta *30*:677 (1985).
277. K. Mishra, et al., J. Electrochem. Soc. *136*:1915 (1989).
278. R. D. Engelken, A. K. Berry, T. P. V. Doren, J. L. Boone, and A. Shahnazary, J. Electrochem. Soc. *133*:581 (1986).
279. L. P. Colletti, S. Thomas, E. M. Willmer, and J. L. Stickney, in *Electrochemical Synthesis and Modification of Materials* (P. C. Searson, T. P. Moffat, P. C. Andricacos, S. G. Corcoran, J. L. Delplancke, eds.), Materials Research Society, Pittsburgh, 1996. Vol. 451, p. 235.
280. T. Mahalingam and C. Sanjeeviraja, Phys. Status Solidi A *129*:K89 (1992).
281. C. D. Lokhande, M. S. Jadhav, and S. H. Pawar, J. Electrochem. Soc. *136*: 2756 (1989).
282. C. Natarajan, M. Sharon, C. Levy-Clement, and M. Neumann-Spallart, Thin Solid Films *237*:118 (1994).
283. J. P. Rai, Solar Energy Mater. *30*:119 (1993).
284. J. P. Rai and K. Singh, Ind. J. Chem. *32*:376 (1993).
285. K. Singh and R. K. Pathak, Ind. J. Chem. *32*:55 (1993).
286. M. Neumann-Spallart and C. Konigstein, Thin Solid Films *265*:33 (1995).
287. C. Wei, K. Rajeshwar, and J. D. Luttmer, J. Electrochem. Soc. *140*:829 (1993).
288. A. J. Panson, Inorg. Chem. *3*:940 (1964).
289. J. Vedel, A. Kampmann, and D. Lincot, 188th meeting of the Electrochemical Society, Chicago, 1995.
290. V. K. Kapur, V. Choudary, and A. K. P. Chu. U.S. Patent #4,581,108, 1986.
291. B. W. Gregory, M. L. Norton, and J. L. Stickney, J. Electroanal. Chem. *293*: 85 (1990).
292. M. P. Soriaga and A. T. Hubbard, JACS *104*:2742 (1982).
293. M. P. Soriaga and A. T. Hubbard, JACS *104*:2735 (1982).
294. I. C. Hamilton and R. Woods, J. Appl. Electrochem. *13*:783 (1983).
295. T. E. Lister, B. M. Huang, R. D. Herrick-II, and J. L. Stickney, J. Vac. Sci. Technol. B *13*:1268 (1995).

296. J. M. M. Droog, C. A. Alderliesten, P. T. Alserliesten, and G. A. Gootsma, J. Electroanal. Chem. *111*:61 (1980).
297. C. B. Ehlers, I. Villegas, and J. L. Stickney, J. Electroanal. Chem. *284*:403 (1990).
298. G. N. Salaita, et al., Langmuir *2*:1986 (1986).
299. N. J. Tao and S. M. Lindsay, J. Phys. Chem. *96*:5213 (1992).
300. H. A. Gasteiger, N. M. Markovic, and P. N. J. Ross, Lang. *12*:1414 (1996).
301. B. G. Bravo, et al., J. Phys. Chem. *95*:5245 (1991).
302. B. M. Ocko, G. M. Watson, and J. Wang, J. Phys. Chem. *98*:897 (1994).
303. S. Sugita, T. Abe, and K. Itaya, J. Phys. Chem. *97*:8780 (1993).
304. R. L. McCarley and A. J. Bard, J. Phys. Chem. *95*:9618 (1991).
305. A. B. Eillis, S. W. Kaiser, J. M. Bolts, and M. S. Wrighton, JACS *99*:2839 (1997).
306. R. D. Herrick, II, and J. L. Stickney, in *New Directions in Electroanalytical Chemistry*, Vol. 96-9 (J. Leddy and M. Wightman, eds.), The Electrochemical Society, Pennington, NJ, 1996, p. 186.
307. D. M. Kolb and J. Schneider, Electrochim. Acta *31*:929 (1986).
308. X. Gao, A. Hamelin, and M. J. Weaver, PRL *67*:618 (1991).
309. L. B. Goetting, B. M. Huang, T. E. Lister, and J. L. Stickney, Electrochem. Acta *40*:143 (1995).
310. A. Nasar and M. Shamsuddin, Thermochim. Acta *205*:157 (1992).
311. L. P. Colletti and J. L. Stickney. J. Electrochem. Soc. *145*:3594 (1998).
312. Q. P. Lei and J. L. Stickney, in *Interface Dynamics and Growth*, Vol. 237 (K. S. Liang, M. P. Anderson, R. F. Bruinsma, and G. Scoles, eds.), Materials Research Society, Pittsburg, 1992, p. 335.
313. T. A. Sorenson, B. K. Wilmer, and J. L. Stickney, in *Solid-Liquid Electrochemical Interfaces*, Vol. 656 (G. Jerkiewicz, K. Uosaki, and A. Wieckowski, eds.), American Chemical Society, Washington, 1997, p. 115.
314. C. A. Ross, L. M. Goldman, and F. Spaepen, J. Electrochem Soc. *140*:91 (1993).
315. C. A. Ross, Ann. Rev. Mater. Sci. *24*:159 (1994).
316. C. E. D. Chidsey, D. N. Loiacono, T. Sleator, and S. Nakahara, Surf. Sci. *200*:45 (1988).
317. E. Holland-Moritz, J. Gordon-II, G. Borges, and R. Sonnenfeld, Langmuir *7*:301 (1991).
318. S. Buchholz, H. Fuchs, and J. P. Rabe, J. Vac. Sci. Technol. B *9*:857 (1991).
319. J. A. DeRose, T. Thundat, L. A. Nagahara, and S. M. Lindsay, Surf. Sci. *256*:102 (1991).
320. D. W. Suggs and J. L. Stickney, J. Phys. Chem. *95*:10056 (1991).
321. L. P. Colletti, B. Flowers, and J. L. Stickney, J. Electrochem. Soc. *145*:1442 (1998).
322. M. P. Soriaga, and J. L. Stickney, in *Modern Techniques in Electroanalytical Chemistry* (P. Vanysek, ed.), Wiley & Sons, New York, 1996, p. 1.

323. G. Binnig, H. Rohrer, C. Gerber, and E. Weibel, Phys. Rev. Lett. *49*:57 (1982).
324. G. Binnig and H. Rohrer, Surf. Sci. *126*:236 (1983).
325. R. Sonnenfeld and B. C. Schardt, Appl. Phys. Lett. *49*:1172 (1986).
326. R. Sonnenfeld, J. Schneir, and P. K. Hansma, in *Modern Aspects of Electrochemistry*, Vol. 12 (R. E. White, J. O. M. Bockris, and B. E. Conway, eds.), Plenum Press, New York, p. 1.
327. T. R. I. Cataldi, et al., J. Electroanal. Chem. *290*:1 (1990).
328. D. A. Bonnell, ed., *Scanning Tunneling Microscopy and Spectroscopy, Theory, Techniques and Applications*, VCH Publishing, New York, 1993.
329. A. J. Stroscio and W. J. Kaiser, eds., *Scanning Tunneling Microscopy*, Vol. 27, Academic Press, New York, 1993.
330. K. S. Kim, N. Winograd, and R. E. Davis, JACS *93*:6296 (1971).
331. B. M. Huang, T. E. Lister, and J. L. Stickney, in *Handbook of Surface Imaging and Visualization* (A. T. Hubbard, ed.), CRC Press, Boca Raton, FL, 1995, p. 77.
332. B. C. Schardt, et al., Surf. Sci. *175*:520 (1986).
333. T. E. Lister and J. L. Stickney, J. Phys. Chem. *100*:19568 (1997).
334. B. M. Huang, T. E. Lister, and J. L. Stickney, Surf. Sci. *392*:27.
335. M. D. Pashley, K. W. Haberern, W. Friday, J. M. Woodall, and P. D. Kirchner, PRL *60*:2176 (1988).
336. B. K. Biegelsen, L.-E. Swartz, and R. D. Brigans, J. Vac. Sci. Technol. A *8*:280 (1990).
337. P. K. Larsen and G. Meyer-Ehmsen, Surf. Sci. *240*:168 (1990).
338. H. H. Farrell, M. C. Tamargo, and J. L. D. Miguel, Appl. Phys. Lett. *58*: 355 (1991).
339. C. Deparis and J. Massies, J. Cryst. Growth *108*:157 (1991).
340. J. F. Moulder, W. F. Stickle, P. E. Sobol, and K. D. Bomben, *Handbook of X-Ray Photoelectron Spectroscopy*, Perkin Elmer Corp., Eden Prairie, 1992.
341. W. J. M. Tegart, *The Electrolytic and Chemical Polishing of Metals in Research and Industry*, Pergamon Press Ltd., Oxford, 1959.
342. A. Hamelin, J. Electroanal. Chem. *142*:299 (1982).
343. J. Schneider and D. M. Kolb, Surf. Sci. *193*:579 (1988)
344. D. M. Kolb, in *Structure of Electrified Interfaces* (J. Lipkowski and P. N. Ross, eds.), VCH Publishers, New York, 1993, p. 65.
345. X. Gao, A. Hamelin, and M. J. Weaver, Phys. Rev. B *46*:7096 (1992).
346. P. Skoluda and D. M. Kolb, Surf. Sci. *260*:229 (1992).
347. X. Gao, G. J. Edens, A. Hamelin, and M. J. Weaver, Surf. Sci. *296*:333 (1993).
348. A. Hamelin, X. Gao, and M. J. Weaver, J. Electroanal. Chem. *323*:361 (1992).
349. O. M. Magunssen, J. Hotlos, R. J. Behm, N. Batina, and D. M. Kolb, Surf. Sci. *296*:310 (1993).

350. X. Gao and M. J. Weaver, J. Phys. Chem. *97*:8685 (1993).
351. I. Villegas, C. B. Ehlers, and J. L. Stickney, J. Electrochem. Soc. *137*:3143 (1990).
352. R. L. McCarley and A. J. Bard, J. Phys. Chem. *95*:9618 (1991).
353. N. J. Tao and S. M. Lindsay, J. Phys. Chem. *96*:5213 (1992).
354. W. Haiss, J. K. Sass, X. Gao, and M. J. Weaver, Surf. Sci. *274*:L593 (1992).
355. X. Gao and M. J. Weaver, Ber. Bunsenges. Phys. Chem. *97*:507 (1993).
356. B. M. Ocko, G. M. Watson, and J. Wang, J. Phys. Chem. *98*:897 (1994).
357. F. T. Wagner and P. N. Ross, Surf. Sci. *160*:305 (1985).
358. X. Gao, Y. Zhang, and M. J. Weaver, J. Phys. Chem. *96*:4156 (1992).
359. R. L. McCarley, Y. Kim, and A. J. Bard, J. Phys. Chem. *97*:211 (1993).
360. L. P. Colletti and J. L. Stickney, unpublished results.
361. H. Siegenthaler, K. Juttner, E. Schmidt, and W. J. Lorenz, Electrochim. Acta *23*:1009 (1978).
362. N. N. Greenwood and A. Earnshaw, *Chemistry of the Elements*, Pergamon Press, Oxford, 1984.
363. Y.-T. Kim, and A. J. Bard, Langmuir *8*:1096 (1992).
364. T. Han and T. P. Beebe, Langmuir *10*:1096 (1994).
365. O. Chailapakul, L. Sun, C. Xu, and R. M. Crooks, JACS *115*:12459 (1993).
366. K. Edinger, A. Golizhauser, K. Demota, C. Woll, and M. Grunze, Langmuir *9*:4 (1993).
367. C. Schonenberger, J. A. M. Sondag-Huethorst, J. Jorritsma, and L. G. J. Fokkink, Langmuir *10*:611 (1994).
368. G. E. Poirier and M. J. Tarlov, Langmuir *9*:2853 (1994).
369. G. E. Poirier, M. J. Tarlov, and H. E. Rushmeier, Langmuir *10*:3383 (1994).
370. C. A. M. Dermott, M. T. M. Dermott, J. B. Green, and M. D. Porter, J. Phys. Chem. *99*:13257 (1995).
371. J. C. Bondos, A. A. Gewirth, and R. G. Nuzzo, J. Phys. Chem. *100*:8617 (1996).
372. J. W. Schultze, F. D. Koppitz, and M. M. Lohrengel, Ber. Buns. Phys. Chem. *78*:693 (1974).
373. D. W. Suggs and A. J. Bard, JACS (1994) *116*:10725..
374. R. F. C. Farrow, G. R. Jones, G. M. Williams, and I. M. Young, Appl. Phys. Lett. *39*:954 (1981).
375. T. H. Myers, Y. Lo, J. F. Schetzina, and S. R. Jost, J. Appl. Phys. *53*:9232 (1982).
376. S. Wood, et al., J. Appl. Phys. *55*:4225 (1984).
377. H. A. Mar and N. Salansky, J. Appl. Phys. *56*:2369 (1984).
378. S. K. Ghandhi and I. Bhat, Appl. Phys. Lett. *45*:678 (1984).
379. Z. C. Feng, et al., Appl. Phys. Lett. *47*:24 (1985).
380. G. M. Williams, C. R. Whitehouse, N. G. Chew, G. W. Blackmore, and A. G. Cullis, J. Vac. Sci. Technol. B *3*:704 (1985).
381. K. J. Mackey, P. M. G. Allen, W. G. Herrender-Harker, and R. H. Williams, Surf. Sci. *178*:124 (1986).

382. T. D. Golding, M. Martinka, and J. H. Dinan, J. Appl. Phys. *64*:1874 (1988).
383. I. B. Bhat, K. Patel, N. R. Taskar, J. E. Ayers, and S. K. Ghandhi, J. Cryst. Growth *88*:23 (1988).
384. N.-H. Cho, B. C. DeCooman, C. B. Carter, R. Fletcher, and D. K. Wagner, Appl. Phys. Lett. *47*:879 (1985).
385. K. Morizane, J. Cryst. Growth *38*:249 (1977).
386. H. Kroemer, J. Cryst. Growth *81*:193 (1987).
387. S. Strite, et al., J. Vac. Sci. Technol. B *8*:1131 (1990)
388. S. B. Quadri, E. F. Skelton, and A. W. Webb, J. Kennedy, Appl. Phys. Lett. *46*:257 (1985).
389. J. L. Bell and S. Sen, J. Vac. Sci. Technol. A *3*:112 (1985).
390. H. F. Wolf, *Semiconductors*, Wiley Interscience, New York, 1971.
391. I. Villegas and J. L. Stickney, J. Electrochem. Soc. *138*:1310 (1991).
392. L. E. Davis, N. C. MacDonald, P. W. Palmberg, G. E. Riach, and R. E. Weber, *Handbook of Auger Electron Spectroscopy*, Physical Electronics Division, Perkin-Elmer Corporation, Eden Prairie, MN, 1978.
393. V. Solzbach and H. J. Richter, Surf. Sci. *97*:191 (1980).
394. Y.-C. Lu, R. S. Feigelson, and R. K. Route, J. Appl. Phys. *67*:2583 (1990).
395. T. Meguro, et al., Appl. Phys. Lett. *56*:1556 (1990).
396. Y. Horiike, et al., J. Vac. Sci. Technol. A *8*:1844 (1990).
397. T. Meguro and Y. Aoyagi, in *Atomic Layer Growth and Processing*, Vol. 222 (T. F. Kuech, P. D. Dapkus, and Y. Aoyagi, eds.), Mater. Res. Soc., Pittsburgh, 1991, p. 121.
398. H. Sakaue, et al., in *Atomic Layer Growth and Processing*, Vol. 222 (T. F. Kuech, P. D. Dapkus, and Y. Aoyagi, eds.), Mater. Res. Soc., Pittsburgh, 1991, p. 195.
399. P. A. Maki and D. J. Ehrlich, Appl. Phys. Lett. *55*:91 (1989).
400. P. Lei, K. R. Choong and J. L. Stickney, University of Georgia, 1995.
401. P. A. Kohl and F. W. J. Ostermayer, Annu. Rev. Mater Sci. *19*:379 (1989).
402. P. A. Kohl, D. B. Harris, and J. Winnick, J. Electrochem. Soc. *137*:3315 (1990).
403. P. A. Kohl, D. B. Harris, and J. Winnick, J. Electrochem. Soc. *138*:608 (1991).
404. E. K. Propst, K. W. Vogt, and P. A. Kohl, J. Electrochem. Soc. *140*:3631 (1993).
405. P. A. Kohl and D. B. Harris, Electrochim. Acta *38*:101 (1993).
406. D. Harris, P. A. Kohl, and J. Winnick, J. Electrochem. Soc. *141*:1274 (1994).
407. L. P. Colletti, R. Slaughter, and J. L. Stickney, J. Soc. Info. Display *5*:87 (1997).

SCANNING TUNNELING MICROSCOPY STUDIES OF METAL ELECTRODES

T. P. Moffat

Materials Science & Engineering Laboratory
National Institute of Standards and Technology
Gaithersburg, Maryland

I. INTRODUCTION

Scanning tunneling microscopy (STM) may be used to directly image the electron density of surfaces with atomic resolution and to follow the dynamics of surface processes in real time. This capability, in combination with the simplicity and accessibility of the method, has contributed to major advances in surface science over the last 15 years [1]. STM was invented by Binnig and Rohrer in 1982 [2,3]. The extraordinary value of the instrument as a real space structural tool was demonstrated early by the rapid assessment of the long unresolved reconstructed Si(111)-(7 × 7) structure [4]. Following the initial reports of Binnig and Rohrer's UHV work, it became apparent that the STM could be operated in an electrolytic environment [5,6], and atomically resolved images of graphite in a variety of electrolytes were reported [5]. A flourish of activity followed examining surfaces beyond easily prepared and unreactive van der Waals solids. In 1988 the first published images of monatomic steps on an immersed noble metal surface appeared [7,8]. Within 2 years papers presenting spectacular atomically resolved images of noble metal electrodes with various ordered metal and anion-based overlayers were published [9,10]. Similarly, the ability to follow the dynamics of surface processes such as surface reconstructions, phase formation, etc. has been demonstrated [11,12]. The synthetic capability of STM has also been explored with a variety of novel structures fabricated via etching, deposition, and/or oxidation reactions occurring within, or in close proximity to, the tunnel gap [13–17]. By the early 1990s the lure of the aesthetic quality and novelty of atomically resolved imaging of electron density began to give way to a more detailed and quantitative assessment of STM images of solid/liquid interfaces. In this sense, perhaps the most unique capability of STM began to emerge, namely, the ability to image localized phenomena on an atomic scale. Defects such as vacancies, dislocation, surface steps, kinks, adatoms, and their dynamics may now be studied in real time and space [18]. More recently, the ability to probe the electronic states of individual adsorbed redox molecules via resonant tunneling was demonstrated [19]. The selectivity associated with the redox potential of a given molecule offers unique opportunities for the studying electron transfer reactions. The close combination of STM with other novel in situ spectroscopic and scattering methods, as well as traditional electrochemical methods, has led to a remarkable expansion of our knowledge of the structure and dynamics of the solid/liquid interface [20–23]. In this light it is interesting to note a remark by Sir Humphrey Davy that "nothing tends so much to the advancement of knowledge as the application of a new instrument. The native intellectual powers of men in different times are not so much the cause of the different success of their labours, as the peculiar nature of the means and artificial resources in their possession'' [24]. The purpose of this chapter is to describe the operation and utility of STM as a technique for probing the structure and dynamics of metal electrode surfaces. This effort builds upon previous reviews [25–30] as well as conference proceedings [1,31–36] on STM in electrochemistry. The discussion centers on metal electrodes since an excellent survey of STM studies of semiconductor electrodes was recently published [37]. Complete coverage of the published literature has not been attempted, rather attention is directed toward developments over the past 5 years which highlight the power and promise of this technique.

The scanning tunneling microscope uses an atomically sharp probe tip to map contours of the local density of electronic states on the surface. This is accomplished by monitoring quantum transmission of electrons between the tip and substrate while piezoelectric devices raster the tip relative to the substrate, as shown schematically in Fig. 1 [38]. The remarkable vertical resolution of the device arises from the exponential dependence of the electron tunneling process on the tip-substrate separation, d. In the simplest approximation, the tunneling current, I, can be simply written in terms of the local density of states (LDOS), $\rho_s(z,E)$, at the Fermi level ($E = E_F$) of the sample, where V is the bias voltage between the tip and substrate

$$I \approx V\rho_s(0, E_F)\exp(-2\kappa d) \tag{1}$$

and where κ is the decay constant. For a semi-classical square potential barrier, ϕ (WKB), the decay constant is given as $\kappa = (\pi(8m_e\,\varphi)^{1/2})/h \approx 0.51(\varphi(\text{eV}))^{1/2}$ where m_e is the electron mass and h is the Planck's constant. An effective or WKB barrier height of 4 eV yields $\kappa \sim 1$ Å^{-1} resulting in an order of magnitude decrease of the tunneling current per Å of electrode separation. Tunneling junctions may also be described in terms of the tunneling conductance, G,

$$G \approx G_0\exp(-2\kappa d) \tag{2}$$

where G_0 is associated with quantum-point contact ($d = 0$, which corresponds to 2–3 Å internuclear separation) ($G_0 = 1/R_0$, where $R_0 = h/2e^2 = 12.9$ kΩ) [39–41]. Point contact has also been treated classically as the Sharvin resistance where the contact aperture has a radius, α, which is much less than the mean free path λ ($\lambda/\alpha \gg 1$), $R_s = 4\rho\lambda/3\pi\alpha^2$, where

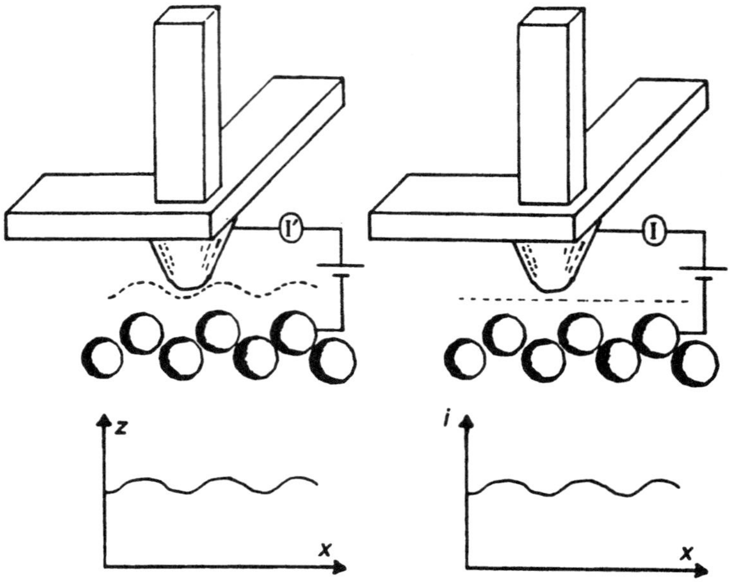

FIG. 1. Schematic representation of the (left) constant current, and (right) constant height modes of operation of an STM. (From Ref. 38.)

ρ is the resistivity of the metal such that a contact 3 Å in diameter corresponds to a resistance of ~10 kΩ [40,41]. Importantly, recent experimental work has demonstrated that the resistance of the fused junction is material dependent, as indicated in Fig. 2 [42].

For imaging purposes, the instrument may be used to follow contours of constant electron density, or alternatively the tunneling current may be monitored while the tip is rastered at a fixed distance from the substrate. The first imaging mode operates using negative feedback to adjust the z-piezo voltage to maintain a set tunneling conductance and is typically referred to as constant current imaging. A plot of the voltage applied to the z-piezo versus the tip raster position yields an image of contours of constant electron density of the surface as indicated in Fig. 1. On the mesoscopic level this technique gives a measure of the surface topography of a bare metal surface, while at the atomic level a more sophisticated description is required that correlates local electron density at the Fermi level with the atomic surface structure. In the second imaging mode the tip is rastered

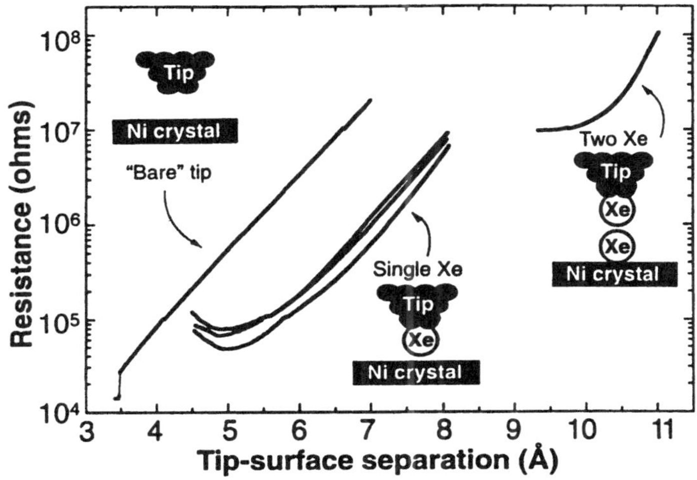

FIG. 2. The characteristics of the tunnel junction are sensitive to the effects of adsorbates. The effect of one or two Xe atoms on the point of contact resistance is clearly shown. Constant current scans over individual adsorbate indicate that the effective diameter of the Xe atom on Ni(110) is ~0.19 nm. (From Ref. 42.)

rapidly over the surface at a fixed height while the tunneling current is monitored. This method is referred to as constant height imaging. In this instance higher scan rates are accessible since the electronics only have to measure the tunneling current fluctuations as opposed to controlling the movement of the z-piezoelectric scanner. This method is effective for imaging flat surfaces, however, interpreting constant height images in terms of topography demands a sophisticated understanding of the relationship between tunnel current and tip-substrate separation.

STM may also be used to characterize the local electronic properties of the surface in terms of the effective tunneling barrier or decay constant (κ), which is derived from the dependence of the tunnel current on the tip-substrate separation. Similarly, the energy distribution of the density of states may be examined via tunneling spectroscopy where the bias dependence of the tunneling current is measured at a fixed tip-substrate distance. This has proven to be particularly useful for examining surfaces that exhibit large changes in the LDOS with bias, such as semiconductors. In a similar vein, the possibility of potential-dependent resonant tunneling through adsorbed molecules may also be studied, thereby offering a unique window for examining electron transfer reactions involving surface-confined species.

Operating a scanning tunneling microscope in an electrolyte adds a new dimension to the tunnel junction. A well-defined experiment in this instance requires the use of a bipotentiostat to independently control the electrochemical potential of the tip and substrate relative to some reference electrode. This configuration is distinct from an ultra high vacuum (UHV) experiment where only the bias between the electrodes needs to be specified. In the electrochemical environment the tip electrode is simultaneously a tunneling probe and an ultramicroelectrode. Consequently, suitable attention must be given to possible faradaic reactions proceeding at the tip as

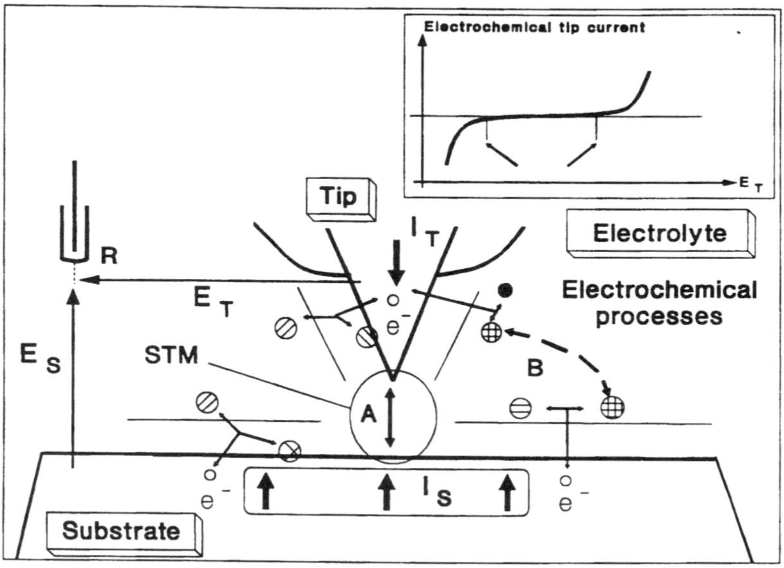

FIG. 3. Schematic presentation of an immersed tunnel junction where in addition to direct tunneling between the tip and substrate there is also the possibility of electrochemical reactions occurring at the tip and substrate. The broken arrow indicates the possibility of coupling between the electrochemical reactions occurring at the tip and substrate, which is the basis of SECM. (From Ref. 26.)

suggested in Fig. 3. These reactions may include redox events as well as deposition and dissolution processes. Under constant current imaging conditions, the set point current is maintained by a combination of electron tunneling and the faradaic process occurring at the tip. Typically, an attempt is made to minimize the faradaic contribution at the tip by coating the probe with an insulating substance, leaving only the apex of tip directly exposed to the electrolyte as indicated in Fig. 3. A typical set point current for atomically resolved STM imaging is on the order of $\sim 1-10$ nA. This corresponds to an extremely large current flux, $\sim 10^6$ A/cm^2, between the apex of the tip and the substrate area being probed, $<10^{-14}$ cm^2. In contrast, any faradaic process would be distributed over the exposed area of the tip, which is often in the range of $\sim 10^{-8}-10^{-10}$cm^2, such that a 10 nA faradaic current would correspond to a current density of 1–100 A/cm^2. Thus, provided the tip electrode is suitably coated, a very large faradaic perturbation is required to destabilize the tunneling-based imaging process. In contrast, the exponential decrease in the tunneling current with increasing tip-substrate separation eventually leads to the limiting case where $i_{faradaic} \gg i_{tunnel}$. Under appropriate conditions, the faradaic current may be used to form images of the electrochemical reactivity of a surface. This is known as scanning electrochemical microscopy (SECM), where the transport and heterogeneous redox activity of species within the junction mediate the tip-substrate interaction. This subject has been thoroughly reviewed [43,44], and an excellent paper demonstrating the transition from STM to SECM is available [45]. The possible contribution of confined redox species to resonant tunneling has also been examined [19,46,47].

Interestingly, electrochemical processes are also evident in certain two-electrode STM experiments performed in air. It is well known that water is absorbed on surfaces exposed to humid environments [48,49]. When such circumstances arise in combination with certain bias conditions, the conventional two-electrode STM exhibits some of the characteristics of a two-electrode electrochemical cell as shown in Fig. 4 [50–53]. This scheme has been used for modifying surfaces and building devices, as will be described in the last section of this chapter. In a similar vein, it has been suggested that a two-electrode STM may be used to perform high-resolution SECM for certain systems that include insulating substrates such as mica [50].

In addition to imaging applications, the sensitivity of the STM may be implemented as a displacement transducer. For example, the small dis-

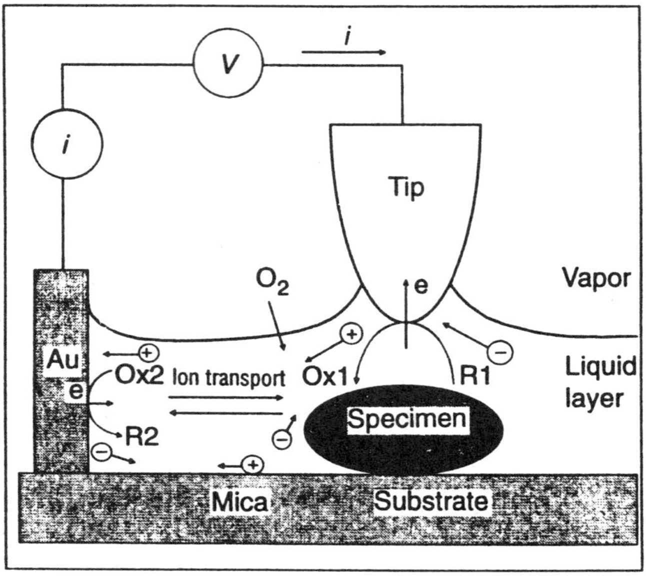

FIG. 4. Many STM experiments performed under ambient conditions incorporate SECM phenomena associated with water adsorption on surface due to the ambient humidity. In some instances the SECM signal may be dominant as in the case of imaging bulk insulators such as mica. (From Ref. 50.)

tortions of a thin film electrode associated with potential dependent changes in surface stress have been quantified in this way [54,55]. Similarly, other forms of scanning probe microscopy, such as atomic force, lateral force, and near-field optical microscopy, have also contributed significantly to electrochemical research [30,33]. The application of these methods is not the subject of this chapter; nevertheless, much of the knowledge gained from these techniques contributes directly to our understanding of the physics of the tunneling junction. Knowledge of the atomic forces that develop between the tip and substrate is particularly relevant for understanding atomically resolved STM imaging as well as qualifying the possibility of surface modification induced by such interactions [56].

II. QUANTUM MECHANICAL TUNNELING

The phenomenon of quantum mechanical tunneling of electrons between two metals separated by vacuum or an insulator is well established. An excellent discussion of electron tunneling, especially as it relates to image formation in STM, is available in a recent monograph [57], textbook [39], and review paper [58]. The most popular theoretical methods rely on perturbation theory, while more sophisticated treatments are available that consider interactions between the tip and sample using scattering theory [57,58]. The Bardeen perturbation formalism yields a description of the tunneling current resulting from the overlap of the wave functions of the individual metal barrier subsystems, Ψ and χ, as shown in Fig. 5 [39]. The probability, p, of electron transfer is given by Fermi's golden rule:

$$p = \frac{2\pi}{\hbar} |M|^2 \delta(E_\Psi - E_\chi) \tag{3}$$

The delta function, δ, limits the analysis to elastic processes. The tunneling matrix element, M, is determined by the overlap of the surface wave functions of the two metal subsystems at a particular separation surface, which also reflects the energy-lowering resonance associated with the interplay of the two states. The tunneling current may be found by summing over

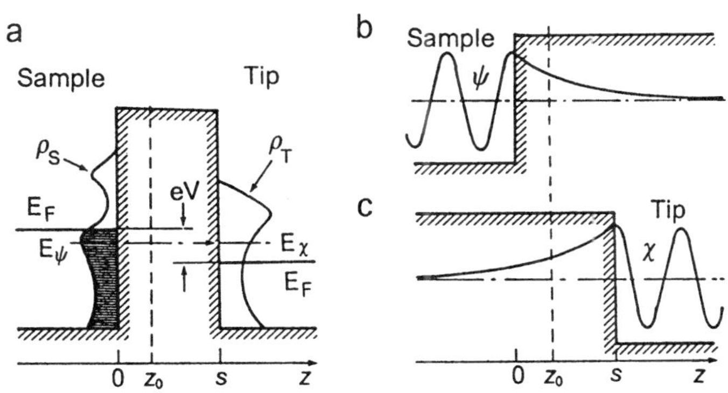

FIG. 5. The tunneling current between the tip and sample is derived from the overlap of the respective wave functions using Fermi's golden rule. (From Ref. 39.)

all relevant states. For a bias voltage, V, the total tunnel current is given by

$$I = \frac{4\pi e}{\bar{h}} \int_{-\infty}^{\infty} [f(E_F - eV + \varepsilon) - f(E_F + \varepsilon)] \times \rho_S(E_F - eV + \varepsilon)\rho_T(E_F + \varepsilon)|M|^2 \partial\varepsilon \quad (4)$$

where $f(E) = \{1 + \exp[E - E_F)/k_BT\}$ is the Fermi-Dirac distribution function and $\rho_S(E)$ and $\rho_T(E)$ are the density of states of the two electrodes [39]. If kT is much smaller than the feature size in the energy spectrum of interest, then

$$I = \frac{4\pi e}{\bar{h}} \int_{0}^{eV} \rho_S(E_F - eV + \varepsilon)\rho_T(E_F + \varepsilon)|M|^2 \partial\varepsilon \quad (5)$$

Furthermore, if the magnitude of the tunneling matrix, M, does not change appreciably in the interval of interest, the tunneling current may be viewed as simply the convolution of the density of states of the respective electrodes.

A. STM Junction in Vacuum

The details of the transport process occurring in atomically resolved STM are significantly different from the tunneling process associated with classical metal-insulator-metal (M-I-M) tunnel junctions. In the latter case the thickness of the insulator between the conducting electrodes is typically 20–30 Å, whereas the ability to resolve the electronic density of individual atoms using STM requires a lateral resolution of ~2 Å. To obtain such resolution requires the tip-sample distance be very short—~3–7 Å (the distance between the nucleus of the apex atom of the tip and nuclei of the top layer of the sample surface). An understanding of the junction requires consideration of the local atomic and electronic structure as well as an assessment of the atomic and electrostatic forces between the tip and the sample [39,57,58]. For example, the measured corrugation amplitude of Al(111) is known to be a strong function of tip-sample separation as well as the electronic state of the tip, as indicated in Fig. 6 [39,59,60]. An exponential dependence of corrugation amplitude with distance was clearly observed, as might be expected. However, the measured amplitude was more than an order of magnitude greater than the corrugation determined by helium-scattering experiments, the latter providing an assessment of the

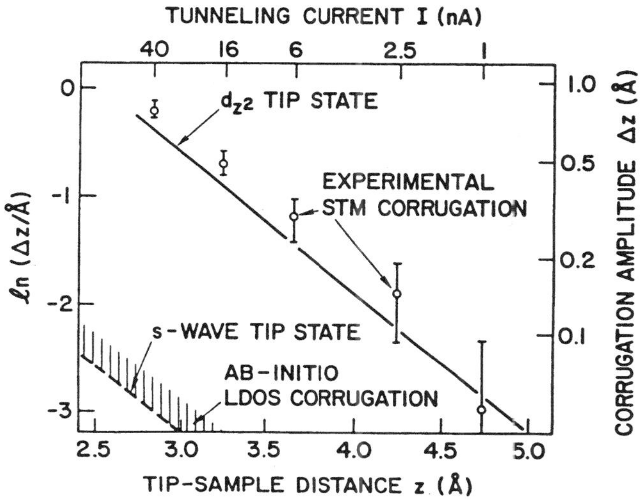

FIG. 6. The corrugation observed while imaging Al(111) is a strong function of tip-sample separation as well as the electronic structure of the tip. Theoretical results for an s and d_z^2 tip state are shown in comparison with experimental data (V_{bias} = −50 mV). (From Ref. 39.)

total charge density integrated over all occupied states while STM probes only states at the Fermi level. The data were explained by considering a d_z^2 tip state as opposed to an s-wave tip state. An s-wave probe would yield a charge density contour reflecting the LDOS derived from the summation of the s-states comprising the substrate atoms. In contrast, a d_z^2 tip state maps the charge density of a fictitious surface with a d_z^2 state on each atom as suggested by the reciprocity principle outlined in Fig. 7 [39,61]. To date atomically resolved imaging of a wide variety of metal surfaces has been reported: Al(111), Au(111), Ag(111), Pt(111), Cu(111), Au(100), Cu(100), Ni(100), Cu(110), Ni(110), Ag(110), Fe(100), etc. [39,60–63].

The imaging process is influenced by interactions between the wave functions of the substrate and tip that result in significant deviation from the idealized square barrier profile. This is particularly true at small tip-substrate separations, where the barrier collapses below the vacuum level, as indicated in Fig. 8 for the Al-Al junction [56,64]. These calculations indicate that for typical STM imaging conditions, the top of the barrier

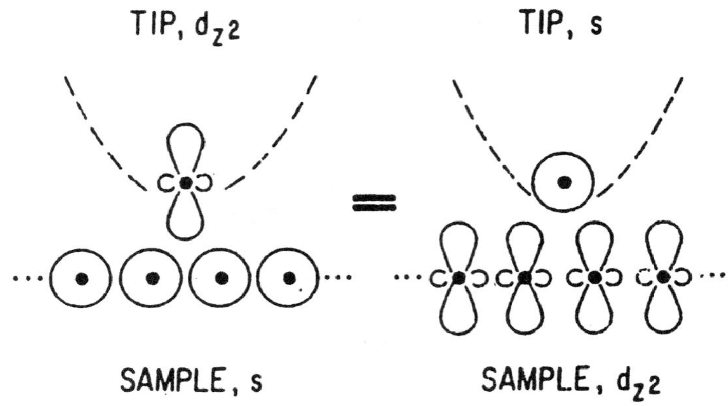

FIG. 7. The importance of the tip state is highlighted by applying the reciprocity principle to imaging a free electron metal surface with a d_z^2 tip. (From Ref. 39.)

between the tip and the sample is either very close to or lower than the Fermi level [56,57,64,65]. From a semiclassical perspective, this condition is referred to as ballistic transport. However, the distinction becomes meaningless for barriers of atomic dimension where the difference between tunneling and ballistic transport becomes indistinguishable, revealing only the unified phenomenon of quantum transmission [39,57].

Significant forces may develop between the tip and sample during atomically resolved imaging. The relationship between these forces and tunneling is closely related to the theory of chemical bonding, whereby the atomic force and tunneling conductance may be correlated to the binding energy gradient and the resonance frequency, respectively [39,57]. The scanning tunneling microscope has been likened to a giant molecule consisting of two component molecules with a controllable intermolecular distance [39]. At tip-sample distance beyond 100 Å the tip-sample interaction is negligible, while between 10 Å $< z <$ 100 Å long-range interactions take place that distort the wave functions and a van der Waals force arises. At short distances, 3 Å $< z <$ 10 Å, electron transfer or exchange may give rise to an attractive interaction corresponding to the resonance condition associated with chemical bonding. At even shorter distances, $z <$ 3 Å, a repulsive force becomes dominant, the onset of which corresponds to mechanical contact. The distance dependence of the atomic forces and tunnel conductance have been described by a variety of models like that shown

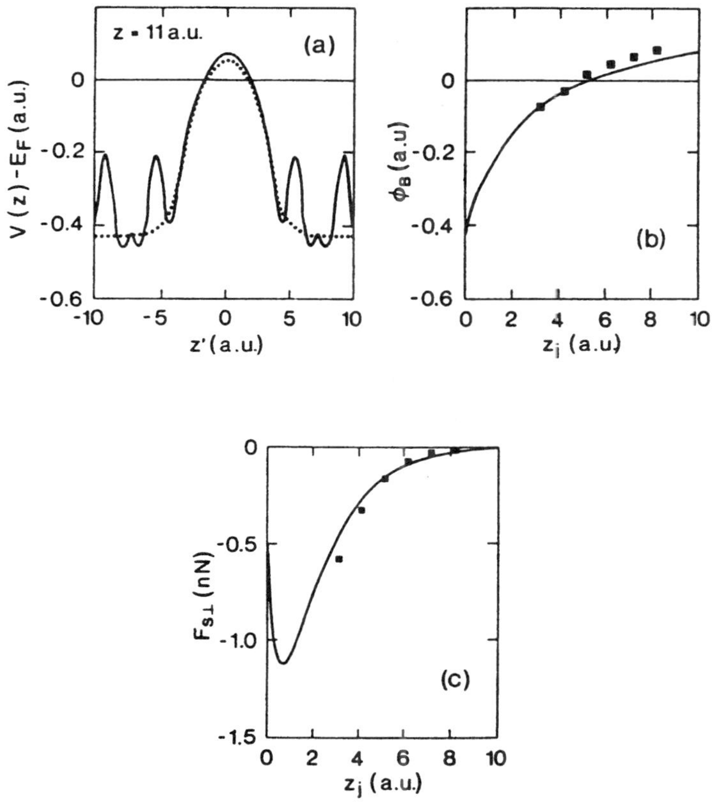

FIG. 8. The potential energy profile between two jellium electrodes separated by ~7.2 a.u. is shown in (a), while the dependence of the barrier height and attractive force on the separation between the two electrodes is given in (b) and (c). (From Ref. 56.)

in Fig. 8 [56]. Importantly, the magnitude of the predicted forces lies in a range, 4–0.01 nN, which is experimentally accessible [57]. The simultaneous measurement of the force gradient and tunneling conductance has been accomplished by monitoring the shift in the thermally induced resonant frequency of a flexible cantilever beam sample as a function of the tunnel conductance [66]. As indicated in Fig. 9, attractive forces operate between

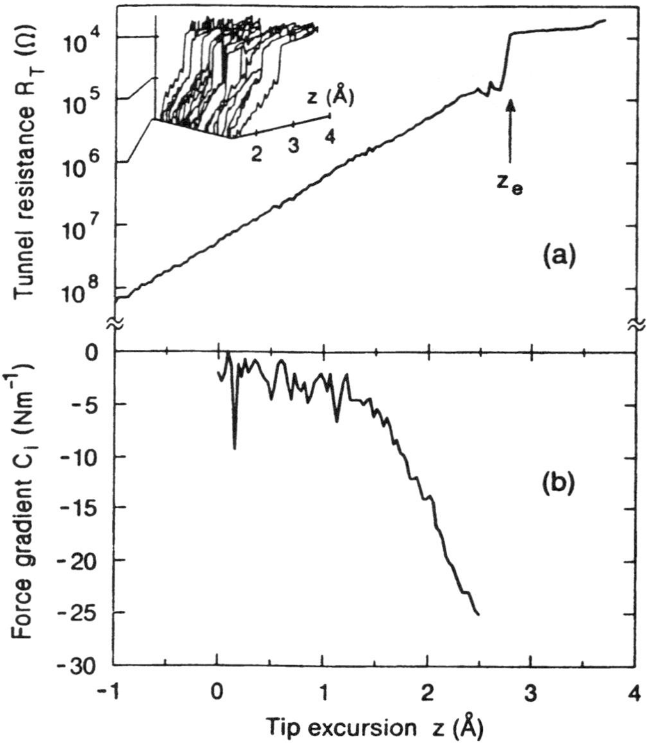

FIG. 9. The tunnel resistance as a function of tip movement is shown in (a). Positive z displacement corresponds to a decreasing tunnel junction width, and the jump in conductance at z_e is associated with point contact. The inset figure shows the degree of reproducibility associated with the experiment which employed an Ir tip and substrate. The gradient associated with the attractive force between the two electrodes was measured simultaneously as shown in (b). (From Ref. 66.)

two metal surfaces at separation distances which are typically used for atomically resolved imaging of metal surfaces [57,66].

The effective tunneling barrier may be quantified by measuring the distance dependence of the tunnel conductance or tunneling current [Eq. (1) or (2), respectively]. Experimentally, the decay constant, κ, may be derived from dc or ac measurements. The more accurate ac modulation method is implemented by applying a small, ~0.05 Å, modulation, Δs, to the z piezo at ~kHz frequency and using a lock-in amplifier to measure the corresponding current modulation (di/ds) where $\kappa = -1/i(di/ds) = -1/2(d(Ln(i))/ds)$. As noted earlier, the decay constant is often converted to yield an effective barrier height, φ (eV), according to the WKB model. However, it is important to note that the square barrier assumed in the WKB model is a physically unsuitable description of the STM junction [39,56,57]. The barrier itself is a function of the tip-substrate separation, as shown in Fig. 8, such that the decay constant measurement incorporates variation of the barrier height with separation distance. Consequently, the WKB-derived barrier height does not reflect the true character of the tunneling barrier. A further complication in assessing the tunnel barrier may result from mechanical strains associated with forces that develop in the junction. In this instance, the small gap displacements that are inferred from the piezo voltage measurement may not necessarily reflect the actual gap displacement [39,56,57]. This has been shown to be particularly important when studying elastically soft materials such as layered van der Waals solids, although the effect is considered to be a minor one for metallic surfaces [39,57,67,68].

Further progress in understanding the mechanism of atomically resolved imaging in STM will require a close coupling between theory and experiment. The power of such comparison in the case of elemental metals has been demonstrated as shown in Fig. 6 [39,57,61], while more recently success has been obtained in describing more complicated systems that involve alloy formation [63,69]. It is noteworthy that the LDOS of many surfaces have been studied by first principle calculations. In contrast, the lack of detailed knowledge of the geometry and electronic structure of probe tips usually limits in-depth evaluation of images and the imaging mechanism. Tip preparation, stability, and characterization remain the central experimental difficulties associated with STM. Nevertheless, significant progress has been made in this area, as suggested in Fig. 2 [42] and elsewhere [58].

The study of adsorbates on metal surfaces is a particularly fruitful area, which has received much attention. Experimentally, atomic adsorbates are known to yield very different images ranging from bumps [S on Pt(111)] to depression [O on Ni(100)] depending on the nature of the interaction with the LDOS of the metal surface [58,70,71]. The phenomenon has been analyzed in terms of the impact of the adsorbate on the local density of states at the substrate Fermi level [57,71–75]. Importantly, even in the cases where the atomic resonance of an adsorbate lies far above the Fermi level, e.g., Xe on Ni(110) [71], the adsorbate may nevertheless contribute to the LDOS due to significant broadening of the resonance upon adsorption. This leads to a long energy tail, which extends to the Fermi level of the substrate as shown in Fig. 10 [71]. If the size of the orbital associated with the resonance is such that it extends considerably further out into the vacuum than the bare substrate wave functions, it will significantly influence image formation [58,70,71]. The tunnel junction has also been analyzed in terms of the electronic coupling between the tip and adsorbate as compared to the tip and substrates [58,70]. The model indicates that the current between the tip and metal surface is attenuated in the vicinity of the adsorbates while the through-atom channel associated with the atomic orbitals of the adsorbate becomes the dominant term in image formation. A comparison between calculated topographic images of several

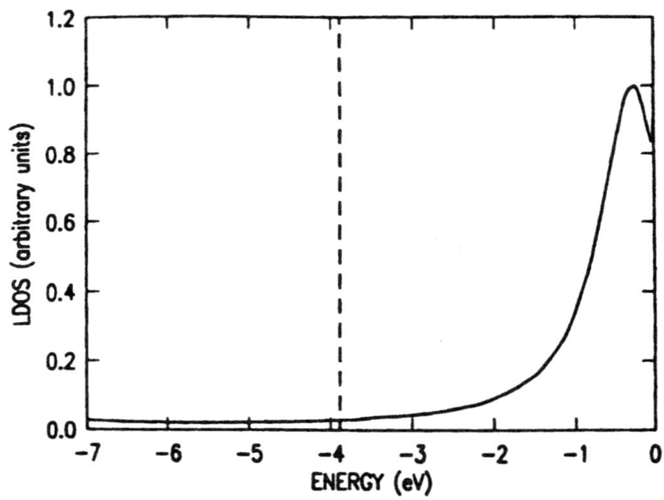

FIG. 10. Theoretical calculations reveal that in the case of adsorption of Xe on Ni the resonance associated with Xe(6s) state is broadened significantly with a long tail that extends to the Ni Fermi level. STM images are determined by the LDOS at the Fermi level. Although the contribution of Xe to the LDOS is small, it significantly extends the spatial distribution of the electronic wave function further away from the surface thereby acting as the central channel for quantum transmission to the probe tip. (From Ref. 71.)

atomic adsorbates on Pt(111) and the electronegativity and polarizability of the adsorbate atom revealed a more favorable correlation with the latter, as shown in Fig. 11 [58,70]. This indicates that the extension in space of the orbitals is more important than the adsorbate's orbital energies [58,70].

B. Resonant Tunneling

Recent studies of molecular species embedded in the M-I-M tunnel junction provide evidence of resonant elastic tunneling associated with the unoccupied orbitals of molecular species [76]. The process has been described in terms of transient reduction of the molecular species. A positive correlation was observed between the orbital-mediated tunneling spectroscopy data and the electrochemical reduction potentials for the solution phase molecular systems as shown in Table 1. The model was also used to explain the spontaneous and permanent reduction of certain compounds within the M-I-M junction. More generally, elastic resonant tunneling and/or through bond tunneling has become a central theme in the interpretation and theoretical development of STM imaging of molecules adsorbed on metal surfaces [58,77,78]. Interestingly, numerous images of molecules adsorbed on metallic surfaces display a likeness to the shape of the highest occupied (HOMO) and lowest unoccupied molecular orbital (LUMO) of the molecules, which simply reflects the participation of the frontier orbitals in the tunneling process [67,79,80]. Initially this was surprising since the HOMO and LUMO of many of these molecules lie several electron volts away from the Fermi level of the tip and substrate [58,81]. However, as noted earlier for atomic adsorbates, despite weak mixing of the molecular HUMO and LUMO with the surface wave functions, the contribution of the adsorbate to the LDOS at the Fermi level can be orders of magnitude larger than the contribution of the bare substrate at distances typically associated with STM probe position. This should not be surprising since the adsorbate sits on top of the surface, which guarantees that its orbitals will extend much farther out from the surface than those associated with the substrate. This scheme is also congruent with the positive correlation found between image brightness, the polarizability of the molecular functional group [58,81], and the ionization potential of the molecule [82]. Since the tunneling probability depends on the LDOS at the Fermi level, the imaging of adsorbate with a large separation between the unoccupied electronic states and the Fermi level, $>\sim 0.5$ eV, should not depend strongly on bias. In contrast, if the molecule's electronic states are within ~ 0.5 eV of the sur-

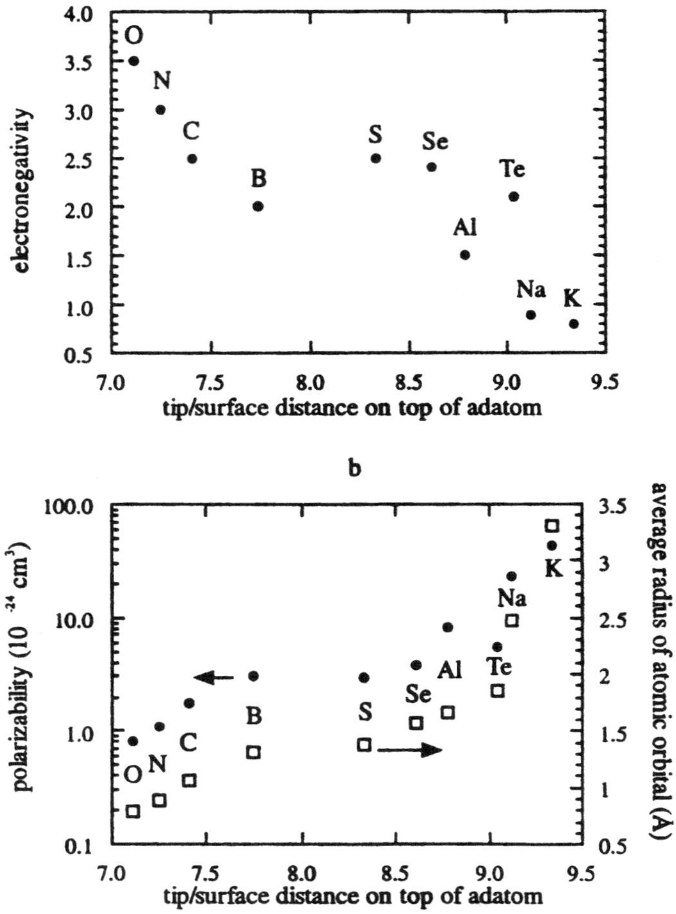

FIG. 11. Correlation between the tip height on top of an adsorbate, the atom electronegativity, and the atomic polarizability, which may be related to the average radius of the appropriate atomic orbital. Far from the adsorbate the tip-substrate separation corresponds to 7.45 Å for a 10 MΩ gap resistance. (From Ref. 70.)

TABLE 1

Standard Reduction Potentials (in Nonaqueous Solvents) and Adjusted First Orbital–Mediated Tunneling (OMT) Band Position (in υ) in a Variety of Compounds

Molecule	$E_{1/2}$	4.2-OMT band max
Coronene	2.62	2.61
Anthracene	2.80	2.84
Perylene	3.00	3.01
Tetracene	3.12	3.21
Zn tetrabenzoporphine	3.24	3.28
Pentacene	3.40	3.57
Cu phthalocyanine		3.90
Cu phthalocyanine tetrasulfonate	3.98	
TCNE	4.50	Spontaneous reduction
TCNQ	4.56	Spontaneous reduction
Ferricyanide	4.80	Spontaneous reduction

Source: Ref. 76.

face Fermi level, strong resonant tunneling may be expected. An elegant example is provided by the observation of potential dependent resonant tunneling associated with the LUMO of a redox center embedded in porphyrin molecules [19]. Related effects have been reported in vacuum studies [83]. Additional resonant imaging effects may be associated with distortion of the electronic structure of molecules due to the strong forces experienced within the tunnel gap [84,85].

C. Inelastic Tunneling

Thus far the discussion has centered on elastic tunneling, but consideration of inelastic processes may offer additional analytical opportunities. An energy scale of the relevant phenomena is presented in Table 2. Inelastic tunneling was first observed in metal-oxide-metal junctions. It was immediately developed as a technique for photon-free vibrational spectroscopy (IETS) where the tunneling electrons dissipate energy by coupling to vibra-

TABLE 2
Energy Scales Over Which Tunneling Experiments Can Be Performed

Type of excitation	Energy
Metal phonons	Up to 50 mV
Molecular or lattice vibrations	50–120 mV
Vibrations of molecules contained within the insulator region	120–500 mV
Electronic transitions	Above 1V

Source: Ref. 87.

tional modes of organic molecules adsorbed at the interface in the metal-oxide-metal tunnel junctions [86]. Typically, the inelastic channel contributes about 1% or less to the tunneling flux and is associated with a threshold bias voltage, V, where $eV = h\nu$, with ν being the vibrational frequency excited in the molecule [87]. The summation of the elastic and inelastic channel results in a change in the slope of the current-voltage response at the bias threshold for inelastic coupling. Thus, spectra are usually presented using derivative methods, i.e., di/dV or d^2i/dV^2. The resolution of IETS is ultimately determined by thermal smearing of the Fermi surfaces of the electrode materials, which corresponds to a FWHM of 5.4 kT. This yields a resolution of 0.136 meV or 1.1 cm^{-1} at 1 K, however, at room temperature the FWHM is on the order of 0.127 eV or 1024 cm^{-1}, which strongly discouraged development of the method for performing vibrational spectroscopy under ambient conditions [87].

In addition to coupling with phonons, photon emission from phonons excited by inelastic tunneling has been reported for STM junctions [88] as well as solid-state devices [87]. Photon emission from metals, semiconductors, and adsorbate-covered surfaces has been observed under ambient conditions as well as UHV [89–91]. A variety of factors including the local dielectric properties, surface geometry, and density of states for inelastic processes affects the probability of inelastic tunneling and photon emission from metal surfaces. Theoretical models have been presented based on tip-induced localized plasmon modes, which are associated with the strong electric field in the tunnel junction. These localized modes are thought to be analogous to those associated with surface-enhanced Raman scattering [90]. Adsorbates may strongly affect the emission intensity by altering the local branching ratio for elastic versus inelastic tunneling. This photoemission process offers the possibility of chemically mapping surfaces by correlating the photon flux with the STM raster. Studies of a monolayer C_{60} film on Au(110) have demonstrated spatial resolution for photon emission of approximately 4 Å, which corresponds to a tunneling path incorporating a single C_{60} molecule [90]. Similarly, the possibility of using photons to perturb the relevant electronic structure within such a junction represents an additional avenue for exploration.

D. The Immersed Tunnel Junction

The influence of an electrolyte on the characteristics of the tunnel junction has been investigated both experimentally and theoretically. Numerous measurements have revealed the anticipated exponential distance dependence of the tunneling current [7,92–94]. The corresponding effective barrier height was found to cover a wide range, from values slightly below that of related vacuum junctions, 0.5 $Å^{-1}$ ~ 2.15 eV, to unusually low values such as 0.12 $Å^{-1}$ ~ 0.2 eV (vacuum junction is typically ~4 eV). In early theoretical work two explanations for the diminished barrier were examined that focused on assessing the role of the solvent, namely, tunneling via the V_0 level or, loosely speaking, the "conduction band" of the solvent [95] or, alternatively, resonant tunneling via a hydrated electron intermediate state [96]. A schematic of the potential energy surface of the respective barriers is given in Fig. 12.

In the first model, the tunneling electron mainly interacts with the electronic polarization of water ($\varepsilon = 1.88$) since tunneling was assumed to be fast in comparison with the orientational response of the dipolar molecules of the liquid. Considering water as a dielectric continuum between a jellium spherical tip and planar substrate yields an effective barrier for tunneling that is about 1 eV lower than that for the vacuum case [95]. This result is consistent with photoemission studies of metal/aqueous interfaces, which reveal electron emission into water at ~1 eV below the vacuum level [95–97]. Similar models have been employed to examine the effect of thermal fluctuations on the tunneling current [98–100]. Likewise, a related model assessing the noise associated with the reorientation of adsorbed molecules has been presented [101].

In the second or resonant tunneling model, the intermediate state is associated with the formation of a solvation cage for the hydrated electron [96]. This scheme has largely been discounted since organizing the state

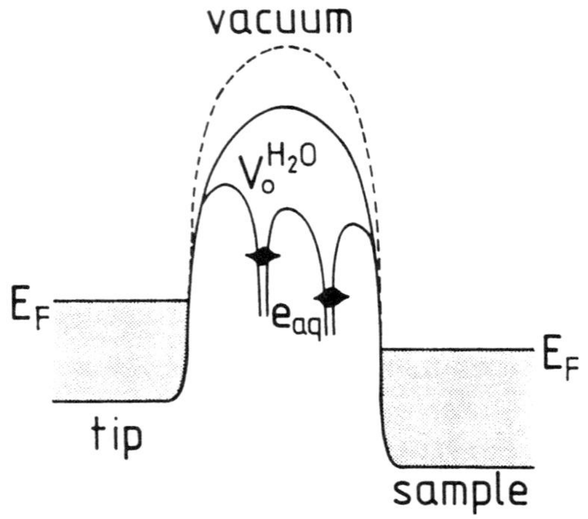

FIG. 12. A schematic representation of the possible role of water on the potential profile within the tunnel junction. Fast electronic polarization of the solvent diminishes the barrier, while the possibility of forming an intermediate hydrated electron resonant state has also been suggested. (From Ref. 96.)

requires an activation energy of 0.5 eV, which makes the process inefficient [102,103]. Furthermore, since hydrated electrons typically have a radius of about 10Å, there would generally not be sufficient space between the tip and the substrates to sustain the intermediate [102,103]. Further analysis reveals that the time spent by an electron tunneling across the gap ($\sim 10^{-15}$ s) is so short that coupling to the orientation polarization of the liquid would be insufficient to significantly lower the barrier [98–100]. Another model of resonant tunneling has been suggested where tunneling is associated with water dipole–induced states near the Fermi level [94]. However, experimental work to date reveals no indication of sharp resonant tunneling at noble metal-tip junctions immersed in simple aqueous electrolytes. For example, the current-voltage characteristics of a Au(111)/0.01 M $HClO_4/Pt_{80}Ir_{20}$ junction of variable dimension exhibits an ohmic character as shown in Fig. 13 [104]. In contrast, a nonlinear response would be a direct indication of resonant tunneling. A degree of caution is warranted at this point, since adsorbates are routinely imaged via resonant tail states, which exhibit a very mild bias dependence. However, no images of adsorbed water (excluding coadsorption effects [105–108]) have been re-

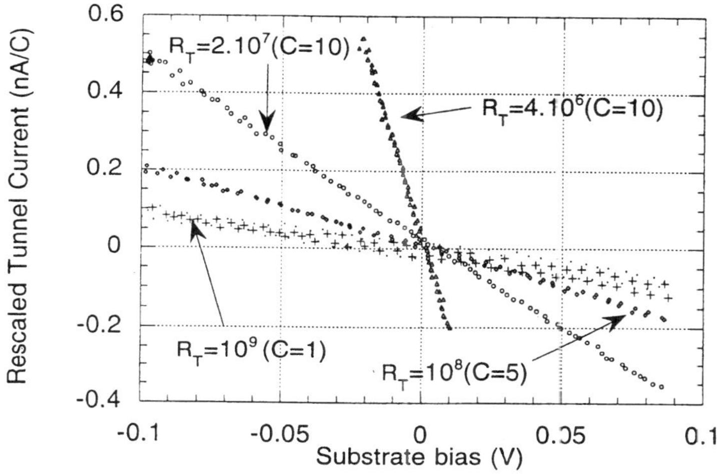

FIG. 13. Current-voltage plots taken at a variety of tunnel gap dimensions. The linear ohmic response suggests the absence of sharp resonant tunneling phenomena occurring in the immersed junction. The gap dimensions were set according to a specific tunnel resistance (R_T), after which the feedback was disengaged to collect the i-V data. The currents were rescaled, (C), for display purposes. (From Ref. 104.)

ported, although several reports of structurally defined double layers exist in the literature [22,109].

The significant variation of the barrier height observed for immersed junctions reflects the experimental difficulties associated with determining the tunneling constant, κ. Two key issues are contamination of the junction and uncertainty as to the structural and electronic character of the tip [104]. Recent data clearly reveal a dependence of the apparent barrier height on tip-substrate separation [7,92–94,104]. Specifically, the effective barrier is observed to diminish for resistance values below $<10^8$ Ω as shown in Fig. 14 [104], the collapse of the barrier being a consequence of the attractive resonance and image forces between the tip and sample [7,39,56]. In Fig. 15 a theoretically derived profile of the distance dependence of the effective tunneling barrier between an unbiased jellium tip and substrate immersed in an electrolyte is compared to the same in vacuum [95,103,110]. From this model the effective barrier height for the immersed junction was shown to be ~1 eV below that for the vacuum case, and the barrier appears to

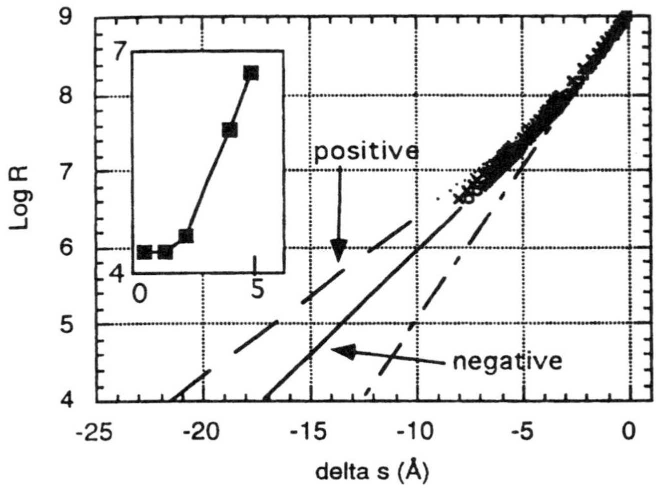

FIG. 14. The dependence of the tunnel resistance on z-piezo voltage or movement for a junction immersed in 0.01 $HClO_4$. Negative displacement corresponds to a diminishing gap between the Au(111) substrate and the Pt-Ir tip. Extrapolation of the data to point contact $\sim R = 10^4 \Omega$ may be used to estimate the size of the tunnel junction (also there are several ambiguities and pitfalls associated with this procedure, as noted in Fig. 2). The long-short dash corresponds to extrapolation from the slope at $10^9\Omega$, while the other lines correspond to extrapolation from $10^7\Omega$. These results reveal a distance dependence on the barrier height and also suggest a dependence on polarity. The inset figure shows data for vacuum tunneling. (From Ref. 104.)

decrease more rapidly as the substrate is approached. The unusual shape of the effective barrier at short separations is due to the fact that both the barrier height and thickness increase simultaneously, which is not accounted for in the WKB square barrier description of the effective barrier. Further work exploring the bias dependence of the effective barrier height indicates that the barrier is noticeably modified only at short tip-substrate separations as shown in Fig. 16 [103]. This prediction is in agreement with measurements that show that the tunneling conductance only becomes sensitive to bias polarity when the tunneling resistance decreases below $10^7\ \Omega$ as shown in Fig. 14 [104].

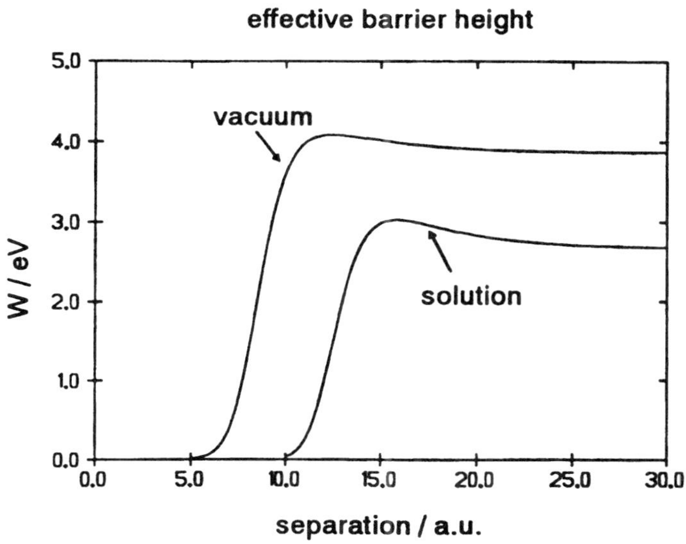

FIG. 15. A comparison between the distance dependence of the tunneling barrier between a jellium tip and substrate immersed in solution versus vacuum under zero bias conditions. The apparent barrier height is derived from the WKB approximation. (From Ref. 110.)

Barrier height measurements, as a function of tunnel conductance, have been used to estimate the dimensions of the tunnel junction via Eqs. (1) and (2). Accordingly, for a given conductance, the observation of a lower effective barrier has been used to infer larger dimensions for an immersed versus a vacuum junction as suggested in Fig. 14 [104]. However, the relevance of parameterizing the immersed junction in this manner is associated with a significant degree of uncertainty. The method often relies on a linear extrapolation of the data to point contact, which, as was shown in Fig. 2, can be very sensitive to the chemistry and structure of the junction. Furthermore, the dimensions derived from Fig. 14 lead to an apparent contradiction in the case of atomically resolved imaging of metals, since vacuum studies indicate that imaging proceeds via interaction of the d-states of the tip and substrate. This overlap would be difficult to sustain at the larger dimensions suggested for the immersed junction. In fact, atomically resolved images of immersed metal surfaces, such as Au(111), are typically

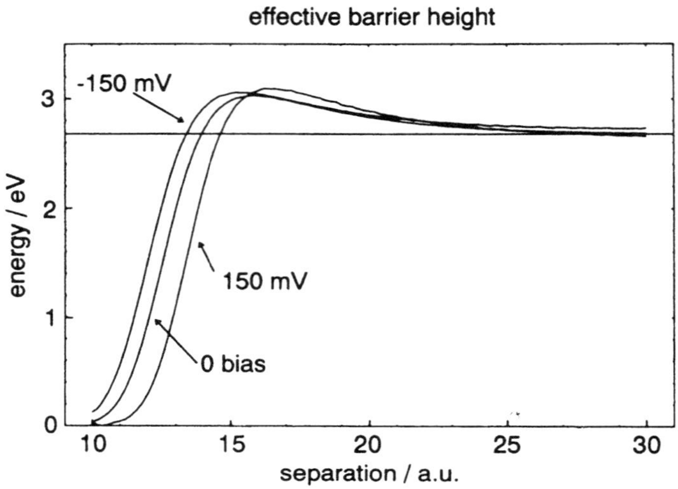

FIG. 16. Bias dependence of the apparent tunneling barrier between a jellium tip and substrate. (From Ref. 103.)

obtained at tunneling resistance values $\leq 10^7\ \Omega$ [111], which is similar to the imaging conditions used in vacuum where the tunnel gap is on the order of 3–5 Å and significant orbital overlap occurs between the localized tip-state and the substrate. Since the diameter of a water molecule is on the order of ~2.5 Å it is quite possible that for atomically resolved imaging the solvent may be largely displaced from the tunneling channel. This would result in physics not unlike that of the vacuum junction. In contrast, at larger tip-substrate separations the solvent must be incorporated into the gap. Analysis of this situation is further complicated by possible electrostriction effects and the distance dependence of the structure and density of water within the double-layer of a single electrode [109,112–115]. Indeed, an understanding of the overlap and interaction of the double layers of the substrate and tip, respectively, is one of the most challenging issues confronting a rigorous electrochemical description of the immersed STM junction. For example, an intriguing experiment has been described where the tunnel conductance was monitored between two mercury electrodes as they were brought into contact by electrodeposition [112]. Discrete changes in junction conductance were observed, which the authors ascribe to changes in water structure when the electrodes were separated by less than about ~1 nm [112]. Additional consideration needs to be given to the magnitude of the adsorption and ordering strength of the solvent relative to the perturbing lateral forces associated with the scanning tip [116]. Lateral force microscopy may prove to be a useful tool for studying this issue [49,117]. Similarly, AFM has been used to examine the double-layer–induced forces between two electrodes, although most studies only deal with the overlap of diffuse double layers at large separations. The results are well described by DLVO theory [57,118–122]. More recently, the structuring of the solvation force between closely spaced surfaces has been explored with AFM [57,116]. In a similar vein the distribution of the electrostatic potential for STM geometry has also been evaluated based on an extension of the Gouy-Chapman theory [57,95,119,124]. For a 0.01 mol/liter 1:1 electrolyte the Debye screening length of the electrolyte is 30.4 Å, while for a 0.1 mol/liter electrolyte it is 9.6 Å. Clearly, the dimensions of the tunnel junction are less than the Debye screening length, and not surprisingly the respective double layers are severely perturbed as suggested in Fig. 17 [124]. A significant perspective on this problem is provided by a simple calculation that indicates that the electron density emanating 2–3 Å from the metal is significantly higher than the ionic density in a 0.1 mol/liter solution [110]. Likewise, strong hydration and resonance forces are known to dominate the junction [57,119,122,123]. The simultaneous measurements of the force and tunnel conductance, in a manner analogous to prior vacuum experiments [85], should help yield insight into the structural and electronic properties of the immersed tunnel junction.

In order to overcome the difficulties outlined above, microscopic models for the STM in water have been recently developed [103,110,125]. In one model a molecular dynamics calculation was used to describe an ensemble of water molecules between two metal plates [103,110]. The potential energy between the plates was calculated as the sum of the one-electron potentials due to the interaction of the plates and the pseudopotentials due to the interaction with the water molecules. The hydrogen atoms were associated with virtual localized states in the gap. The tunneling process was envisaged as the scattering of electrons from a three-dimensional potential energy surface with several maxima and minima [103,110]. The tunneling probability was calculated for a range of static barriers in order to obtain statistical data on the effect of solvent fluctuations [103]. The discrete nature of the model resulted in "thermal fluctuations" of the tun-

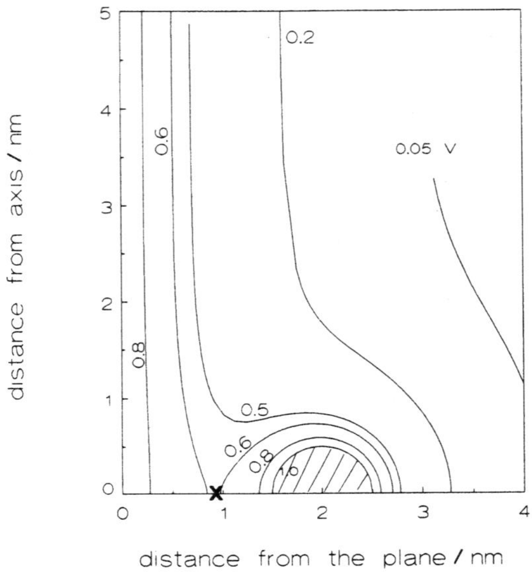
distance from axis / nm
distance from the plane / nm
0.2
0.05 V
0.6
0.8
0.5
0.6
0.8
1.0
0
1
2
3
4
5

(A)

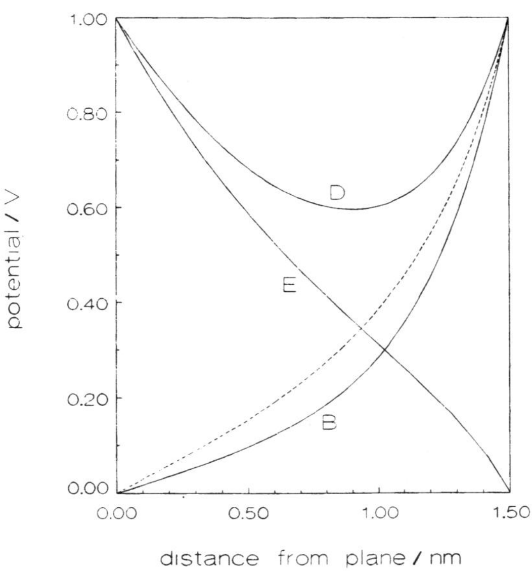
potential / V
distance from plane / nm
1.00
0.80
0.60
0.40
0.20
0.00
0.50
1.00
1.50
D
E
B

(B)

neling probability spanning an order of magnitude. The influence of the fluctuations was predicted to decrease as the dimensions of the junction were increased. The calculations yield a barrier height of about 2 eV for a gap containing roughly three layers of water (~9.6 Å). Imposing a surface charge of 0.2 C/m^2 exhibited no significant effect on the tunneling probabilities [103]. These results are in reasonable agreement with the experimental results shown in Fig. 14 [104].

A general understanding of the influence of the tip upon the electrochemical potential of the substrate within the tunnel junction is central to understanding the appropriate conditions for imaging as well as rational development of various tip-induced surface modification and synthesis schemes [12–17]. To date, the successful observation of intelligible potential-dependent phenomena (e.g. potential-induced surface reconstruction) at small tip-sample bias provides much optimism as to the relevance of analytical in situ STM studies [8–11,27]. In contrast, studies of metal deposition, dissolution, and oxidation have revealed a strong influence of tip potential, i.e., inhibition or catalysis, on the respective reaction [126–132]. In addition to the vertical junction gradients, the geometry of the overlapping tip-substrate double layers also leads to a variety of lateral gradients. For example, in the case of imaging liquid metal surfaces, a tip-induced surface tension gradient has been associated with destabilization of the surface [133]. The gradient leads to "waving" of the liquid metal surface, which has unfortunately limited the use of STM to study the otherwise well-characterized mercury electrolyte interface [134]. For solid metals, perturbation of the surface stress may likewise become important, as suggested by the recent demonstration of tip-induced lifting of the surface reconstruction of gold under certain bias conditions [135,136].

The role of ions in the tunneling process has received limited attention. By virtue of the limited size of the tunneling gap during atomically resolved imaging, completely solvated ions are unlikely to exist in the junction. In contrast, ions that are specifically absorbed on either the tip or

FIG. 17. (A) Contour plot of potential between a spherical tip and planar substrate separated by 1.5 nm in a 0.1 M electrolyte while both electrodes are posed at 1 V relative to a counter located some distance away. (B) Potential profiles along the tip-substrate normal for various bias conditions. The dashed line represents the analytical result in vacuum. (From Ref. 124.)

substrate will change the LDOS and the surface dipole, thereby altering the distribution of the electrostatic potential within the gap in a manner analogous to the vacuum junctions [71,137]. Similarly, ions with electronic states near the Fermi level may act as resonance centers for the tunneling electron, thereby significantly enhancing the current.

Enhanced tunneling associated with resonant states has been extensively studied in the case of M-I-M junctions [87] as well as for electron transfer reaction between oxidized metal electrodes and soluble redox couples [138]. In both instances, the resonant state is ascribed to localized defect sites within the barrier oxide. The development of modified electrodes, such as adsorbed redox metalloproteins with energy levels accessible at low bias voltages, represents an excellent medium for examining the issue of resonant tunneling [19,46,139–142]. In this case interactions with the tip and substrate are expected to be minimal, while the adsorbate levels may be coupled to the environment, in terms of nuclear motion. An elegant demonstration of resonant tunneling associated with transport through the LUMO of redox species immobilized at an electrode surface was recently reported [19]. Aligning the Fermi level of the substrate and tip with the Fe LUMO in Fe(III)–protoporphyrin IX (FePP) resulted in a 10-fold increase in the tunneling current over that associated with protoporphyrin IX (PP), as shown in Fig. 18. This demonstrates that structurally similar molecules can be identified based on their differing redox properties. The sharpness of the resonant tunneling is revealed by the apparent height of FePP versus the neighboring PP molecules as a function of the substrate potential. The broad nature of the resonant state ~0.3 eV shown in Fig. 19a may be due

FIG. 18. A demonstration of resonant tunneling is provided by studying the effect of bias on the imaging of a mixture of surface confined species such as protoporphyrin IX (PP) and Fe-protophorphyrin IX (Fe-PP). The energy diagram shows the position of the HOMO and LUMO of the protophorphyrin (PP) relative to the Fermi level of the tip and substrate. The addition of Fe(III) to the protophorphyrin (Fe-PP) causes little change in the protophorphyrin (PP) energy levels. STM images, along with the corresponding cyclic voltammograms of a monolayer of FePP/PP adsorbed on a graphite substrate from 0.05 M $Na_2B_4O_7$ solutions, are shown for FePP:PP monolayers in the ratio of 0:1 (A), 1:4 (B), 4:1 (C), and 1:0 (D). The images were taken with a substrate potential of −0.41 V, a tunneling current of 30 pA, and a tip-substrate bias of −0.1 V. (From Ref. 19.)

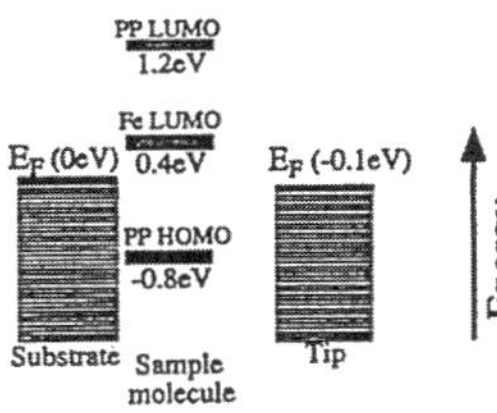

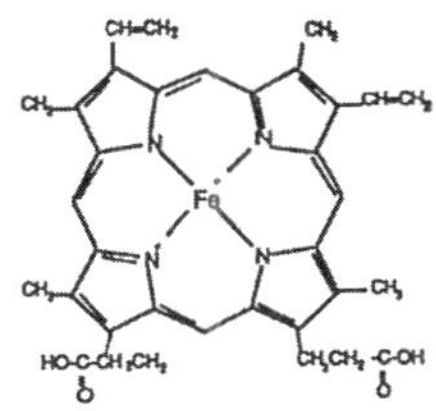

1. Fe-Protoporphyrin IX

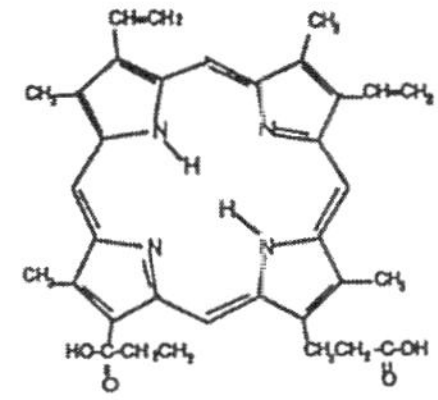

2. Protoporphyrin IX

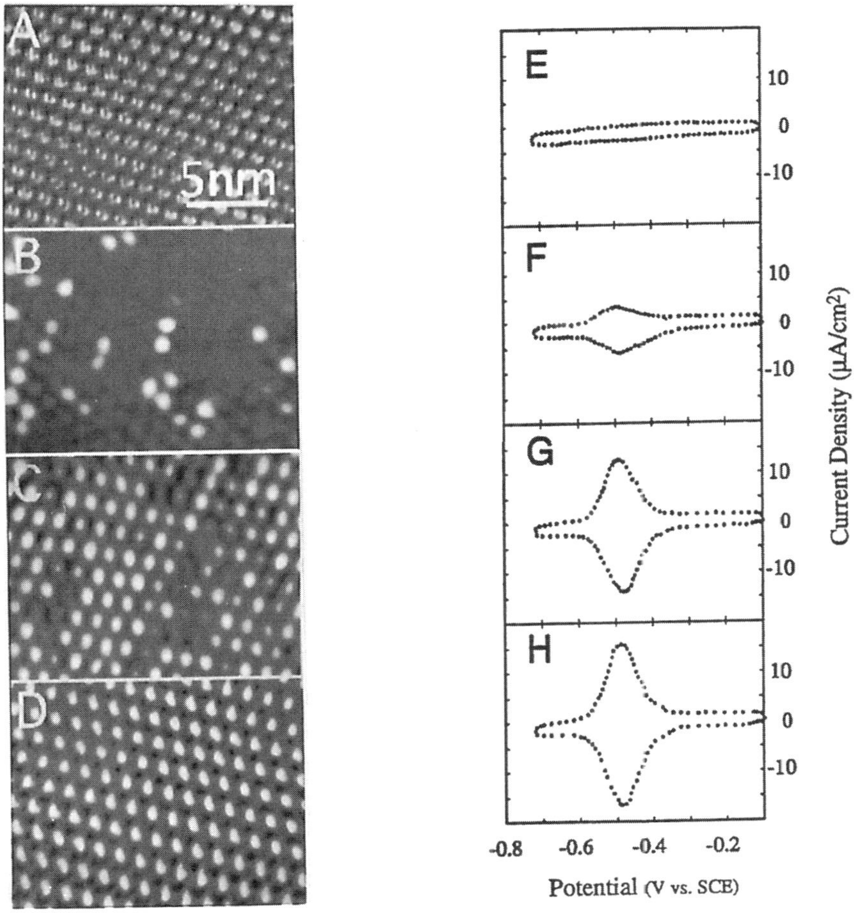

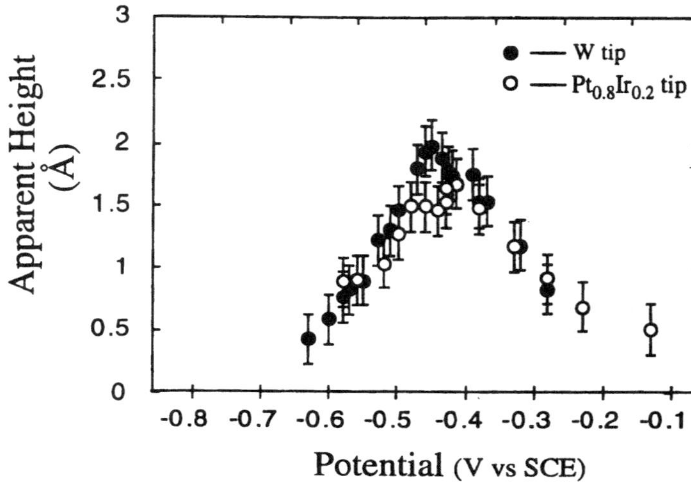
W tip
$Pt_{0.8}Ir_{0.2}$ tip
Apparent Height (Å)
Potential (V vs SCE)

(a)

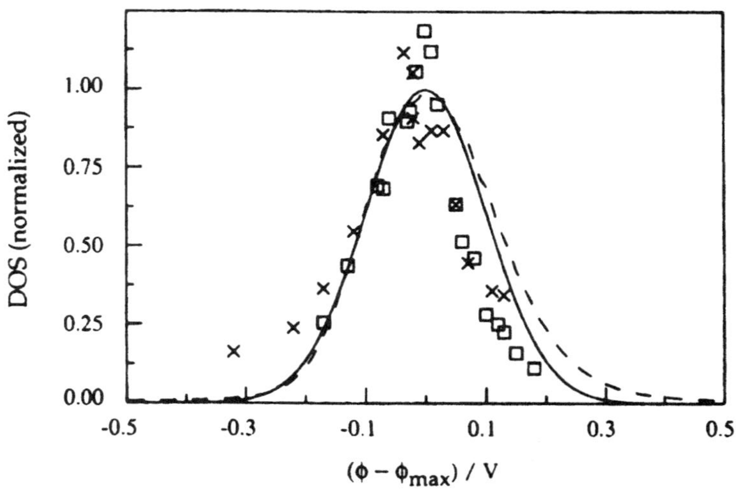
DOS (normalized)
$(\phi - \phi_{max})$ / V

(b)

to a combination of interactions of the redox center with the surrounding molecules as well as the substrate. Both a semiclassical and quantum model are available for assessing the density of states of the resonant channel as shown in Fig. 19b [19,143]. The semiclassical model of resonant elastic tunneling considers coupling of the redox species to the surrounding molecules [140]. In this case further experiments investigating the effects of adsorbate-substrate separation and varying the dielectric character of the solvent should help clarify the relative contribution of solvent reorganization and redox center-substrate coupling to broadening of the resonant tunneling band. More recently, a detailed quantum mechanical model based on scattering theory indicates that when the bias between the tip and substrate is sufficiently large, inelastic transitions associated with vibrationally excited species may contribute significantly to the net current [144]. However, there is no compelling evidence for inelastic scattering in Fig. 19b. Nevertheless, it is important to note that coupling between the tunneling electron and the vibrational modes can be much stronger than in ordinary inelastic tunneling spectroscopy due to the resonant nature of the transition. In conventional inelastic tunneling spectroscopy, the inelastic channels typically contribute 0.1–1% to the total current, while in the case of resonant tunneling contributions on the order of 10% are anticipated [144]. Thus, measurement of vibrational spectra of inner sphere modes associated with electron transfer may be plausible. Nonaqueous electrolytes, which exhibit a large electrochemical window as well as low freezing point to reduce the thermal broadening associated with the Fermi-Dirac statistics, represent the most promising environment for further examination of this issue [143].

FIG. 19. (a) A quantitative measure of the contribution of various resonant tunneling channels may be obtained by examining the apparent height of the Fe-protoporphyrin relative to protoporphyrin as a function of the substrate potential. (b) By consideration of the distance dependence of the tunneling current, it is possible to convert the apparent height data to the tunneling current that would be measured at constant height, thereby allowing comparison with theoretical density of states calculations as shown. The full line is for a semiclassical model, while the dashed line is from a quantum model. However, the energy of reorganization is a sensitive function of the decay constant used for conversion of the data. The squares and crosses denote values obtained using a tungsten and Pt-Ir tip, respectively. (From Refs. 19, 143)

III. EXPERIMENTAL CONSIDERATIONS

The relative simplicity and low cost of STM instrumentation has contributed significantly to the rapid increase in the number of in situ electrochemical studies performed over the last decade. An excellent discussion of the general aspects of STM design and construction is available in a recent textbook [39]. Beyond instrumentation, insightful experiments depend on the preparation of a flat, well-defined substrate and the formation of a stable tip capable of atomically resolved imaging. In this sense, the ability to reliably produce high-quality noble metal electrodes outside UHV has been central to the success of many STM studies [145–148]. In contrast, our knowledge of the structure, chemistry, and operation of the probe tip may be more aptly viewed as an art form.

A. Electronics

For electrochemical studies, two-, three-, and four-electrode systems have been utilized. The two-electrode configuration involves simple immersion of the tunnel junction. This approach was used in early studies, where the ability to operate STM in liquids and electrolytes was first demonstrated (see Ref. 5). More recently, the two-electrode configuration has been used for surface modification where the tip counterelectrode localizes the spatial extent of the reaction occurring on the working electrode substrate [13,15,16]. However, the lack of independent control of the potential of the immersed electrodes severely limits possible applications of the device. In contrast, the four-electrode system, based on the bipotentiostatic principle, allows independent control of the potential of the tip and substrate relative to some reference electrode. A schematic diagram of this configuration is shown in Fig. 20. A variety of different circuit designs have been examined, with the chief distinction among them being electrical grounding considerations that ensure compatibility between potentiostat and STM electronics [26].

B. Electrochemical Cell

The design and construction of an electrochemical cell derives from consideration of the system being examined. Potential sources of contamination must be carefully evaluated. Cell components are typically made of inert materials such as Teflon or Kel-F. Alternatively, electrolyte contact with confining materials may be avoided altogether by letting the cell be defined by the geometry of a hanging meniscus. The latter method has been incor-

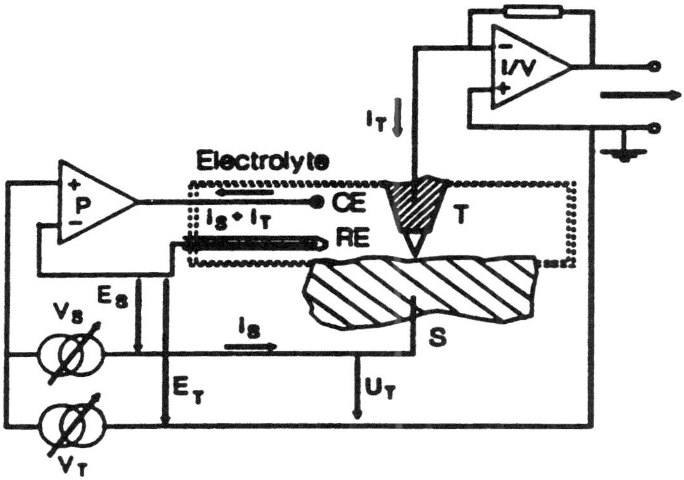

FIG. 20. A bipotentiostat allows independent control of the tip (E_T) and substrate (E_S) potential relative to a reference electrode (RE). (From Ref. 26.)

porated into a UHV system where conventional surface analytical methods such as low energy electron diffraction (LEED), Auger electron spectroscopy (AES), x-ray photoelectron spectroscopy (XPS), etc., may be used for complementary analysis of the immersed electrode [149]. This instrument will also enable the surface to be examined during synthesis of a double layer by dosing the surface with water, HCl, etc., in vacuum [150].

Experiments may be performed in either a separated or unseparated cell, which typically has an electrolyte volume in the range of 0.1–1 ml. In the case of the four-electrode configuration, an unseparated cell is usually comprised of a platinum wire counterelectrode and a simple reference electrode, such as metal/metal ion, metal/metal hydride, metal/metal oxide, and/or $Ag/AgCl/Cl^-$. It is noteworthy that a sufficiently unpolarizable electrode, such as $Ag/AgCl/Cl^-$, Ag/Ag^+, etc., may be simultaneously used as counterelectrode and reference electrode in a three-electrode configuration. To date, unseparated cells have proven to be remarkably effective, although great care must be given to possible interference and/or contamination effects between the various electrodes. When a quasi-reference electrode is used, it is usual practice to compare the voltammetric results obtained in the unseparated STM cell with that obtained in a conventional cell in order to verify the accuracy and precision of the potential scale. In general, these uncertainties may be avoided by using a separated STM cell, which incorporates a robust reference electrode such as $Ag/AgCl/Cl^-$, $Hg/Hg_2Cl_2/Cl^-$, or $Hg/Hg_2SO_4/SO_4^{2-}$. In this instance, it is important to consider the possible impact of ions, such as Cl^-, leaking from the reference electrode through the separator, i.e., vycor, ZrO_2, etc., into the working electrode compartment. Similar consideration must be given to the reactions occurring at the counterelectrode.

Good electrochemical engineering practice requires that proper consideration be given to the current distribution in the electrochemical cell. The theory for treating this problem is well developed, although the important screening effects associated with tip-substrate geometry have just begun to be examined [126–131]. In a conventional cell, heterogeneity of the primary current distribution becomes important when the ratio of the polarization resistance to the electrolyte resistance is low. Due to the limited temporal resolution of existing microscopes, atomically resolved studies have been limited to examining dynamics at relatively slow rates. Fortuitously, in this instance the placement of the counterelectrode should be relatively unimportant provided the electrolyte is well supported. Nonetheless, early designs centered on maintaining a symmetrical arrangement of the counterelectrode around the working electrode, giving a radial symmetric primary current distribution [26]. In contrast, the counterelectrode in several commercially available cells is simply a wire placed randomly within the cell. As the STM is adapted to study processes, such as the formation of three-dimensional phases at industrially relevant rates, i.e., traditional electroplating applications, the homogeneity of primary current distribution will unquestionably become a significant issue. The poor primary current distribution associated with a conventional atomic force microscopy (AFM) cell has already been demonstrated by finite element numerical analysis [151]. Control of tertiary current distribution is also an important issue. The ability to perform AFM under conditions of forced fluid flow was recently demonstrated [152,153].

In many STM studies little effort has been made to control the atmosphere within the electrochemical cell. Yet oxygen is known to exert a major role in the chemistry and corrosion of many transition metals. For example, several STM studies have used the copper/copper ion reference electrode, yet the electrode is known to be polarized from its reversible condition by oxygen, leading to significant dissolution [154]. These effects become particularly significant in the study of metal deposition and dissolution reactions at small overpotentials, where precise knowledge of the supersaturation is quite important for detailed analysis. Oxygen reduction also complicates voltammetric analysis and may obscure other adsorption reactions of interest, as well as alter the surface chemistry of the system under investigation. It is thus desirable to deaerate electrolytes prior to use and blanket the cell with inert gas during the experiment. Similarly, isolation from water is critical to the stable operation of many nonaqueous and molten salt experiments.

Accurate atomically resolved STM studies require stable thermal conditions, otherwise differential thermal expansion and contraction of the various components of the device can give rise to significant image distortion or ''drift.'' Thus, it is common practice to isolate the microscope from convective air currents associated with most laboratory environments. Furthermore, following assembly or sample exchange it is important to allow the device to thermally equilibrate prior to performing an experiment. Another potential source of image drift may be associated with slow relaxation of the mechanical devices used to position the sample and/or scanner. For example, coarse positioning in many microscopes incorporates fine-pitch screws, which have a certain degree of backlash associated with their operation. The slow relaxation of the tip/sample positioning devices along with thermal perturbations account for much of the commonly observed drift as well as its diminution with time following device assembly.

C. Tip Selection and Preparation

The tip is the source of the greatest uncertainty in STM due to a combination of its ill-defined structure, electronic properties, and instability associated with interactions with the sample or surface contaminants. In the case of UHV systems field ion microscopy (FIM) studies of STM tips have provided detailed information on the effect of various tip-preparation methods [39]. FIM also provides a mechanism for characterizing the tip before and after use. In electrochemical STM systems the uncertainties are even greater by virtue of possible interaction between the tip and the electrolyte, i.e., adsorption, corrosion, etc. Bipotentiostatic operation enables the tip to be biased in a potential regime, which minimizes faradaic processes occurring at the tip. The faradaic component of the tip current is further minimized by coating the tip with an insulating coating, such that the residual background current in inert electrolytes is at least below 0.1 nA. Currently, the figure of merit defining a suitable tip is simply the demonstrated ability to obtain atomically resolved images of a graphitic or noble metal surface. An amusing perspective on the problem is provided by some of the early reports of atomically resolved images of noble metal. These results were obtained using remarkably simple and unsophisticated means for forming Pt-Ir tips, such as cutting Pt-Ir wire with pliers, followed by coating with nail polish—a particularly popular variety going by the brand name ''Wet 'n Wild.'' Since that time some effort has been made to formalize the engineering of tips, although a scientific understanding of the etching, coating, and characterization of the tips has hardly progressed in the past 5 years [26].

Almost any conductive material may be used as a tip, although electrochemical studies have largely focused on Pt, Pt-Ir, Ir, and W wire, which may be easily fashioned into tips by etching or mechanical fracture. The relatively inert noble metals were the first materials to be considered since the potential can be posed in the ''double-layer'' regime where no net faradaic reactions proceed. Platinum-iridium was chosen instead of gold because of its superior mechanical properties as well as its historically successful use in FIM experiments. Tungsten electrodes have also received extensive use as a consequence of the ease and rapidity of fabrication. The success of tungsten is somewhat surprising in light of its reactive nature in aqueous environments. Interestingly, it has been noted that the imaging quality of W tips can be greatly improved by depositing small amounts of metal from solution (e.g., Cu) [28,148,155]. Looking to the future, carbon nanotube STM and AFM tips have recently been demonstrated [156–158]. The stability and well-defined structural and electronic characteristics of these materials may enable rigorous modeling of STM images of immersed electrodes [57].

Conventional W or Pt-Ir tips are usually made by either a mechanical or an electrochemical etching process. In the mechanical processes the tip is formed by either cutting the wire at 45° with pliers or straining the wire to failure in tension. Tips have also been prepared by mechanical polishing [130,159]. In the case of etching, a wide variety of electrochemical methods have been used ranging from ac to dc processes. A summary of the appropriate etching conditions for a variety of materials is available [26,39,160]. Etching of W in KOH and Pt-Ir in $CaCl_2$ or cyanide solutions appear to be the most extensively used. Importantly, the final apex may be formed by chemo-mechanical fracture of the narrowing ligament that forms during etching. The precision of the process is dependent on the design of the electrochemical cell and sample geometry. In addition to the issue of current distribution, the length of wire suspended below the etching ligament determines the mechanical loading on the gage section. In certain protocols the fractured apex receives a final polishing step.

Once the tip has been formed, it is coated with an insulating layer. A wide variety of coatings have been developed ranging from inorganic glasses to polymers [26]. However, the pace of development has slowed, with most papers reporting the use of either Apiezon wax or polyethylene. Typically, the coating is applied by immersion of the tip in the liquid phase followed by controlled removal while monitoring the solidification or glass transition process [26,161,162]. In another procedure the tip is rotated about its axis while molten wax is transferred to the tip from an adjacent heated wire. An alternative scheme involves electrophoretic painting of the coating followed by subsequent polymerization [163]. The effectiveness of any coating may be examined by monitoring the voltammetric response of the ultramicroelectrode tip in an electrolyte containing a well-defined redox couple. Estimates of the exposed or electroactive area of the coated tip range from 10 μm down to ~10 nm. [43,47,161–165]. Finally, in some studies substrate contamination due to degradation of tip-coating material has been noted [107], and this possibility must always be carefully considered, particularly as experiments move beyond simple aqueous electrolytes.

In UHV experiments a variety of post-etch treatments are used to ensure that the tip is well formed as well as oxide- or contamination-free. In the case of immersed Pt and Pt-Ir alloys, it is possible to characterize the tip by examining oxide formation and reduction by voltammetry, although this leads to changes in microstructure which are ill-defined and difficult to characterize. In contrast, exposure of the commonly used tungsten tip to an electrolyte and air results in the formation of a hydrated WO_3-based oxide. It is worthwhile to note that freshly etched W tips typically do not provide atomically resolved imaging when first engaged in UHV studies [39]. Rather, effective imaging usually begins spontaneously after repeated tunneling and/or scanning, although the onset occurs in an unpredictable way. A variety of phenomena ranging from oxide removal to transfer of material from the substrate have been used to rationalize this observation. As noted earlier, optimal image resolution in some of the earliest in situ STM studies was associated with copper deposition on the W tip [28,148,155]. From a different perspective, pretreating an etched W tip in a hot I_2/N_2 stream reportedly led to a strong potential dependence during imaging. Under low bias conditions the platinum substrate was imaged, while at higher bias the overlying adlayer was revealed [166]. This effect remains unexplained, but related work suggests that controlled alteration of the surface chemistry of the tip may be a profitable way to probe the electronic structure of the tunnel junction [42,58,167]. This is analogous to the chemical imaging schemes that have been developed for lateral force microscopy [168].

D. Substrate Preparation

Atomically resolved STM studies require preparation of a flat surface with well-defined crystallography. Studies to date have focused on either single-crystal or highly textured thin film noble metal electrodes. The traditional approach to single crystal preparation involves growing an ingot or boule by solidification from a melt using a seed crystal to control the orientation. The boule is then sliced into specimens by a variety of means ranging from abrasive to chemically assisted cutting processes. The specimens are subsequently treated to produce a smooth, deformation-free surface. This may be accomplished by annealing in vacuum or in an inert or reactive gas, where the partial oxygen pressure is maintained below that of the relevant oxide, or alternatively chemical or electrochemical polishing may be utilized.

There exist several other methods for producing single-crystal electrodes. A remarkably simple and inexpensive technique involves using a well-controlled H_2-O_2 flame to melt the end of a polycrystalline noble metal wire and allowing a grain at the end of the wire to act as a seed crystal during solidification of the molten sphere [145–147]. By allowing the bead to solidify slowly, followed by remelting and/or annealing, symmetrical facets develop that reveal the formation and orientation of a single crystal. This method has been widely used for producing gold and platinum electrodes and more recently has been extended to other noble metals such as Rh, Pd [169]. Thin film deposition methods, such as e-beam evaporation and sputtering, provide another avenue for producing very flat, highly oriented thin films. The substrate used for deposition exerts a major influence on the evolution of roughness and texture during film growth. For example, highly textured Au(111) [170], Ag(111) [171], and Cu(111) [172] may be easily deposited on freshly cleaved mica, while the (100) textured films may be obtained by evaporation on a variety of substrate with cubic symmetry, etc. [173]. Bulk single crystals as well as thin films may also be prepared by the electrolytic deposition [174,175]. In particular, a capillary growth technique has been used extensively in the study of the deposition and dissolution of silver [176,177].

Electrochemical studies of solid electrodes require that the surface be routinely restored to its original conditions. Flame annealing, where the noble metal surface is annealed in a hydrogen-oxygen flame, represents a major advancement in convenience that allows a crystal to be rapidly refurbished for repetitive experiments. This method has been used for thin film electrodes [148] as well as bulk crystals [145–148,178,179]. Crystal quality is a sensitive function of the cooling procedure with slow cooling in air or inert gas being recommended over rapid quenching in water [180]. Alternatively, furnace annealing under a controlled atmosphere may be used, although this is less convenient than flame annealing. However, for more reactive materials such as nickel, control of the partial pressure of oxygen is central to providing a clean surface via thermal annealing.

Another approach to protecting reactive surfaces involves using a strongly chemisorbed species to displace impurities and passivate the surface against further contamination during preparation and transfer to electrochemical cell. Iodine-covered surfaces have received particular attention in the case of Pt and Rh electrodes [166,181–183]. Typically, the procedure involves annealing in a hydrogen flame followed by cooling in a glass dosing cell containing iodine crystals under a flux of inert gas. The iodine layer may be displaced by bubbling CO through the solution while holding the potential at a negative value. Subsequently the CO layer may be removed by oxidation, e.g., CO absorbed on Pt or Ni [183,184]. This method is particularly useful for studying systems, such as I on Pt(111), however, the difficulty encountered in trying to electrochemically desorb such layers without altering the underlying metal surface limits the general applicability of the technique. It is noteworthy that many successful STM experiments have dealt with systems that exhibit strong adsorption reactions, which tend to minimize the effects of a variety of less strongly adsorbing impurities. The corollary of this situation is that examining systems that involve weak adsorption phenomena is likely to be very demanding from the perspective of cleanliness.

An alternative or additional step to flame annealing is electrochemical or chemical polishing. The fundamental aspects of electropolishing were reviewed recently [185], and a list of polishing procedures and parameters is available [185,186]. This method has been successfully applied to the preparation of gold, silver, and copper electrodes for STM studies [177,180,188]. It is important to note that different mesoscopic structures may arise according to the specific preparation procedures. For example, electropolishing a mechanically prepared Au(100) surface followed by flame annealing generates a surface with atomically flat terraces ranging between 0.1 and 0.5 μm wide, which were separated by step bunches up to 6 nm in height. In contrast, flame annealing the sample immediately after mechanically polishing produced a surface with terraces tens of nanometers in dimension, separated by monatomic steps [180]. An understanding of such phenomena remains to be developed.

In addition to removing deformed material associated with mechanical preparation, electropolishing has also been used as a final preparation step in several studies [26]. To date this method has proved to be most useful for preparing copper electrodes, which cannot be flame annealed in the conventional sense. Many of the details concerning electropolishing of copper have received attention, although the nature of the interface following oxidation at high potentials remains unclear [187]. Nevertheless, other studies indicate that electropolishing does not influence subsequent electrochemical experiments [188], particularly those involving strongly adsorbed species such as halides in acid media [188,189]. Similarly, the effectiveness of electropolishing silver has been explored to a limited extent [190].

IV. APPLICATIONS

The ability of the STM to generate atomically resolved images of the solid/electrolyte interfaces has generated tremendous excitement in the electrochemical community. While STM may reveal structures that exhibit long-range order, this capability should be viewed as complementary to the more precise and well-developed scattering techniques (SXS, LEED, etc). In contrast, STM is a very effective and efficient tool for real-time, in situ characterization of mesoscopic structures, such as island density, step geometry, etc., which are difficult to study by scattering methods. This capability has been amply demonstrated in studies of phase transformation, which involve nucleation and growth processes associated with surface reconstruction, adsorption, deposition, dissolution, or passivation of metals and alloys. The STM has also been implemented as a synthesis tool for building novel structures by etching, deposition, or oxidation, where the spatial extent of the reaction is more or less defined by the tunnel gap. The following pages will present some striking examples of the utility of STM for examining the rich variety of phenomena that occur at solid/liquid interfaces.

A. Imaging Surface Dynamics

In principle, quantitative evaluation of the rate parameters associated with various surface processes in combination with simulation and theory should allow a direct correlation to be established between microscopic mechanisms and traditional macroscopic electrochemical measurements. The dynamics of individual surface atoms, i.e., terrace and step edge diffusion, have been monitored with STM, although for many processes of technical interest the rate of individual atomic events often exceeds the response time of conventional instruments. Images often represent a convolution of the dynamic changes that occur during the raster [191,192]. Nonetheless, a variety of noteworthy schemes have been implemented to examine the individual atomic events. For example, vacuum STM studies have been performed under variable temperature control, whereby the time constants of the various relaxation processes may be matched to the capabilities of the probing device [193]. Alternative approaches have focused on increasing the instrumental time resolution. For example, reducing the dimensionality of the experiment by disabling the y-raster allows the one-dimensional movement of well-defined surface features, e.g., an array of steps, to be followed [194]. The development of faster microscopes, as well as novel imaging modes, should allow the direct observation of individual atomic events to become routine. An encouraging example is provided by a novel "atom tracker" technique, which employs lateral-positioning feedback to lock the STM probe tip into position above a selected atom with sub-Angstrom precision [195]. This device has been used to follow the random walk of a Si addimer. By tracking individual species, temporal resolution is increased by a factor of $\sim 10^3$ over conventional STM imaging techniques.

Despite the limitations of conventional instrumentation, significant advances in the understanding of surface dynamics have occurred using STM to follow the mesoscopic evolution of surfaces. Phenomena ranging from equilibrium step fluctuations [192,193,196–201], to roughness evolution during film growth [191,202,203] to surface alloy formation [204], etc., have been successfully examined from both a theoretical and experimental perspective. In particular, comparison between scaling and spectral analysis of experimental images and simulations has been used to infer the underlying atomic mechanisms. This approach has been used extensively in vacuum studies of metal surfaces (e.g. see Ref. 205), while there is an expanding interest in examining the metal/electrolyte interface [206]. For example, thermally driven equilibrium step fluctuations have been observed for both Ag and Cu surfaces in vacuum [192,193,197,200,201]. The steps actual appear to be "frizzy" due to kink motion, which is rapid compared to the tip raster speed with the rms of the fluctuations increasing with temperature. The impact of the potential on equilibrium step fluctuations of immersed Ag(111) electrode has also been studied [18,207,208]. In this instance, the step fluctuations increase sharply as the potential is moved towards the reversible value of the Ag/Ag^+ electrode. This reflects the high exchange current or standard rate constant of the Ag/Ag^+ reaction. Time correlation methods have enabled the origin of step fluctuation to be evaluated [207,208]. The potential dependence of the fluctuations is shown in Fig. 21. At negative potentials the fluctuations follow a $t^{0.25}$ power law independent of step spacing, which is similar to that observed in vacuum studies. This suggests that the electrolyte exerts negligible influence on mass transport along step edges. In contrast, at higher potentials, a $t^{0.5}$ power law is observed, with the fluctuations being dependent on step separation. This is rationalized in terms of exchange of adatoms between steps and terraces as well as transfer between the terraces and electrolyte. These fluctuations amplify sharply as the reversible potential for Ag/Ag^+ is ap-

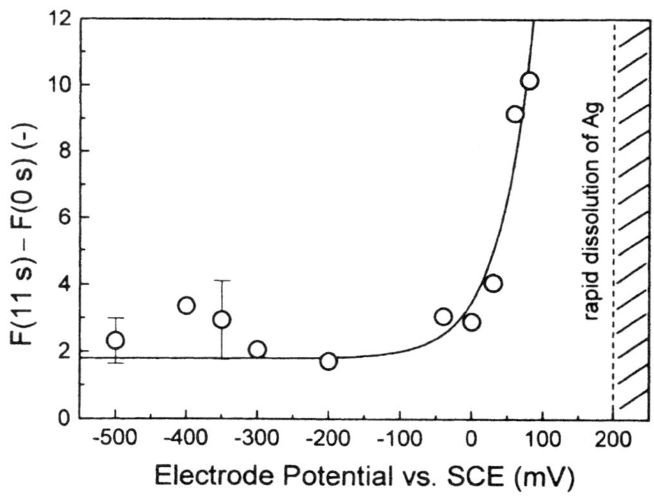

FIG. 21. The influence of potential on step fluctuations, x(t), may be described by means of a time correlation function $F(t) = \langle (x(t) - x(0)^2 \rangle$. At negative potentials, fluctuations are due solely to mass transport along the steps, while at more positive potentials the magnitude of the fluctuations increases rapidly. This is attributed to the onset of adatom exchange with terraces as well as the electrolyte, which occurs even at the potential well below the reversible value for Ag/Ag^+. (From Refs. 207, 208.)

proached. In other experiments, larger-scale coarsening effects have been observed and correlated with the loss of SERS activity of intentionally roughened electrodes [209]. In subsequent sections, other studies will be described that deal with potential-induced reconstruction, adsorption, and other phenomena where mechanistic insights have been revealed that help rationalize numerous results of prior optical and electrochemical studies. Likewise, the impact of surface heterogeneity on a variety of chemical and electrochemical reactions has been examined. From a larger perspective, the convergence of theory, simulation, and experiment in assessing the impact of individual rate processes on the evolution of the surfaces represents a timely opportunity for electrochemists to examine many longstanding questions concerning the interplay between electrode structure and electrochemical kinetics.

B. Reconstruction Phenomena

Termination of a periodic solid leads to an imbalance of forces, which results in the structure of the free surface deviating from that of an ideally terminated lattice [210–212]. The simplest case involves minor relaxation or contraction in the bond lengths normal to the interface, while a more dramatic effect, surface reconstruction, involves lateral rearrangement (i.e., bond breaking) of the atoms leading to a structure that differs markedly from that of the underlying lattice. The electron density associated with the interatomic bond at the metal surface exerts a crucial role in determining the stability of such reconstructed surfaces. This makes research using electrochemical methods particularly fruitful since the excess charge may be conveniently altered via the electrode potential. Negative surface charge induced at lower potentials usually favors the reconstructed state, while adsorption of anions at more positive potentials results in a ''lifting'' of the reconstruction [210–213]. There is limited experimental evidence that distinguishes between lifting driven by anion adsorption versus charging the electrode to a positive value [214]. The reconstruction of gold has received the most attention, while limited studies of platinum electrodes have been reported [210]. It is also important to realize that reconstruction phenomena, whereby the surface structure becomes potential dependent, are of fundamental importance in studies of electrochemical reactivity. An example of this is the structural dependence of the activity of Au(100) towards O_2 reduction [215]. A listing of STM studies of surface relaxation and/or reconstruction for a variety of crystal faces is given in Table 3.

TABLE 3

STM Studies of Surface Reconstruction and Relaxation

Surface	Phase transition	Ref.
Au(111)	(1 × 1) ↔ (1 × 23)	27,54,210,211,216,225–227
Au(100)	(1 × 1) ↔ (hex) or c(26 × 68) for perfect reconstruction	27,54,210,211,215,216,221–223
Au(110)	(1 × 1) ↔ (1 × 2), (1 × n)	27,214,228–230
Au(331)	Edge-atom depression and row buckling ↔ (1 × 2)	231
Au(221)	Edge-atom depression and row buckling	231,232
Au(533)	(1 × 1) ↔ edge-atom depression and row buckling	231,232
Au(311)	(1 × 1) ↔ edge-atom depression and row buckling	231,232
Au(210)	Close to bulk termination	231
Au(410)	Close to bulk termination	231
Pt(110)	(1 × 1) ↔ (1 × 2)	233
Pt(100)	(1 × 1) ↔ hex	234,235

In addition to examining the surface structure, the STM may also be used as a transducer to monitor deformation associated with potential dependence of the surface stress and associated electrocapillary phenomena at solid electrodes. In this instance, the high z-sensitivity of the STM is used to follow the minute displacements of a surface, which is supported in a cantilever geometry [54,55].

It is noteworthy that prior to the advent of scanning probe microscopy electrochemically driven reconstruction phenomena had been identified and studied using traditional macroscopic electrochemical measurements [210,211]. However, STM studies have provided insight as to the various atomistic processes involved in the phase transition between the reconstructed and unreconstructed state and promise to provide an understanding of the macroscopically observed kinetics. An excellent example is provided by the structural evolution of the Au(100) surface as a function of potential and sample history [210,211,216–223]. Flame annealing of a freshly electropolished surface results in the thermally induced formation of a dense hexagonal close-packed reconstructed phase referred to as Au(100)-(hex). For carefully annealed crystals a single domain of the reconstructed phase can extend across micrometer-sized terraces [217]. As shown in Fig. 22, the surface exhibits a one-directional buckling with a period of 14.5 Å due to the structural misfit between the hexagonal close-packed reconstructed surface and the bulk lattice. In some instances an additional corrugation is observed associated with a slight rotation of the hexagonal structure with respect to the substrate. This yields a large unit cell, c(26 × 68), for a well-annealed reconstructed Au(100) surface. In this discussion the hexagonal close-packed variants will simply be referred to as Au(100)-(hex). The transition between Au(100)-(1 × 1) and Au(100)-(hex) is associated with a sharp peak in the voltammogram, which arises from the difference in the point of zero charge (pzc) of the two phases [210,216]. The peak potential is a strong function of the electrolyte composition, particularly the anion. The Au(100)-(hex) surface contains about 25% more atoms than the unreconstructed Au(100)-(1 × 1). Thus, adsorbate-induced deconstruction or "lifting" of the Au(100)-(hex) → (1 × 1) results in the excess atoms being expelled and subsequently coalescing into islands by surface diffusion as indicated in Fig. 22 [217,218]. The islands coarsen with time at a rate that is highly sensitive to the nature of the electrolyte, i.e., anion. Specifically adsorbed anions have been reported to enhance the mobility of steps with the effect being larger, the more positive the potential and/or stronger the anion-gold interaction ($SO_4^{2-} < Cl^- < Br^- < I^-$ [210]. For instance, if the phase transition is performed in the presence of iodide, only a few large Au islands are observed due to rapid coarsening [219]. The lifting process has been investigated in some detail in sulfate media where at low overpotentials heterogeneous nucleation of Au(100) − (1 × 1) occurs at surface defects such as steps. Subsequent growth proceeds via sequential transformation of reconstructed rows reflecting the anisotropy of the Au(100)-(hex) phase. Lateral growth across the reconstructed rows also occurs, but at a much slower rate under these conditions.

When the potential is swept in the negative direction, charge-induced Au(100)-(1 × 1) → (hex) reconstruction occurs. The transformation proceeds by the anisotropic propagation of the reconstructed rows along the two main directions of the square lattice [217,218]. The alignment of the hexagonal reconstructed layer on top of the square substrate exhibits certain deviations such as curved rows associated with growth around various surface defects [220]. Surface steps and island edges supply the adatoms, which are necessary to form the more densely packed hexagonal phase. The simultaneous formation of domains rotated by 90° restricts the size of the reconstructed regions since the collision of a propagating hex row with an orthogonal hex boundary terminates further growth, as shown in

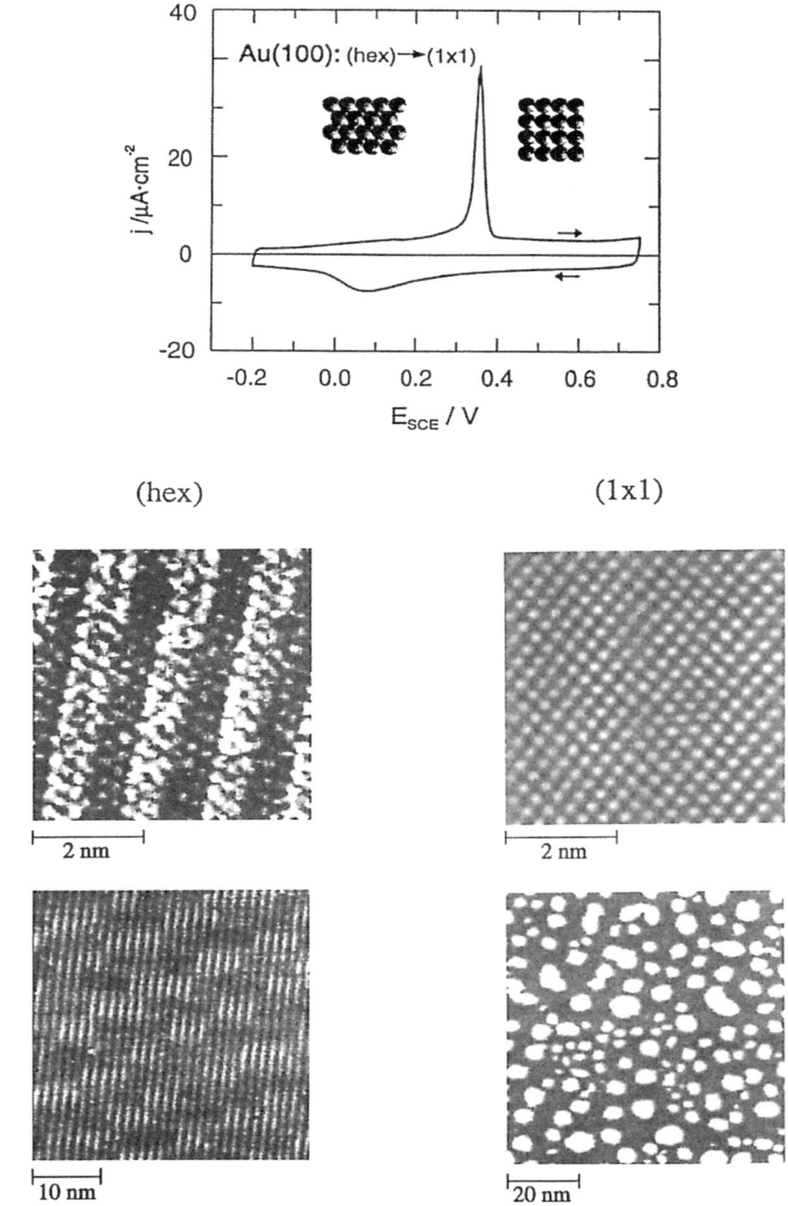

Au(100): (hex)→(1x1)
j /μA·cm⁻²
40
20
0
-20
-0.2
0.0
0.2
0.4
0.6
0.8
E_SCE / V
(hex)
(1x1)
2 nm
2 nm
10 nm
20 nm

Fig. 23. Significant coarsening of the domain structure did not occur, at least for the time scale investigated. Thus, charge-induced reconstruction in sulfuric acid results in a much smaller domain size than that associated with a freshly flame-annealed crystal. The domain boundaries act as preferred nucleation sites for subsequent (hex) $\rightarrow$ (1×1) transformation, which occurs much more rapidly than for a freshly flame-annealed sample. Studies of this nature have provided keen insight into aging effects that would be difficult to assess by other means.

The important influence of anion adsorption on the rate of the various surface processes involved in surface reconstruction has been noted [210]. It is also important to realize that anion adsorption can lead to changes in the dominant mechanism operating during potential-induced reconstruction [219]. For example, a disordered iodide adlayer catalyzes the Au(100)-(1×1) $\rightarrow$ (hex) transition such that single hexagonal domains tend to cover a large portion of each terrace in contrast to the nucleation and growth of narrow (hex) strands associated with the transformation in the presence of sulfate. In the case of iodide adsorption, the additional atoms required to form the Au(100)-(hex) are extracted directly from terraces of the (1×1) phase resulting in the formation of holes in the terraces, which then undergo rapid coarsening. Weakening of both the interlayer and intralayer metal-metal substrate bonds by adsorbed iodide is thought to facilitate the surface processes involved in the reconstruction [219]. More generally the coverage and chemical dependence of adsorbate-induced changes in the nature of metallic surfaces has been the focus of much theoretical and experimental activity [224].

C. Oxide Formation on Metal Electrodes

Thin oxide films formed on metal electrodes are of widespread technical importance for passivation and/or catalysis of certain electrode reactions. For example, the stability of most engineered metallic structures towards environmental degradation, i.e., corrosion, is largely dependent upon the formation of thin protective oxide overlayers. Alternatively, electrosynthe-

---

FIG. 22. Atomic and mesoscopic structure of the reconstructed and unreconstructed Au(100) surface. The sharp peak in the voltammogram demarks the phase transition and is physically associated with charging the double layer due to the difference in the pzc of the respective phases. (From Ref. 216.)

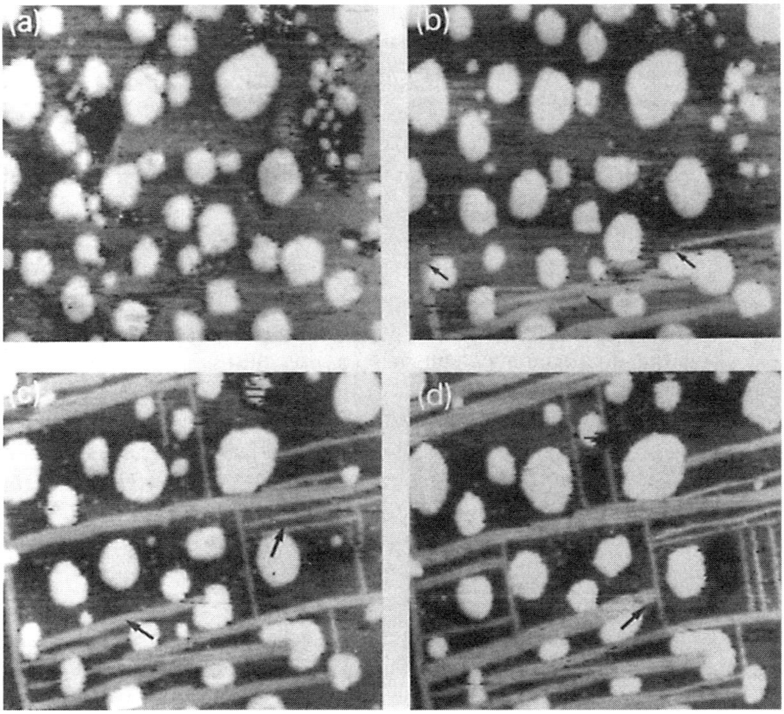

FIG. 23. Selected STM images of the (1 × 1) → hex transition on Au(100) in 0.01 M H_2SO_4. (a) Image of (1 × 1) at 0.6 V (b) the potential was swept in the negative direction at 10 mV/s and at −0.05 V reconstructed domains begin to form (i.e. light gray strings). Subsequent images were collected at −0.15 V after (c) 125 s and (d) 250 s. (From Ref. 217.)

sis at oxidizing potentials usually proceeds on oxide surfaces, which exhibit selective catalytic activity towards a desired reaction. From another perspective, the preparation of noble metal electrodes for electroanalytical experiments often entails repetitive formation and reduction of the oxide films in order to generate a "reproducible" and clean metal surface. STM is providing key insight into the structure, morphology, and kinetics associated with oxidized metallic electrodes.

Oxide formation on noble metal electrodes usually proceeds by two-dimensional adsorption followed by place exchange between the metal and oxide species to form a three-dimensional overlayer. Subsequent reduction of the overlayer leads to roughening of the surface due to the asymmetrical nature of the place-exchange process. An excellent review article on many aspects of oxide formation on gold and platinum electrodes is available [236]. A direct assessment of the morphological evolution and dynamics of repetitive oxidation and reduction was provided by some of the earliest high-resolution STM studies of Pt [237], and Au [8,238]. Subsequently, more detailed studies of oxide formation and reduction on Au(111) [127,216,239], Au(100) [240], Pt(111) [169], Rh(111) [169], and Pd(111) [169] have been published. An example of the roughness introduced during an oxidation-reduction cycle of a Au(111) electrode is shown in Fig. 24. These images were collected during a linear potential sweep such that the y-axis data represents a superposition of the topographic raster and the potential ramp as noted in the figure caption [127]. Formation of the oxide results in the terrace becoming somewhat rough, ~0.1 nm, and the terrace edges shift several nanometres and become diffuse. In other work, growth of the oxide on Au(111) was observed to initiate at step edges and propagate across the terraces [239]. During the reverse sweep minor changes were observed until the onset of oxide reduction. Upon reduction the surface roughens with the formation of small gold islands and monolayer-deep pits [127,216,239], which coarsen rapidly. Consequently, the observed morphology is highly dependent upon the history of the sample, i.e., the potential waveform. Cyclic oxidation and reduction of Pt, Rh, and Pd (111) [169] also results in roughening, but the characteristic length scale is significantly different, presumably due to the slower rate of surface diffusion compared to Au. Purposefully roughened electrodes produced by repetitive oxidation-reduction cycles have also been examined by in situ and ex situ STM [241–245].

STM studies have also begun addressing several longstanding questions concerning the structure and dynamics of passivating films formed on reactive metals and alloys. For example, the complex nature of the earliest stages of room temperature oxidation of Al(111) [246], Ni(100) [247], and Ni(110) [248] has been examined by UHV-STM. Less sophisticated studies of the anodic oxidation of transition metals in aqueous electrolytes have revealed atomically resolved images of the protective crystalline oxide overlayers formed on polycrystalline Fe [249], Ni(111) [250–253], Ni(100) [253–255], Cr(110) [256,257], Cr(100) [257], $Fe_{78}Cr_{22}(110)$ [258,259], polycrystalline $Fe_{85-75}Cr_{15-25}$ [260,261], and $Fe_{69}Cr_{18}Ni_{13}(100)$ [262]. The potential and temporal dependence of the passive film structure has also been investigated. Examination of the defect structure and mesoscopic surface morphology has generated new insights into the dynamics of passiv-

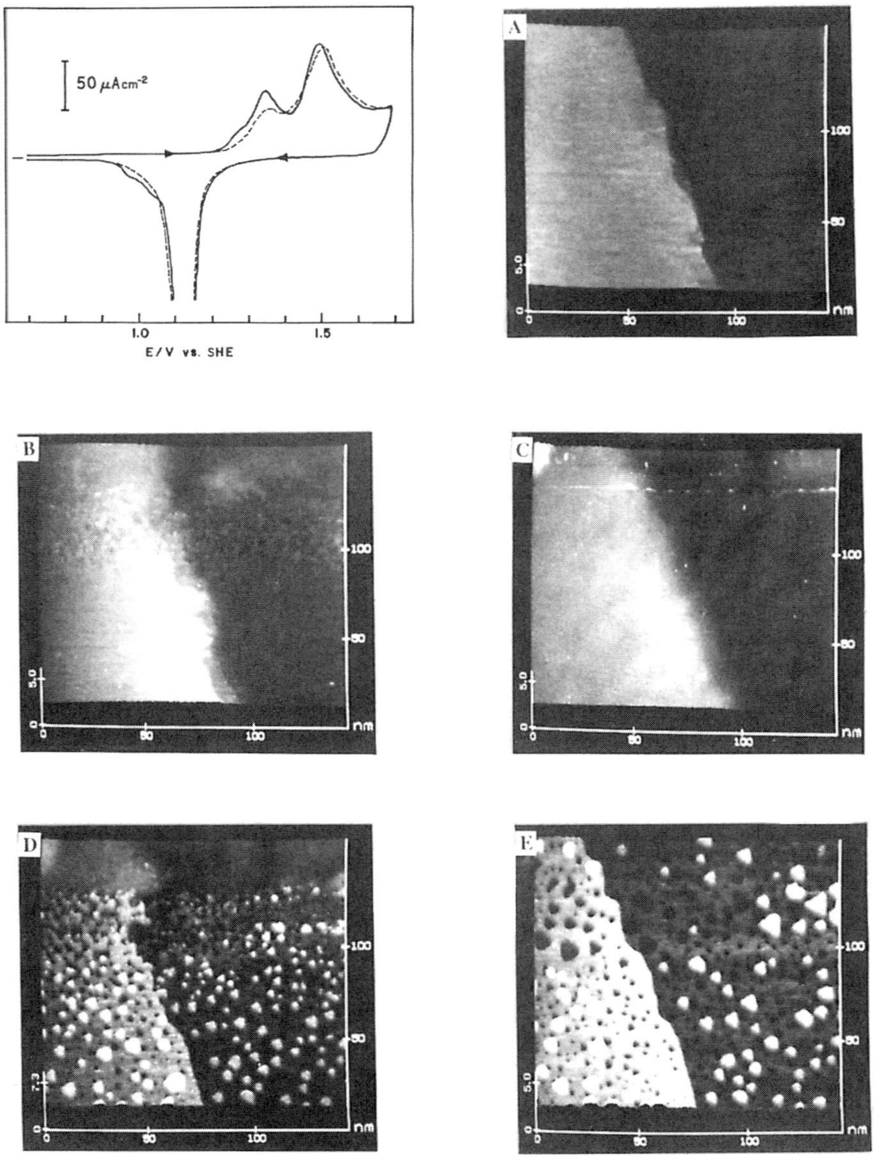
50 μAcm⁻²
1.0
1.5
E/V vs. SHE
A
B
C
D
E
nm

ation. For example, strain accommodation during passivation of Ni(111) is reported to proceed via 8° tilting of the NiO(111) film [250–252]. The potential dependence of the crystal structure of the passive film formed on Ni(100) has also been demonstrated [254]. Likewise, structural changes associated with the growth and aging (dehydration) of the nanocrystalline passive film formed on chromium [256], ferritic (FeCr) [258–261], and austenitic (FeCrNi) stainless steel [262] were noted along with the effects of alloy composition [260,261]. Barrier height and tunneling spectroscopy measurements may be useful for characterizing the evolution of oxide films. For example, in an early study the nucleation and growth of graphitic oxide on highly oriented pyrolytic graphite (HOPG) was mapped by following the spatial variation of the effective barrier height [263]. Nevertheless, interpretation of the atomically resolved STM images of thin oxide overlayers on metals will remain somewhat murky until a more rigorous correlation is established between the imaged electron density and its chemical nature. This is particularly true for nanocrystalline films where complementary information from scattering experiments may not be available. Questions such as whether oxide, bound water, or hydroxide states are imaged or determining the contribution of mixed valent cation states to image formation remain to be addressed. Tunneling spectroscopy in combination with barrier height measurements should be helpful towards characterizing the various transport channels that contribute to image formation. Theoretical prediction on the influence of thin oxide overlayers on the tunneling conductivity of NiO/Ni(100) have been presented [264]. Under bias conditions where the electrons tunnel from the conduction band of the oxide, the tunneling probability should be independent of the film thickness, while in contrast the probability of electrons tunneling from the underlying metal decreases exponentially with the thickness of the insulating oxide overlayer. In fact, the variety of conductance channels available for tunneling in metal/metal oxide systems may result in a marked bias dependence during imaging as noted for epitaxial Al_2O_3 grown on NiAl(110) [265]. Under low bias conditions the dominant contribution is thought to arise from states

FIG. 24. A sequence of STM images of the cyclic oxidation and reduction of Au(111) in 0.1 M $HClO_4$. Each image required 20 s to acquire: (A) downward tip rastering, 0.8–1.2 V; (B) upward, 1.2–1.6 V; (C) upward, 1.6–1.2 V; (D) downward, 1.2–0.8 V; (E) upward 0.8–0.4 V. Tunneling current was 1 nA, while the tip potential varied between 0.65 and 1.0 V SHE. (From Ref. 127.)

at the metal/metal oxide interface, while at larger bias the electronic states of the oxide film contribute. Similarly, the successful imaging of thin MgO films (bulk MgO E_{gap} = 7.8 eV) grown on Mo(001) has been reported for films with thicknesses <25 Å [266]. This conclusion is based on the assumption that the electronic structure of the film is essentially independent of film thickness. However, several studies indicate that this is rarely the case. Barrier height measurements on passivated iron reveal a decreasing barrier height with increasing passive film thickness or film growth potential. A correlation between changes in the barrier height and the donor concentration of the mixed valent (Fe^{2+}/Fe^{3+}) film was suggested [267]. Tunneling spectroscopy of the surface density of states (SDOS) of native and anodically grown oxide films on titanium reveals a large SDOS for the native oxide film at energies corresponding to mid-gap states of TiO_2 [268]. These states were found to facilitate heterogeneous electron transfer reactions, while subsequent anodic oxidation and film thickening resulted in a SDOS similar to that of a rutile single crystal. Recently, a variant of tunneling spectroscopy was used to investigate the influence of the space charge layer on the tunneling response of the passive film [269].

In addition to examining the structure and growth of thin passivating oxide films, the process of film reduction and breakdown promises to be a fruitful area of research. Some preliminary results involving SECM as well as STM have been published [257,270,271].

D. Anion Adsorption

Recent STM and surface x-ray scattering (SXS) studies have provided remarkable insight into the structural arrangement of chemisorbed species at the solid/electrolyte interface [272]. Table 4 presents a list of some of the systems that have been examined to date by STM, with the adsorption of halides on noble metal surfaces having received the most attention. The strength of the interaction increases in the order $Cl^- < Br^- < I^-$, with the adlayer coverage being potential dependent. At low coverage the adlayer appears to be disordered since SXS and STM reveal only the ordered electron density associated with the underlying metal lattice. This observation may be a result of the high mobility of adsorbate adatoms and/or the disordered nature of the adlayer, i.e., a two-dimensional gas. At more positive potentials a critical value is reached where the adlayer transforms into an ordered structure. This typically corresponds to anion coverage of ~70–80% of the saturation value. The ordered adlayer may form either commen-

TABLE 4

STM Studies of Anion Adsorption

Surface	Anion	Ref.
Au(111)	$HSO_4^-/SO_4^{2-}/H_30^+$	105,106
	I^-	27,288,293–296
	polyiodide	288,289
	Br^-	272,290–298
	S_2^-	275,299,300
	AuCN	301
	α-$SiW_{12}O_{40}^{4-}$	302
Au(100)	I^-	289,303,304
Au(110)	I^-	27,289,291,303–305
	Br^-	323
Ag(111)	I^-	306–310
	Br^-	306–309
	Cl^-	306–308
	F^-	306
	S_2^-	311
	α-$SiW_{12}O_{40}^{4-}$	302
Ag(100)	I^-	283
Pt(111)	$HSO_4^-/SO_4^{2-}/H_30^+$	312,313
	I^-	10,290,314.315
	Br^-	316,317
	CN^-	292,318
	CO	319–321
Pt(100)	I^-	182,235,314,322–326
	Br^-	316,327
	S^{2-}	328,329
	CO	323
Pt(110)	I^-	232
	Br^-	316
	SO_4^{2-}	330
Rh(111)	I^-	225
	$HSO_4^-/SO_4^{2-}/H_30^+$	332
	CO	333
Rh(110)	CO	334
Pd(111)	I^-	284,285
Pd(100)	I^-	284
Pd(110)	I^-	284,285
Cu(111)	Cl^-	276,334,336
	SO_4^{2-}	337
Cu(100)	Cl^-	277–280
Cu(110)	I^-	338
	Br^-	338
	Cl^-	338
Ni(100)	O^{2-}	339
	S^{2-}	287,339

surate, incommensurate, or higher-order commensurate structures. Incommensurate structures usually exhibit potential-dependent in-plane dimensions as a consequence of electrocompression [272,273]. In comparing SXS and STM, the precision of lattice parameter measurements by SXS greatly exceeds that associated with STM due to thermal drift and imperfect calibration of the piezoelectric drives. However, the mismatch between the substrate and the incommensurate adlattice usually creates a long-range modulation or Moire' pattern, like that shown in Fig. 25 for bromide adsorption on Au(111), which may be used to provide a reasonably accurate view of in-plane dimensions [272]. The relationship between the superlattice period, Λ, of the Moiré pattern, the lattice spacing of the overlayer, a, and substrate, b, and the rotation angle between them, Ω [274], is given by,

$$\Lambda = \frac{ab}{\sqrt{a^2 + b^2 - 2ab\cos\Omega}} \tag{6}$$

Images may be analyzed by approximating the adlattice structure as a high-order commensurate overlayer and comparing the result with simulated Moiré patterns. Reasonable agreement with the x-ray data for electrocompression has been found for the case of bromide adsorption on Au(111) [272]. The potential dependence of adlayer structure may also be followed by performing potential step or sweep experiments during imaging where discontinuities in the image dimensions and symmetry become readily apparent (''composite-domain'' images [27]), as shown in Fig. 25b. This method may be particularly helpful for determining adsorbate-substrate registry [27,275].

Chemisorbed species are also known to exert a strong effect on the step dynamics of vicinal surfaces. An interesting example is provided by the chloride adsorption on Cu(100) as shown in Fig. 26 [276–280]. Immersion in hydrochloric acid results in the oxidative adsorption of chloride forming a $(\sqrt{2} \times \sqrt{2})R45°$ adlayer, which stabilizes the surface steps in the ⟨100⟩ direction as opposed to the close-packed ⟨110⟩ orientation associated with an adsorbate-free copper surfaces in UHV systems. The ⟨100⟩ oriented step edge corresponds to the close-packed direction of the commensurate chlorine adlattice yielding kink saturated ⟨100⟩ metal steps beneath the overlayer. At more negative potentials, the chloride layer is partially desorbed and the remaining adlayer becomes disordered. Interestingly, x-ray studies of the analogous Ag(100)-Br system suggest that the second-order

FIG. 25. STM images revealing the formation of a sequence of ordered hexagonal bromide adlayers on Au(111) in 1 mM NaBr + 0.1 M HClO4 (a, c, d), 8.0 nm × 8.0 nm; (b), 6.5 nm × 6.5 nm). (a) At 0.44 V only the Au substrate is visible; upon increasing the potential from 0.48 V (upper edge of image) to 0.59 V (lower edge). (b) A rotated-hexagonal bromide adlayer is formed. (c) A Moiré pattern is evident due to the mismatch between the adlattice and the substrate. The dimensions of the pattern change as the potential (coverage) is increased from 0.59 to 0.74 V (c, d). (From Ref. 272.)

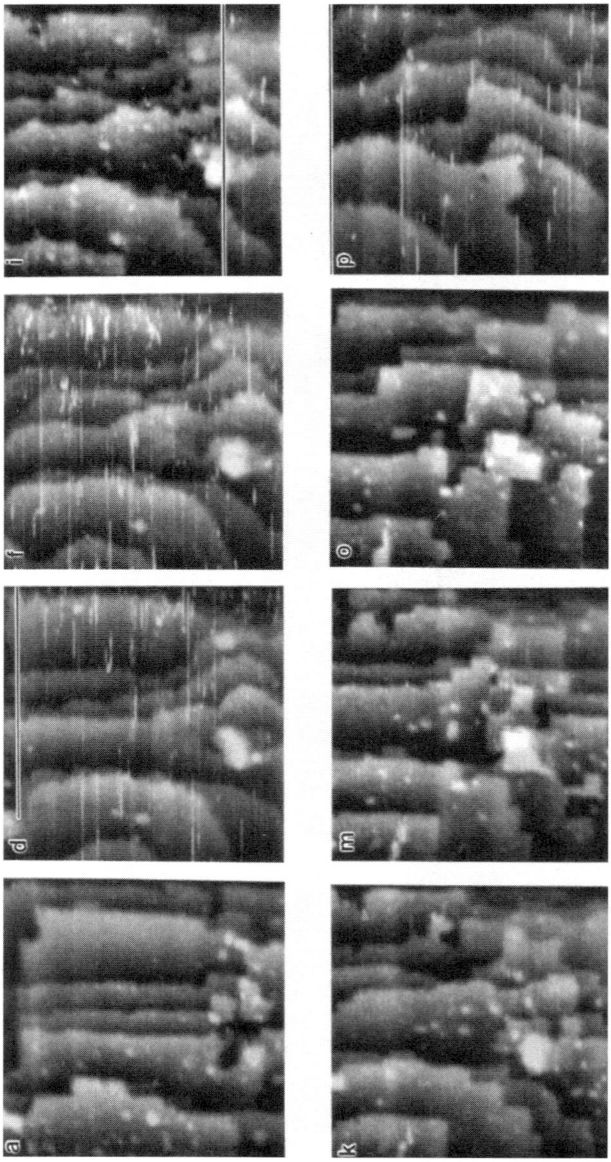

transition follows a two-dimensional Ising model [281], and recent efforts to describe the nature of the lateral interactions are available [282]. Coincident with the disruption of the ordered chloride adlayer, the ⟨100⟩ oriented steps immediately become frizzy due to the rapid movement of the destabilized kinks. Similarly, a variety of relatively small islands appear to dissolve to the adatom state. This effect in combination with the development of the curved steps correlates to an elevated adatom activity on the terraces. When the potential is stepped back in the positive direction, the ordered adlayer reforms and the steps rapidly adopt the ⟨100⟩ orientation. Simultaneously, the elevated adatom concentration associated with the previously curved steps is quenched by nucleating a variety of small islands on the terraces. The surface structure then coarsens rapidly with time to reduce the total number of kinks in the halide adlayer. Similar absorbate-induced step dynamics have been observed on other surfaces [283–287].

The STM has also been used to follow the evolution of surface-confined reactions such as the oxidation of adsorbed sulfide to form adsorbed S_8 and iodide to polyiodide [275,288,289]. The substrate exerts a strong influence on the dimensions and ordering of the adsorbed molecules, particularly the formation of the first monolayer. In a similar manner, studies of the impact of different adlayer structures on the electron transfer kinetics of various soluble redox species have been initiated [290].

In addition to anion adsorption, there exists the possibility of adsorption of cations at negative potentials along with coadsorption phenomena. For example, mixed layers of alkali cations with iodine on Au(110) [291] or cyanide on Pt(111) [292] have been reported. In fact, coadsorption has proven to be quite common among numerous underpotential metal deposition reactions as described below.

E. Underpotential Deposition of Metals

Underpotential deposition (upd) originates from an adatom-substrate bond being formed using less energy than that required to form subsequent ada-

FIG. 26. Step faceting associated with formation of a saturated ordered chloride adlayer on Cu(100). A saturated ($\sqrt{2} \times \sqrt{2}$) R45° chloride adlayer covers the surface at −0.25 V (a, k, m, o) while stepping to more negative potential, −0.65 V, leads to partial desorption of the adlayer (d, f, p). The black lines in d, i, and p represent the time at which the potential was stepped between the two potentials. The abscissa and ordinate correspond to the ⟨100⟩ direction. (From Ref. 278.)

tom-adatom bonds associated with bulk deposition. In the last few years significant advances in the understanding of upd of metals have been derived from a combination STM, AFM, SXS, quartz crystal gravimetry as well as chronocoulometry experiments [340,341]. STM images of ordered upd structures have been reported, and experiments are underway probing the dynamics of the deposition process in several systems. Table 5 lists some of the systems examined to date. Not surprisingly, STM images of ordered metal adlayers exhibit many of the characteristics already noted for anion adsorption. For example, Pb upd on Ag(111) exhibits a Moiré pattern associated with the incommensurate rotated-hexagonal Pb overlayer, which is formed at high coverages [342,343]. The adlayer is electrocompressive, and STM measurements of the dimensions of the Moiré pattern reveal favorable agreement with the potential dependence originally determined by more precise SXS measurements as shown in Fig. 27 [274,343–345].

For many ordered upd systems structural assignment of the imaged electron density can be particularly difficult due to coadsorption. This has been most clearly demonstrated for Cu upd on Au(111) in H_2SO_4. At low coverages random absorption leads to a gaslike adlayer, which is not imaged in the STM presumably due to the mobility of the adatoms. This is followed by formation of a $(\sqrt{3} \times \sqrt{3})$ R30° structure, which subsequently gives way to a close-packed monolayer at more negative potentials. Initially, the $(\sqrt{3} \times \sqrt{3})$ R30° structure was attributed to 1/3 coverage of copper [9,148,346,347]. However, subsequent electrochemical measurements suggest a copper coverage of two thirds and a sulfate coverage of one third [348]. This finding was consistent with the interpretation of prior ex situ LEED measurements [349], model calculations [350], as well as a recent SXS study [351], which resulted in the structural model of the coadsorbed Cu/SO_4^{2-} adlayer shown in Fig. 28. A comparison with the STM data reveals that the maxima in the tunneling response correspond to the protruding $(\sqrt{3} \times \sqrt{3})$ R30° sulfate species. Additional studies demonstrate that the intermediate open structures associated with submonolayer metal coverage are dependent on the identity of the anion [9,148,347]. Thus, submonolayer upd structures are notably different from the pseudomorphic overlayers that are often associated with vacuum deposition. In contrast, upd reactions involve balancing the interactions between the electrolyte anion (cation), upd metal atom(s), and the substrate surface [352,353]. Depending on the relative strength of these interactions, anion coadsorption may yield structures ranging from an ordered halide adlayer adsorped upon a psuedo-

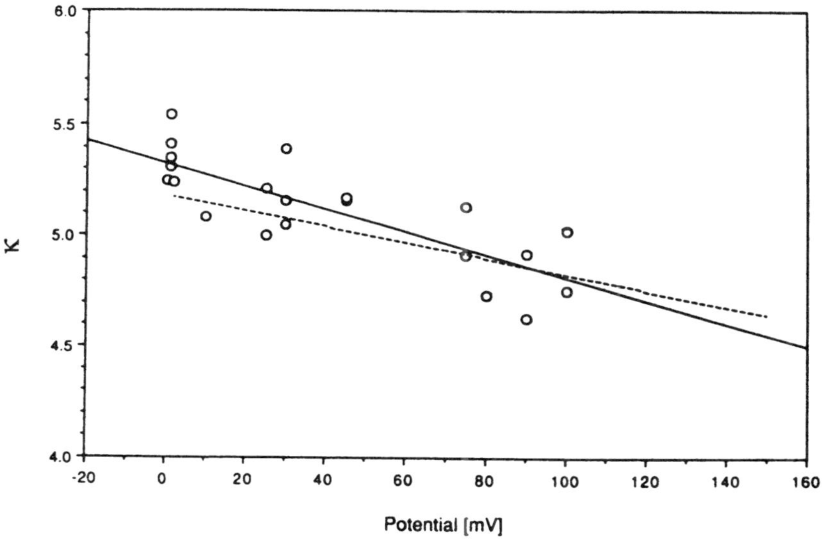

FIG. 27. A comparison between the in-plane dimensions, determined by SXS and STM, of the electrocompressive Pb upd adlayer formed on Ag(111). The open circles are the STM data and the solid line is a least-mean-square fit to the data, while the dotted line is derived from SXS measurements ($\kappa = \Lambda/a$ when Λ is the period of the Moiré pattern while a is the lattice spacing of the overlayer). (From Ref. 343.)

morphic copper upd layer formed on Pt surfaces, to more complex crystal-like structures such as CuX (X = Cl^-, Br^-, I^-, SO_4^{2-}), which are formed on Au(111) [352]. These results highlight the difficulty in interpreting STM images of multicomponent structures and demonstrate the need for corroborating evidence when assigning tunneling contrast to atomic species.

The dynamics of upd reactions have also been examined by STM. The formation of the ordered copper/sulfate layer [354] and copper chloride layer [355] on Au(111) was examined in a dilute solution of Cu^{2+} where the reaction was under diffusion control so that growth proceeded on a time scale compatible with STM measurements [354]. In another study, the importance of step density on nucleation was examined and the voltammetric and chronoamperometric response for Cu upd on vicinal Au(111) was shown to be a sensitive function of the crystal miscut, as

TABLE 5
STM Studies of Upd

Substrate	Metal	Anion	Ref.
Au(111)	Ag	ClO_4^-	372–377
		SO_4^{2-}	374,377–379
		$ClO_4^- + I^-$	380
	Bi	ClO_4^-	381
	Cd	CH_3CO_2H, SO_4^{2-}	299,365
		SO_4^{2-}	366,367,382
	Cu	SO_4^{2-}	9,148,346,347, 354,356, 383–385
		$SO_4^{2-} + Cl^-$	9,347,352, 384,385
		$SO_4^{2-} + Br^-$	352
		$SO_4^{2-} + I^-$	352
		Cl^-	383
		$ClO_4^- + Cl^-$	355
	Hg	$SO_4^{2-} + ClO_4^-$	386
	Ni	$H_2NSO_3^-$	387
	Pb	ClO_4^-	341,357,358–360,362
	Pt	$ClO_4^-/PtCl_6^{2-}$	388
	S	$S^{2-}/CH_3CO_2Na/KOH$	299
		S^{2-}	366
	Se	SO_4^{2-}	365
	Te	SO_4^{2-}	364,367
	Tl	NaOH	298,389
	Zn	H_xPO_4	409
Au(100)	Ag	ClO_4^-	375,390
	Cd	CH_3CO_2H, SO_4^{2-}	364,367
	Cu	$SO_4^{2-} + Cl^-$	391
		SO_4^{2-}	9
	Pb	ClO_4^-	357,392
	Se	$HSeO_3^-$	369
	Te	SO_4^{2-}	304,364,367
Au(110)	Cd	CH_3CO_2H, SO_4^{2-}	364,367
	Cu	$SO_4^{2-} + Cl^-$	394
	Se	SO_4^{2-}	365
	Te	SO_4^{2-}	364,367

TABLE 5 Continued

Substrate	Metal	Anion	Ref.
Ag(111)	Cd	SO_4^{2-}, $P_2O_7^{4-}$	368
	Pb	ClO_4^-	341–343,358, 362,375,376, 393,395–401
	Cd	SO_4^{2-}, $P_2O_7^{4-}$	368
	Ti	ClO_4^-	362,400
Ag(100)	Pb	ClO_4^-	341,359,365, 376,387,392, 395,397–402
Pt(111)	Ag	SO_4^{2-}	403
		I^-	310
	Bi	ClO_4^-/CO/Bi	319
	Cu	SO_4^{2-}	353,371,404, 405
		SO_4^{2-} + Cl^-	353,384
		SO_4^{2-} + Br^-	353,384
		SO_4^{2-} + I^-	315,353,384
	Pb	ClO_4^-	406
	Ti	SO_4^{2-}	353
Pt(110)	Cu	SO_4^{2-}	330
Pt(100)	Ag	ClO_4^-	410
	Bi	SO_4^{2-}	326
	Cu	SO_4^{2-}	407
		SO_4^{2-} + Cl^-	9,347,352
		SO_4^{2-} + Br^-	352
		SO_4^{2-} + I^-	352
	Pb	ClO_4^-	406
Cu(111)	Pb	Cl^-	276
Cu(100)	Pb	Cl^-	408

revealed by STM [356]. The potential dependence of two-dimensional nucleation has also been investigated directly with the STM. At low supersaturation, Pb upd on Au(111), Ag(111), Au(100), and Ag(100) proceeds exclusively at steps, while at a higher supersaturation nucleation may also proceed on terraces [341,357–359] as indicated in Fig. 29. The correlation between the microscopic mechanism and the degree of supersaturation was in good agreement with theoretical considerations [341,358]. The influence

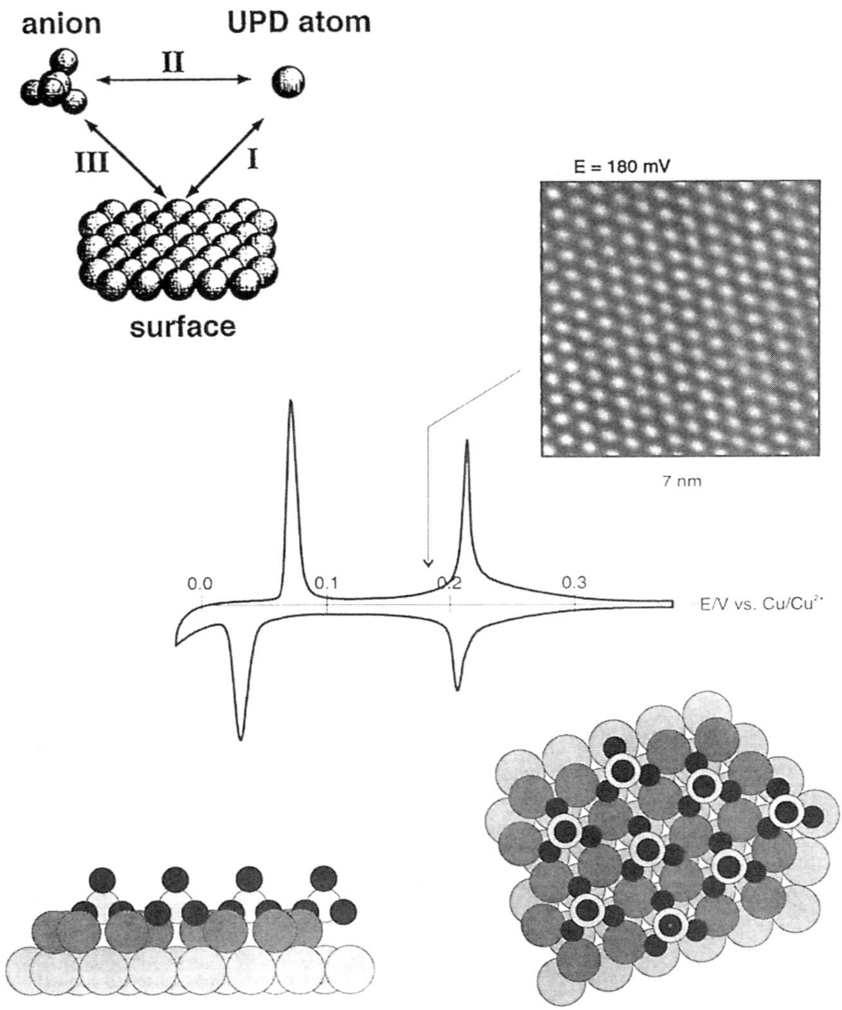

FIG. 28. The formation of upd layers often involves complex interactions between the anion, upd metal, and substrate. An example is provided by the ($\sqrt{3} \times$ v$\sqrt{3}$) R30° STM image of the copper/sulfate upd layer formed on Au(111) and the SXS of the interfacial structure. (Adapted from Refs. 148, 351, 353.)

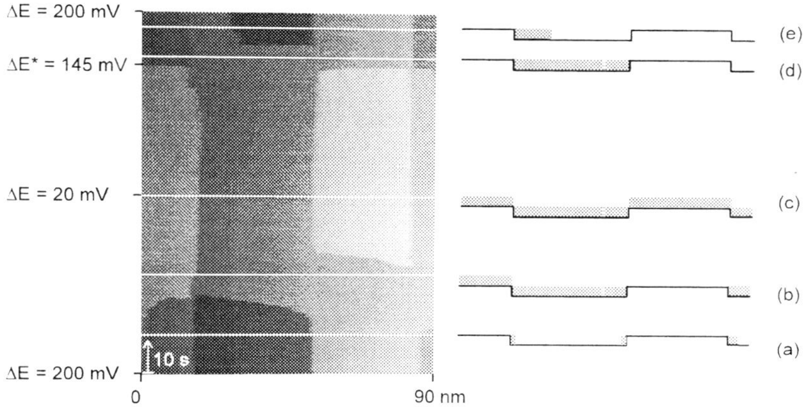

FIG. 29. In situ STM line scan plot showing the formation and dissolution of a Pb upd overlayer on Ag(111) during a cyclic potential scan between $20 < \Delta E < 200$ mV in 4 mM Pb $(ClO_4)_2$ + 10 mM $HClO_4$, where ΔE^* represents the underpotential of a condensed Pb phase at $\theta \sim 0.5$. The cross sections reveal (a) step decoration at $\Delta E > \Delta E^*$; (b) followed by completion of a 2d Pb_{ads} overlayer on top of substrate terraces and by growth starting from the monatomic steps ($\Delta E < \Delta E^*$); (c) nucleation and growth of 2D Pb islands on top of substrate islands starting at more negative potentials ($\Delta E < \Delta E^*$); (d) dissolution of a condensed 2D Pb_{ads} overlayer proceeds from island and terrace step edges at $\Delta E > \Delta E^*$; (e) dissolution of the condensed 2D Pb_{ads} overlayer from substrate pits occurs at more positive potentials $\Delta E > \Delta E^*$. (From Ref. 341.)

of surface reconstruction on the initial stages of deposition has also been examined. In the case of Pb upd on Au(111), Pb clusters were found to preferentially nucleate and grow along the rows of the $(23 \times \sqrt{3})$ reconstruction of Au(111) in a manner analogous to vacuum deposition experiments [360].

In many instances upd is found to be strongly dependent on sample history. This is usually an indication of alloying between the upd layer and the substrate. Repetitive alloying and dealloying usually leads to surface roughening [361] in a manner somewhat analogous to that associated with cyclic oxidation and reduction of noble metals. STM studies have revealed that the step density of the substrate exerts a strong influence on the rate of alloy formation [359,361,362]. Simulations of two-dimensional alloying via vacancy generation at steps have been published and promise to contribute towards a more quantitative description of upd alloying phenomenon [204]. Interestingly, alloying has been observed in many systems that are immiscible as bulk alloys. The lack of bulk solubility in certain cases stems from the strain associated with the mismatch between the size of the substrate and solute atoms. However, the free surfaces offer a means for partial relaxation of the strain, thereby allowing the overall free energy of the system to be diminished by alloy formation [363]. In this case it is likely that the energetics will still restrict alloying to two dimensions, i.e., one or two monolayers.

Implementation of upd in material synthesis has also been explored. A particularly interesting effort has focused on the production of II–VI compounds by successive upd reactions performed in two different electrolytes. Importantly, process development has been tightly coupled with STM studies of both upd and overpotential deposition (opd) of the constituents [299,304,365–369]. Similarly, the influence of upd on catalytic activity towards certain reactions is well known [370]. An STM study of the inhibition of four-electron oxygen reduction on Pt(111) by upd Cu clearly demonstrates the importance of upd structure on reactivity [371].

F. Overpotential Deposition of Metals

STM has already generated significant insight into electrocrystallization phenomena and promises to be a central tool in the further development of the subject. The following discussion will focus on studies of homo- and heteroepitaxial deposition on single-crystal electrodes, although it should be mentioned that SPM has also found application in monitoring the evolution of film structure and roughness during electrolytic and electroless deposition of and on polycrystalline electrodes. To date most studies have dealt with heteroepitaxial deposition of metals onto easily prepared noble metal surfaces and (0001) HOPG. In large measure these studies have focused on characterizing the role of substrate defect structure on the nucleation and growth of the new phase. In contrast, homoepitaxial growth processes have received somewhat limited attention. A significant obstacle to rigorous development of STM in electrodeposition studies is the unresolved influence of the probe on the electrochemical conditions within the tunnel junction [126–130]. As described earlier, electrostatic shielding by the tip can result in a nonuniform current distribution around the junction. The tip may also limit access of metal ions in solution to the growth centers being studied [126,132]. Similarly, rastering the tip may also perturb mass transport of the reactant. However, experience with SECM suggests that the latter effect is minimal, and recent theoretical work for the STM geometry supports this contention [132]. The combination of these effects demonstrates the need for caution when studying bulk deposition processes with scanning probe methods. A particularly easy check for possible imaging-related artifacts is to make significant changes in the area being investigated in order reveal the presence or absence of surface modification. For example, under certain conditions a marked shielding effect has been noted for copper deposition on Au(111), as shown in Fig. 30 [126]. Similar effects have also been observed for copper deposition on gold at a macroscopic current density of 30 μA/cm^2, while no such effects were observed for

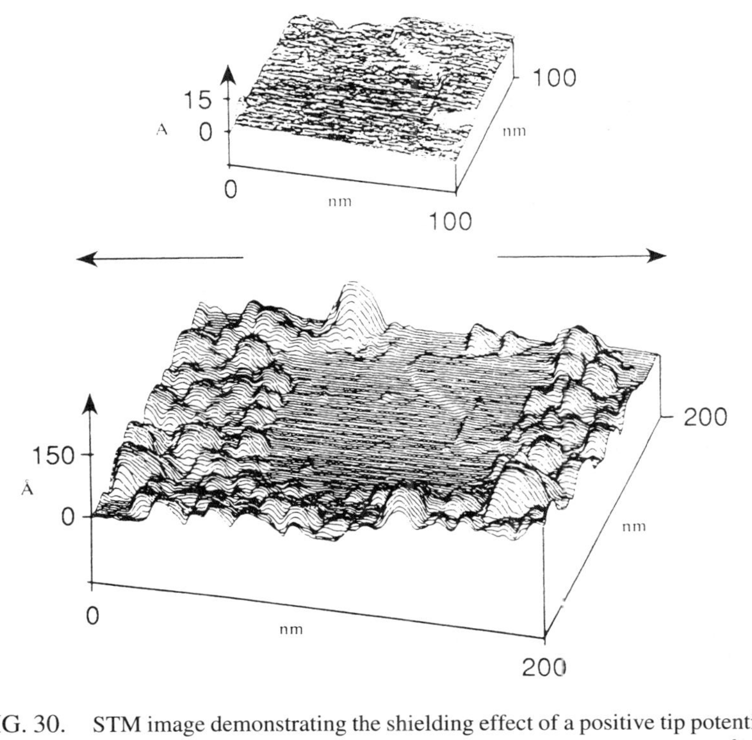

FIG. 30. STM image demonstrating the shielding effect of a positive tip potential on bulk copper deposition on Au(111). Substrate potential ~−375 vs. Cu/Cu^{2+} in 50 μM $CuSO_4$ and 5 mM H_2SO_4. (From Ref. 126.)

silver deposition and dissolution at 2 μA/cm^2 [128], although in the latter case the tip was shown to influence the morphological evolution of the electrodeposited silver under open circuit conditions. Clearly, further investigation of these important effects will be central to the successful implementation of STM in examining phase-formation reactions. From an alternative perspective, it should be noted that such phenomena also represent an opportunity for spatial resolved synthesis, as will be described in a later section.

1. Homoepitaxial Metal Deposition

A recent textbook summarizes much of what is known about electrocrystallization [341]. Metal-on-metal homoepitaxial deposition is a subject that incorporates a variety of different processing schemes ranging from electrocrystallization to physical vapor deposition to chemical vapor deposition, etc. Recent developments in the characterization of vapor-deposited films by surface science methods coupled with the ability to perform realistic simulations are driving the rapid growth in our knowledge of metal deposition [191,202,205,411–413]. Importantly, homoepitaxial systems can be used to study the kinetic processes affecting film growth without the complicating effects of differential surface energy and misfit strain that accompany heteroepitaxial growth [202,205,411–414]. Consequently, provided appropriate consideration is given to the site bias dependence of the electrochemical reduction reaction (see Ref. 341), it is likely that realistic simulations of the electrocrystallization process will become a reality in the near future. STM will be an essential tool for obtaining information on individual atomic processes as well as for characterizing the evolution of surface structure on the mesoscopic scale. Papers dealing with scaling and spectral analysis of electrodeposited films have been published, although these studies pertain to complicated systems [206,415]. Nevertheless, rapid progress in this area is anticipated.

Reports of homoepitaxial deposition proceeding by step flow have been presented for Ag(111) [173], Cu(111), and Cu(100) [276–279]. In the case of copper deposition on Cu(100), a $\sqrt{2} \times \sqrt{2}$ R45° Cl adlattice has been shown to exert a strong influence on film growth by acting as a template guiding step flow in the $\langle 100 \rangle$ direction [277–279]. At high overpotentials growth proceeds by multinuclear multilayer growth. In many commercial applications of metal electroplating, a variety of organic and inorganic additives are used to control the morphology and grain structure of the resulting films. STM studies of the influence of various organic additives on copper deposition have been initiated, although the majority of these are complicated by phenomena associated with heteroepitaxy growth [148,347,416–422]. A noteworthy exception to this is a recent study of the copper deposition on Cu(100) in the presence of benzotriazole (BTA) [423]. Relative to growth in an additive-free sulfuric acid solution, adsorption of BTA was found to suppress island nucleation while favoring deposition via step flow. This resulted in leveling of the substrate roughness, which had been induced by prior dissolution. Slow metal deposition via isotropic step flow has also been observed for gold deposition in solutions containing tetramethylthiourea [424].

2. *Heteroepitaxial Metal Deposition*

Traditionally, three different heteroepitaxial growth modes have been identified: Volmer-Weber, Stranski-Krastanov, and Frank-van der Merwe growth [330,406,407]. The origin of this classification scheme is based on thermodynamics, namely, the free energy balance between the bulk phases, interfaces, and misfit strain. The potential dependence of the surface stress and surface free energy associated with an immersed electrode represents an additional variable. Nevertheless, it is well established that kinetic phenomena exert a decisive role in determining the mode of film growth as well as the mechanism of strain relief [413,425,426]. An example of the complexity involved is given by the observation of three-dimensional growth in the ~ zero misfit system, Ag on Au [372,373]. This was attributed to hindered interlayer transport as a result of a step edge barrier. Such barriers are often responsible for the multinuclear multilayer growth mode, as opposed to layer-by-layer growth, observed during homoepitaxial growth [406–414,427,428]. In addition to the classical growth modes outlined above, both vacuum and electrochemical studies have revealed a fourth growth mode involving rapid two-dimensional alloying. In this case the substrate surface atoms are displaced by alloying while the adatoms so generated coalesce with the incoming deposition flux, which results in a mixed and roughened interphase region. Interestingly, this growth mode has been shown to occur even for constituents that are essentially immiscible in the bulk form [429].

The first in situ STM study of metal deposition was for Ag on HOPG [430], which is representative of the Volmer-Weber growth mode. Since that time several other studies of metal deposition on HOPG have been reported: Ag [393,431–433], Pb [129], Pt [434,435], Au [436], and Ni [437]. In these studies examining the small metal particles proved to be difficult. In several cases no deposits were observed in the probed region or significant migration of small clusters was noted [431,435]. The movement of the clusters was attributed to repulsive force interactions between the tip and the weakly bound clusters [435]. According to a UHV-AFM study, forces below ~1 nN are required to prevent displacement of such clusters [435,438]. More recently, noncontact AFM has been shown to be a particularly effective tool for characterizing such systems [439,440]. It remains to be determined if there is a preferred in-plane orientational relationship between the electrodeposited metal cluster and the HOPG substrate. In the case of Au deposition, the particle morphology was found to be a sensitive function of potential [436]. Interestingly, a recent TEM study [441] revealed that the shape and structure of gold particles grown from a $HAuCl_4$ solution were potential dependent. This observation was rationalized in terms of potential-dependent reconstruction phenomena.

Several systems that exhibit Stranski-Krastanov growth have been investigated. Copper deposition on Au [12,148,347,356,421,422] and Pb deposition on Ag [341,393,395] have received the most attention. The important role of defect structure in the nucleation of three-dimensional growth centers has been demonstrated for Cu deposition on Au(111) [12,148,347,421,422]. The contribution of surface step density has also been explored using a series of miscut as well as roughened (ORC) Au(111) crystals [356]. In contrast, the initial deposition of copper on Ag(111) was observed to occur in a layer-by-layer manner with nucleation occurring exclusively on the terraces [148,178]. It is noteworthy that Au and Ag have practically the same lattice parameter, however, unlike gold, no copper upd phenomena are observed on Ag. Similarly, unlike Au-Cu, alloys of Ag and Cu are immiscible at room temperature. It was suggested that copper nucleation on the terraces of Ag(111) may be due to preferential adsorption of anions at the steps due to the positively charged silver surface [i.e., Epzc ~ -0.7 V SCE for Ag(111) versus 0.23 V SCE for Au(111), with copper deposition occurring at ~0 V SCE] [148,178]. At higher overpotentials or after deposition of a number of copper layers, three-dimensional growth begins. In another variation, Pd deposition on Au(111) proceeds with upd of the first layer, followed by nucleation at steps and two-dimensional growth of the second layer with the onset of three-dimensional growth occurring during deposition of the third layer [442,443]. Stranski-Krastanov growth has also been reported for copper deposition on Pt(100) [255]. Yet another interesting variation is Ag deposition on Pt(111), where following the deposition of two upd layers smooth epitaxial silver films have subsequently been grown to a thickness of 25 monolayer equivalents [402].

Studies of the influence of organic additives on nucleation and growth have also been reported for copper deposition on Au(111) [148,347,416–422]. Chloride salts of benzothiazolium derivatives [417,418] and crystal violet [148,347,417,421,422] were found to favor lateral growth of the copper nuclei, with the former additive being more effective in this capacity. No atomically resolved images of adsorbate structures were obtained in these studies. However, more recently, small ordered regions have been observed in the BTA adlayer on Cu(100) [423], and several other studies of ordered aromatic molecules adsorbed on noble metal electrodes have been published (see next section). These studies highlight the importance of the interactions of adsorbates with surface defects, such as steps, for both inorganic [148,178] and organic species [419]. For example, in an electrolyte containing crystal violet, copper is found to nucleate preferentially around the perimeter, or rim, of the gold islands that are formed as a consequence of the lifting of the reconstructed Au(100) surface [419]. Copper deposition in the presence of thiourea has also been examined, and the adsorbate is found to prevent the formation of a uniform upd monolayer of copper [148,444]. The resulting island structure leads to a much higher nucleation density during bulk deposition, which demonstrates the impact of upd phenomena on bulk deposition. It is also worth noting that older studies suggest that thiourea adsorption on copper leads to deviations from epitaxial growth, which are associated with incorporation of sulfur in the deposit. However, the mechanistic details of this important process remain to be resolved, particularly with respect to the dependence on the deposition rate [445]. Significant insight into the role of organic adsorbates may also be obtained from studies of reasonably well-defined, self-assembled monolayer films [446–450].

Another demonstration of the impact of upd on bulk deposition is provided by Pb and Tl deposition on Ag(111) and Ag(100), where the orientation of the three-dimensional crystallites reflects the epitaxially relationship established by the upd layer [341]. For example, in the case of Pb deposition on Ag(111) [395], a two-dimensional layer, Ag(111)[110]$\|$ compressed 2D hcp Pb [110] R 4.5°, is initially formed followed by nucleation of a three-dimensional cluster having the same orientational relationship, Ag(111)[110]$\|$ 3DPb(111)[110] R4.5°.

Similarly, the defect structure associated with surface-reconstruction phenomena is known to exert an influence on heteroepitaxial deposition. This has been demonstrated for both lead and nickel deposition on reconstructed Au(111) [353,360,451]. For nickel deposition, nucleation was observed to proceed in three distinct, potential-dependent steps [354,451]. At low overpotentials, place exchange of Ni and Au atoms occurred at the elbows of the herringbone $(23 \times \sqrt{3})$ reconstruction followed by nucleation on top of these sites when the overpotential exceeds 80 mV, while nucleation at step edges occurred when the overpotential was greater than 100 mV. The decorated elbows correspond to the position where the two-dimensional lattice of the reconstructed Au surface layer is dislocated. The lateral dimensions of the exchange site are in good agreement with the size of the distorted atmosphere or zone surrounding the dislocation [452]. Similar effects have also been seen for vacuum phase deposition on reconstructed Au(111) [453]. The sensitivity of the nickel deposition reaction to overpotential can be used to alter the dynamics of film growth and also offers the possibility of generating two-dimensional magnetic nanostructures. For example, growth of the first nickel monolayer on the reconstructed Au(111) surface was highly anisotropic, resulting in the formation of two-dimensional needle-like islands [452]. The orientation of the needles was determined by the weak anisotropy of the reconstructed surface, while the needle-like shape of the islands was ascribed to structural anisotropy of the nickel deposit due to uniaxial contraction perpendicular to the needle, which improves the packing density while the width is constrained to avoid occupation of energetically unfavorable top sites [452]. Another interesting result was the observation of layer-by-layer growth at least for the first seven monolayers of nickel deposition on Au(111), as shown in Fig. 31 [454]. Nickel deposition has also been examined on Au(100) and Cu(100) [454,455], where the films were found to be much rougher than those grown on Au(111) under similar conditions. In the case of nickel deposition on Cu(100), nucleation of second-layer islands was observed at a total coverage as low as 0.1 monolayer [454,455]. Even rougher films were grown on Au(100) [454,455]. Interestingly, these observations are congruent with the results of vacuum studies of nickel deposition.

Strain relief during heteroepitaxial film growth remains a topic of central importance to the understanding of microstructural evolution during film growth. In vacuum studies on hexagonal and quadratic surfaces, a variety of intermediate structures have been observed, associated with relaxation of the misfit strain [456–460]. Importantly, the high z-resolution of the STM permits detection of single misfit dislocation glide lines, making STM a highly precise tool for studying dislocation emission during the early stages of strain relaxation [461]. The first images of strained overlayer structures and misfit accommodation during electrodeposition have been reported [130,462,463]. In the case of Cu deposition on Ag(100), a smooth

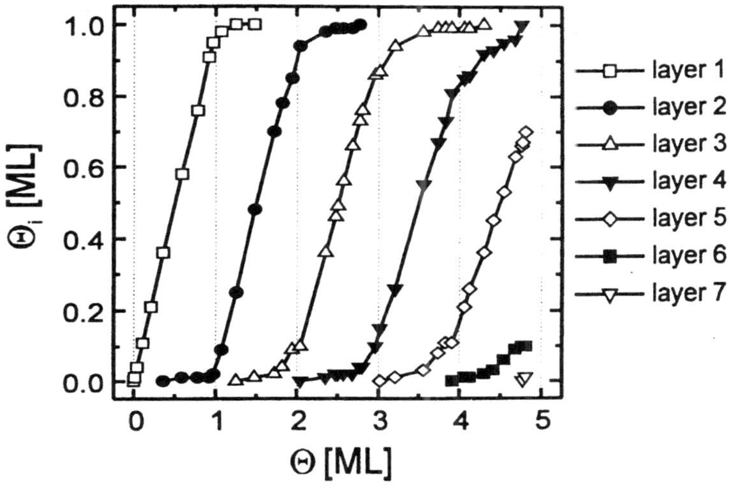

FIG. 31. Layer-by-layer growth is observed at least for the first seven layers of nickel deposition on Au(111). Specifically, the occupation, θ_i, of the individual layers i ($1 \leq i \leq 7$) as a function of total Ni coverage, θ, is shown. The data were obtained by quantitative evaluation of a series of STM images recorded in 10^{-2} M H_3BO_3 + 10^{-4} M HCl + 10^{-3} M $NiSO_4$. (From Ref. 454.)

bcc overlayer is formed up to the first 8 monolayers, after which a one-dimensional wavy structure develops associated with the onset of the transformation to the fcc phase [463]. The same structural evolution is observed for Cu deposition on Au(100), except that the shear transformation occurs after 10 monolayers are deposited, as shown in Fig. 32 [462]. The reason for this difference is unknown. Strain accommodation has also been observed during copper deposition on Pt(100) [130]. After deposition of 5–10 monolayers, the overlayer is observed to relax to a lattice-spacing characteristic of bulk copper. This resulted in a square Moiré pattern reflecting the symmetry of the substrate and the misfit of the overlayer [130]. Subsequent growth proceeds via formation of three-dimensional clusters. This result differs significantly from that observed for films produced by physical vapor deposition [130]. It was argued that this may be a thermodynamic effect associated with changes in surface stress, which is a function of electrolyte composition and applied potential.

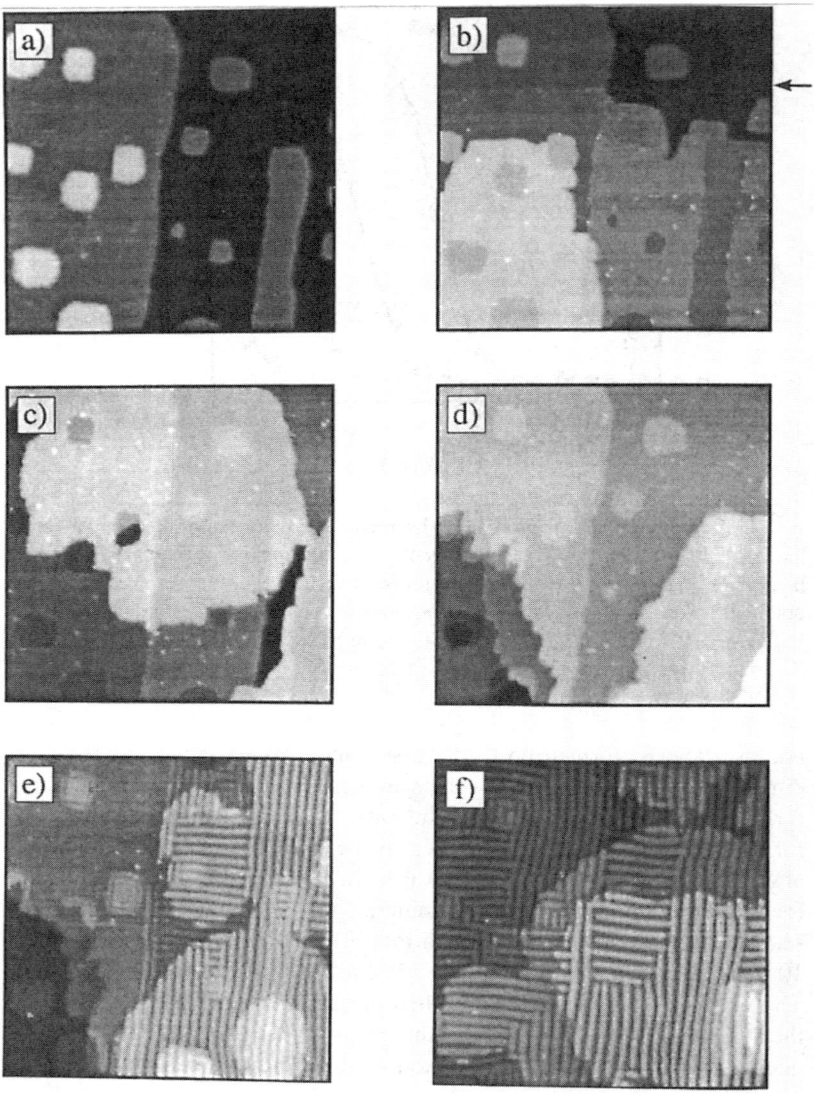

FIG. 32. The first 10 layers of copper deposition on Au(100) proceeds in a smooth layer-by-layer mode with the formation of b.c.c. copper. During the deposition of the eleventh layer, a striped structure appears due to the onset of strain relief via a shear transformation. (From Ref. 462.)

In addition to imaging the structural rearrangements associated with strain relief, the STM has been used to study stress development during film growth. This is accomplished by monitoring the bending of a cantilever sample with the STM. The technique has been used to reveal the strong influence of overpotential on stress development during the deposition of thin copper films on Au(111) [464].

G. Adsorption of Molecules

The ability to image individual molecules, monolayer films, as well as the dynamics of adsorption and reaction is an extremely exciting prospect for chemists, biologists and engineers [58,80,465,466]. Recently, in situ images were acquired for the prototypical aromatic molecule, benzene, adsorbed on Pt(111) [467], Rh(111) [467], and Cu(111) [468]. High-resolution images revealed the internal structure of the adsorbed species along with the site dependence, as shown in Fig. 33 [467]. Theoretical calculations provide support for the interpretation of such images [58,469,470]. Ordered adlayers of slightly larger molecules such as naphthalene on Rh(111) [471] and Cu(111) [468] have been reported, while a disordered arrangement was found on Pt(111) [467]. Images of adsorbed anthracene and napthoquinones were also presented. Similarly, molecularly resolved images of ordered adlayers of purine and pyrimidine bases [472–476], 2-2′ bipyridine [477], pyridine [478], cysteine [479], phenanthraquinone [480], and tetramethylthiourea [424,481] on gold have been reported. More recently, uracil adsorption on Ag(111) has been studied [482]. Individual fullerenes as well as films of C_{60} and C_{70} have also been imaged on Au(111) and Au(110) [483]. Images of even larger molecules such as crystal violet, porphyrin (TMPyP), and methylpyridinium-phenylenedivinylene (PV) have also been reported [484–487]. Interestingly, a hydrophobic iodine adlayer was found to be a necessary precursor for the formation of an ordered overlayer of porphyrin molecules [484,485] on Au(111). Similar structures were also observed on iodine-modified Pt(111) [485], Rh(111) [485], and Ag(111) [486], while a fourfold symmetrical arrangement was found for iodine-modified Pt(100) [487]. The relatively weak van der Waals interaction between the hydrophobic halogen adlayer and the organic molecules is thought to be a key factor promoting the ordering of these large molecules [484–487]. This is supported by similar results observed on HOPG and other van der Waals surfaces [19,80]. Nonetheless, it is clear that the geometry of the halide layer also exerts a significant influence on the packing of the molecules [487].

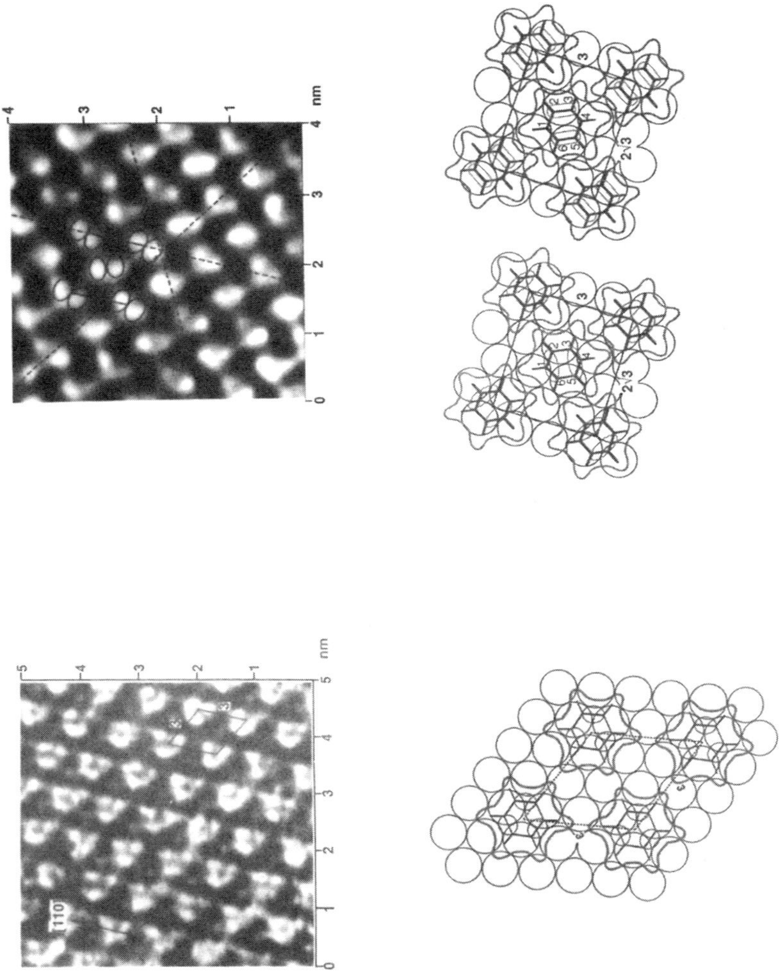

FIG. 33.

STM images of molecules are often a sensitive function of the tunnel conductance. For example, when examining TMPyP on Au(111)-I, the iodine underlayer was imaged at a tunnel resistance of 5×10^6 ohms while the electron density of TMPyP was observed at 8×10^8 ohms [484]. Similar effects have been noted for protoporphyrins [488] and purines [489,490] adsorbed on graphite.

STM has also been used to examine the dynamics of potential-dependent ordering of adsorbed molecules [475–478]. For example, the reversible, charge-induced order-disorder transition of a 2-2′ bipyridine monolayer on Au(111) has been studied [477]. At positive charges, the planar molecule stands vertically on the surface forming polymeric chains. The chains are randomly oriented at low surface charge but at higher potentials organize into a parallel array of chains, which follow the threefold symmetry of the Au(111) substrate as shown in Fig. 34. Similar results were found for uracil adsorption on Au(111) and Au(100) [475,476].

The interesting class of monolayer films derived from spontaneous adsorption of alkanethiols on Group 1B metals has also been examined by STM. Several atomically resolved studies have been performed in both air and vacuum (see, for example, Refs. 466, 491–497). Imaging of extended amphiphile monolayers in a low perturbation mode requires a tunneling impedance in the GΩ range [466], and the defect structure of these films is a sensitive function of preparation conditions. Studies of the formation of alkanethiolate films under potential control have been reported [496,497]. Similarly, the reductive desorption of three different alkanethiolates has been examined [498]. In general terms the bonding state between the molecule and the metal is thought to be in the form of Au-thiolate. However, a recent study of 4-mercaptopyridine adsorbed on Au(111) suggests that the molecules are dimerized to form disulfide [499]. The STM has also been used to study metal deposition on derivatized electrodes in order to probe both the nature and distribution of the defect sites [446–450] as well as the effect of coadsorbates, e.g., Ag upd, on alkanethiolate-derivitized Au(111) [500]. Metal deposition on top of organic monolayer films has

---

FIG. 33. Images of benzene adsorbed on Rh(111) in a 0.01 M HF solution containing 0.05 mM C_6H_6. (left) At 0.25 V RHE a (3×3)-adlattice is observed with C_6H_6 occupying threefold hollow sites. (right) At 0.45 V RHE a $c(2\sqrt{3} \times \sqrt{3})$rect adlattice is imaged with the molecules occupying twofold bridging sites. (From Ref. 467.)

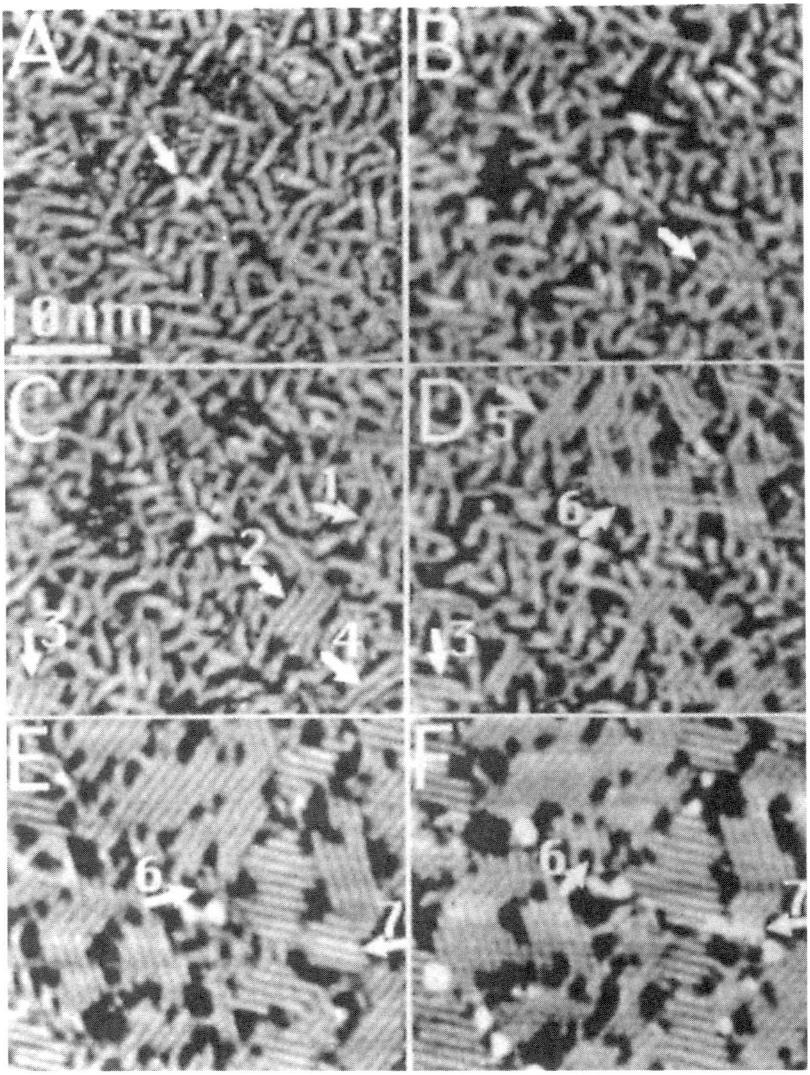

FIG. 34. A sequence of STM images of a surface charge–induced transition from the disordered to the ordered phase of 2-2′ bipyridine adsorbed on Au(111). At positive charges the molecules stack into polymeric chains, which are initially disordered as shown in (A). As the potential is increased, small ordered domains begin to form and expand as the potential is increased from 0.14 to 0.25 V (B–E). The domains follow the symmetry of the Au(111) substrate. (From Ref. 467.)

also been explored [501]. In this instance deposition was restricted to the film surface by using an electroless deposition process whereby the position of the palladium catalysis was restrained by the terminal amine functionality of the aminothiolate film.

H. Dissolution of Elements and Alloys

The etching or dissolution of metals is a subject of great technical importance ranging from the fabrication of devices to their destruction by corrosion processes. Several atomically resolved STM studies have now been published on the defect and adsorbate sensitivity of etching reactions occurring on Au [131,272,424,479,481,502,503], Cu [277,280,335,423,504, 507], Pd [284–286,505], Ni [287,339], and Ag [283,287,506]. Dissolution of vicinal surfaces at low potentials usually proceeds by step propagation, while etching at higher potentials may involve additional dissolution from terrace sites, e.g., Au(111) in CN solutions [502]. Pit formation as a result of dissolution directly from terrace sites has also been observed for dissolution of Au(111) in tetramethylthiourea solutions [424,481] and Cu(100) in the presence of a benzotriazole/sulfuric acid solution [423]. Less effective inhibition of Cu(100) dissolution by benzotriazole in hydrochloric acid solutions has also been examined [507]. Adsorption of molecules and anions may significantly alter the morphological evolution of the surface during etching. This was powerfully demonstrated by the effect of halide ions on the dissolution of Cu [277,280,335], Pd [284–286], and Ag [283,287], and S^{2-} on Ni [287,339], respectively. In the case of the (100) surface, step propagation occurs predominantly in the ⟨100⟩ direction, which corresponds to the migration of kink-saturated metal steps, which are stabilized in the close-packed direction of the floating $\sqrt{2} \times \sqrt{2}R45°$ halide or sulfide adlayer [280,283,284,287,504]. Adsorption may inhibit or catalyze the dissolution process. An example of the former is provided by thiolate adsorption on Au [131,449,450,502] and Cu [508]. In contrast, the catalytic nature of such adlayers towards metal dissolution was clearly demonstrated in a series of experiments using iodine-modified Pd(111) [284,286,505], Pd(100) [284,505], Pd (110) [284,286] and Ag(100) [283], and sulfur-modified Ni(100) [287,339]. The surface was initially covered with a chemisorbed adlayer, which prevented passivation by oxide species during subsequent experiments performed in adsorbate-free electrolyte. These findings are likely to have a significant impact on the study of localized corrosion processes where sulfide and halide-induced breakdown of the passive state of many transition metals is well known.

Studies of alloy dissolution have largely focused on two systems; Ag-Au [509,510] and Cu-Au [510–512]. The film-free dissolution characteristics of Ag and Cu in combination with the large difference between the reversible potential of these elements and gold make these materials attractive for studying the limiting case of selective dissolution. From a metallurgical point of view, Ag-Au represents a close approximation to an ideal solid solution, while Cu-Au includes the possibility of investigating ordered alloys as well as strain effects associated with dealloying due to the compositional dependence of the lattice parameter. Analysis of these systems is simplified because under certain conditions the oxidation current may be ascribed to selective removal of the more electrochemically active component, while the concurrent enrichment and rearrangement of the noble species may be monitored. At low overpotentials this layer passivates the surface, while above a so-called ''critical potential'' catastrophic global breakdown of the passive surface is observed. While much interest has focused on understanding the nature of this phase transition, STM studies are somewhat limited since the reaction front rapidly becomes inaccessible during the transition from two-dimensional to three-dimensional dealloying. Nevertheless, the dynamics associated with two-dimensional dealloying and passivation are open to study. The observation of alloy formation and dealloying during certain upd reactions represents an opportunity in this respect since a great variety of systems may be studied including some for which the equivalent bulk phase alloys are unknown [341,359,361, 362,408]. In an important development, UHV-STM studies have demonstrated the ability to identify the individual alloy constituent both in the case of ordered and disordered surfaces [63,69,513–515].

I. Surface Modification

During the last decade STM has proven to be a unique tool for the synthesis of novel structures. Perhaps the most elegant demonstration of this was the positioning of individual Xe atoms on Ni(110) with atomic precision in a low-temperature UHV experiment [516]. A variety of structures that exhibit the physics of quantum confinement have been produced in this manner [517], and more recently, the manipulation of individual molecules at room temperature has been demonstrated [518,519]. It is now clear that there are several possible mechanisms for atomic and/or molecular manipulation [520]. Similarly, two reviews of various related schemes for sub-μm surface modification are also available [521,522]. In addition to published studies focused on surface modification, it is worth noting a statement made in a recent overview of STM that "anyone who has performed an scanning probe microscopy experiment has, at one time or another, modified a surface" [523]. This comment reflects the delicate balance between imaging for analytical purposes versus surface modification. However, what might be viewed as an imaging artifact in an analytical study may alternatively represent a potential opportunity for controlled surface modification.

In vacuum studies tip-sample interactions have generally been attributed to the influence of atomic forces, the electric field, and/or the tunneling current [520]. In the first instance, modification results from atomic force interactions between the tip and sample even in the absence of a voltage bias. In the second case, the dominant field-dependent forces derive from the polarizability and dipole moment of the adsorbate. Under conditions of a high junction bias, -10^8 V/cm, the field becomes comparable to that experienced by the valence electrons in atoms and molecules. This results in the redistribution of electrons and thereby changes in bonding, which may lead to charge transfer, field desorption, and/or other alterations. The spatial resolution of the modification method may be optimized by combining chemical and electrostatic forces, whereby the tip is brought close to a species of interest, which reduces the activation energy for desorption. This technique has been coined "chemically assisted field desorption" [520]. Alternatively, surface modification may also result from the very high current density associated with the STM. In this instance modification may be a consequence of vibrational or electronic excitation, which activates local chemical reactions via inelastic tunneling processes. There exist several STM reports of the dependence of electrochemical etching and deposition, as well as surface diffusion, on the imaging conditions, e.g., bias potential [127–131], the explanation for which may lie with one of the mechanisms outlined above.

In addition to the "physical" interactions described above, the tip may also be used to alter the local chemical conditions within the tunnel junction. For example, catalytic rehydrogenation of carbonaceous fragments on Pt(111) by tip-directed production of atomized hydrogen in vacuum at the Pt-Ir tip has been described [524]. Similar modification schemes may also be envisioned based on limiting the transport of reactants and products into or away from the partly occluded tunnel junction. As noted earlier, such effects may be important in the study of electrodeposition and etching process [126–131]. Nonetheless, much remains to be understood about the detailed physics and chemistry of the immersed tunnel junction.

The following discussion will be limited to a brief summary of modification schemes that clearly involve electrochemical reactions, thereby effectively combining concepts associated with STM and SECM [43–45].

In the case of a tunnel junction immersed in an electrolytic solution, two successful tip-directed schemes have been reported for producing small clusters and nanometer-sized wires of metal. In the first method a four-electrode STM is utilized [14,28,525–528]. The material of interest is first concentrated by overpotential deposition on the tip and then subsequently transferred to the substrate by bringing the tip into close proximity to the substrate where a ''jump to contact'' between the tip and sample can occur as outlined in Fig. 35. When the tip is retracted the connecting neck breaks leaving a cluster behind [28,525–530]. The method has been extended to the point that large cluster arrays may be formed, as shown in Fig. 36. Arrays as large as 10^4 Cu clusters on Au(111) have been reported [528]. The amount of material transferred between the tip and substrate is found to be a function of the minimal tip-substrate distance attained [525]. The smaller the tip-substrate separation, the greater the amount of material transferred. However, if the tip is brought too close, holes appear in the substrate. The precise operating conditions for performing the exchange procedure are a function of the feedback characteristic of the instrument

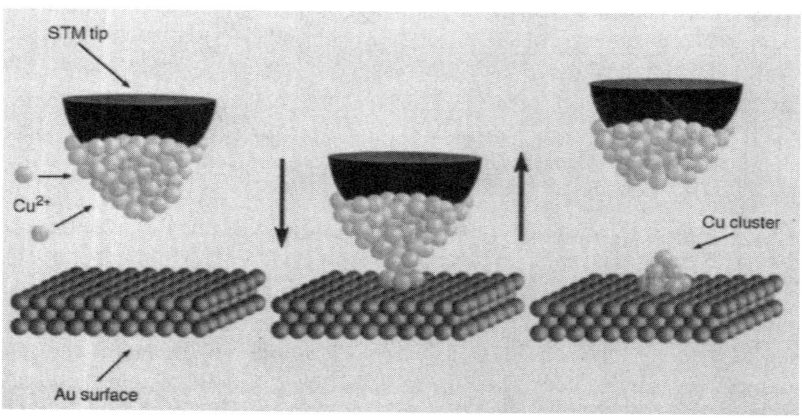

FIG. 35. Schematic diagram of the proposed mechanism of material transfer from a Cu-covered STM tip to the Au substrate induced by an appropriate tip approach toward the substrate. (From Ref. 525.)

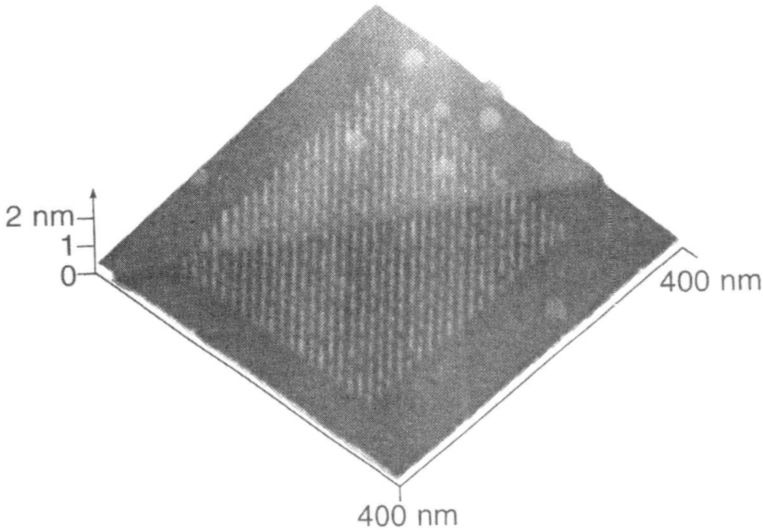

FIG. 36. An array of 400 copper clusters produced on Au(111) in 50 mM H_2SO_4 + 1 mM $CuSO_4$. The sample was held at +10 mV and the tip at −30 mV versus a Cu/Cu^{2+} reference electrode. The clusters were formed by periodically imposing a voltage pulse on the z-piezo, d_z = 0.8 nm for 2.5 ms, while the tip was scanning over the surface using a tunneling current of 2 nA. (From Ref. 525.)

[14,525,526]. The as-deposited clusters were found to be remarkably stable, which was attributed to screening of the substrate by the tip, although stabilization associated with alloying is also a possibility [525,526]. Thus far, deposition of Pb, Pd, Cu, and Ag clusters on Au(111) and Ag(111) have been reported [28,526]. No deposits were observed on HOPG due to the weak interaction between graphite and the selected metals. An interesting variation, or derivative, of this method involves fabricating a narrow metallic constriction between the tip and substrate by electrochemical deposition. By appropriate control of the respective electrodes, nanowires can be formed that exhibit quantum transport characteristics [531].

A second method of tip-directed synthesis involves a two-electrode STM configuration to form small clusters of metals, polymers, and semiconductors on graphite surfaces immersed in a dilute electrolyte [13,532–535]. Initially, the material to be deposited (i.e., Ag) is concentrated by upd on the tip. A positive potential pulse is then applied to the tip such that the desired material is released and diffuses across the tunnel junction where it deposits on the substrate. The resolution of the patterning method is further enhanced by using a short potential pulse to produce a small circular pit in the graphite basal plane, which acts as a catalytic site for electron transfer and cluster nucleation. Pit formation, although not well understood, occurs only in the presence of water reflecting the electrochemical nature of the process [535–537]. In a typical potential pulse program a 6 V:5 μsec pulse is used to generate the circular pit followed by stepping to a lower potential, for tens of μsec, to sustain the metal deposition reaction. The length of the pulse may be used to control the size of the deposited cluster. The silver particles produced by this method are strongly adherent and electrically connected with the graphite substrate [13,532–534]. More recently, the two-electrode method has been used to create holes ~20 Å in diameter using short voltage pulses (−2 V:60 nsec) [535]. The confinement of the modification is thought to be caused by depletion of the ionic concentration within the tunnel junction for which restoration by diffusion takes longer than the duration of the pulse [535].

A third electrochemical modification scheme involves a two-electrode STM operating under humid ambient conditions where a thin layer of water may be adsorbed on the surface, thereby creating a two-electrode electrochemical cell. In fact, a liquid bridge between the tip and substrate can even form spontaneously due to capillary condensation of vapor [439]. In the early years of ambient STM the influence of water adsorbed on surfaces in humid environments [538] was not widely appreciated, although the consequence of electrochemical reactions soon became evident with reports of unanticipated surface modification, particularly during operation under high bias conditions. In comparison to operation of a conventional SECM immersed in an electrolytic solution, the geometry of the cell formed by a thin layer of water under humid ambient conditions further constrains transport phenomena associated with any soluble species and also limits the effective electrochemically active area of the tip and substrate, thereby confining surface modification to the tens of nm range. To date this scheme has been largely used to perform etching and anodization [15,539–546], while the ability to deposit material has received less attention. For example, monolayer deep etching of van der Waals solids like graphite [15], orthorhombic phosphorus [539], gold [540,541], and self-assembled monolayer films on gold [542–544] has been described. The etching process has been shown to occur only above a critical humidity value [540–543]. In the case of gold the critical value of 18 $\pm$ 4% corresponds to a 2-nm-thick layer of water, which is the same order of magnitude as the tip-substrate separation [540]. Thin titanium films have been successfully patterned by anodization [545] where the resolution was found to be a function of the humidity. Similarly, oxidation of chromium has also been examined [546]. Depending on the specific condition, either an insoluble Cr_2O_3 film or a soluble CrO_3 film is grown. Features down to 25-nm linewidths were formed in chromium [546], while even finer dimensions ($\sim$10 nm) have been obtained for anodized lines of titanium [547]. The robust nature of the titanium system has enabled interesting electronic devices to be fabricated. For example, a single electron transistor produced by STM-directed anodization has demonstrated coulomb staircase behavior during operation at room temperature [547]. Similar devices have been fabricated via essentially the same mechanism using an electrically addressable AFM probe [548]. In closing, it is worthwhile drawing attention to recent discussions of the relative merit of STM, in either a serial or parallel processing mode, as a synthesis tool for prototyping versus actual production of devices [522,549].

REFERENCES

1. Papers from the Eighth International Conference on STM/Spectroscopy and Related Techniques, J. Vac. Sci. Technol. B *14*(2): (1996).
2. G. Binnig and H. Rohrer, Helv. Phys. Acta *55*:726 (1982).
3. G. Binnig and H. Rohrer, Rev. Mod. Phys. *56*:615 (1987).
4. G. Binnig, H. Rohrer, C. Gerber, and E. Weibel, Phys. Rev. Lett. *50*:120 (1983).
5. R. Sonnenfeld and P. K. Hansma, Science *232*:211 (1986).
6. H-Y. Liu, F-R. F. Fan, C. W. Lin, and A. J. Bard, J. Am. Chem. Soc. *108*: 3838 (1986).
7. J. Wiechers, T. Twomey, D. M. Kolb, and R. J. Behm, J. Electroanal. Chem. *248*:451 (1988).
8. D. J. Trevor, C. E. D. Chidsey, and D. N. Loiacono, Phys. Rev. Lett. *62*: 929 (1989).
9. O. M. Magnussen, J. Hotlos, R. J. Nichols, D. M. Kolb, and R. J. Behm, Phys. Rev. Lett. *64*:2929 (1990).
10. S. L. Yau, C. M. Vitus, and B. C. Schardt, J. Am. Chem. Soc. *112*:3677 (1990).
11. X. Gao, A. Hamelin, and M. J. Weaver, Phys. Rev. B *44*:10983 (1991).

12. R. J. Nichols, D. M. Kolb, and R. J. Behm, J. Electroanal. Chem. *313*:109 (1991).
13. W. Li, J. A. Virtanen, and R. M. Penner, J. Phys. Chem. *96*:6529 (1992).
14. R. Ullmann, T. Will, and D. M. Kolb, Chem. Phys. Lett. *209*:238 (1993).
15. R. L. McCarley, S. A. Hendricks, and A. J. Bard, J. Phys. Chem. *96*:10089 (1992).
16. H. Sugimura, T. Uchida, N. Kitamura, and H. Masuhara, Jpn. J. Appl. Phys. *32*:L553 (1993).
17. J. A. Dagata, J. Schneir, H. H. Harary, C. J. Evans, M. T. Postek, and J. Bennett, Appl. Phys. Lett. *56*:2001 (1990).
18. M. Dietterle, T. Will, D. M. Kolb, Surf. Sci. *327*:L495 (1995).
19. N. J. Tao, Phys. Rev. Lett. *76*:4066 (1996).
20. A. J. Bard, H. D. Abruna, C. E. Chidsey, L. R. Faulkner, S. W. Feldberg, K. Itaya, M. Majda, W. O. Melroy, R. W. Murray, M. D. Porter, M. P. Soriaga, and H. S. White, J. Phys. Chem. *97*:7147 (1993).
21. *Electrochemical Interfaces, Modern Techniques for In Situ Interface Characterization* (H. D. Abruna, ed.), VCH Publishers, Inc., New York, 1991.
22. *Structure of Electrified Interfaces* (J. Lipkowski and P. N. Ross, eds.), VCH Publishers, Inc., New York, 1993.
23. *Adsorption of Molecules at Metal Electrodes*, (J. Lipkowski and P. Ross, eds.), VCH Publishers, Inc., New York, 1992.
24. T. Hager, *Forces of Nature*, Simon and Schuster, New York, 1995, p. 86.
25. T. R. I. Cataldi, I. G. Blackham, G. A. D. Briggs, J. Pethica, and H. A. O. Hill, J. Electroanal. Chem. *290*:1 (1990).
26. H. Siegenthaler, STM in electrochemistry, in *Scanning Tunneling Microscopy II* (R. Wiesendanger and H.J. Gunterodt, eds.), Springer-Verlag, New York, 1992.
27. M. J. Weaver and X. Gao, Annu. Rev. Phys. Chem. *44*:459 (1993).
28. A. S. Dakkouri, M. Dietterle, and D. M. Kolb, Festkörperprobleme/Advances in Solid State Physics, Vol. 36, ed. by R. Helhig (Vieweg, Braunschweig/Wiesbaden) 1997.
29. A. A. Gewirth and B. K. Niece, Chem. Rev *97*:1129 (1997).
30. *In-Situ Local Probe Techniques for Studies of Electrochemical Interfaces* (W. J. Lorenz and W. Plieth, eds.), VCH-Wiley, New York, 1998.
31. Proceedings of the IUVSTA Workshop on Surface Science and Electrochemistry, Surf. Sci. *335*: (1995).
32. *The Liquid/Solid Interface at High Resolution*, The Royal Society of Chemistry, London, 1992.
33. *Nanoscale Probes of the Solid/Liquid Interface* (A. A. Gewirth and H. Siegenthaler, eds.), NATO ASI Series E, Vol. 288, Kluwer Academic Publishers, Boston, 1995.

34. *Scanning Tunneling Microscopy and Related Methods* (R. J. Behm, N. Garcia, and H. Rohrer, eds.), NATO ASI Series E, Vol. 184, Kluwer Academic Publishers, Boston, 1990.
35. Proceedings of 6th Int'l Fischer Symp. on Nanotechniques in Electrochemistry, Electrochim Acta *40*:10 (1995).
36. Electrolytic Metal Deposition: Fundamental Aspects and Applications (A. S. Dakkouri and D. M. Kolb, eds.) 5. Ulmer Electrochemische Tage, R. Uldenbourg Verlag, München (1999). This is reprinted from Zeitschrift für Physikalische Chemie, Bd. 208 (1999) by R. Oldenbourg Verlag, München, 1999; Surface structure and electrochemical reactivity, Electrochim. Acta *40*:1 (1995).
37. P. Allongue, *Advances in Electrochemical Science and Engineering* (H. Gersischer and C. W. Tobias, eds.), VCH Publishers, New York, 1995.
38. P. K. Hansma and J. Tersoff, J. Appl. Phys. *61*:R1 (1987).
39. C. J. Chen, *Introduction to Scanning Tunneling Microscopy*, Oxford University Press, New York, 1993; R. Wiesendanger, Scanning Probe Microscopy and Spectroscopy: Methods and Applications, Cambridge University Press (1994).
40. J. K. Gimzewski and R. Moller, Phys. Rev. B *36*:1284 (1987).
41. L. Escapa and N. Garcia, in *Scanning Tunneling Microscopy and Related Methods* (R. J. Behm, N. Garcia, and H. Rohrer, eds.), NATO ASI Series E, Vol. 184, Kluwer Academic Publishers, Boston, 1990.
42. A. Yazadani, D. M. Eigler, and N. D. Lang, Science *272* 1921 (1996).
43. A. J. Bard, F-R. F. Fan, and M. V. Mirkin, *Electroanalytical Chemistry*, Vol. 18 (A. J. Bard, ed.), 1994. Marcel Dekker, Inc., N.Y., N.Y.
44. A. J. Bard, F-R. F. Fan, and M. V. Mirkin, *Physical Electrochemistry* (I. Rubinstein, ed.), Marcel Dekker, New York, 1995.
45. M. V. Mirkin, F-R. F. Fan, and A. J. Bard, Science *257*:364 (1992).
46. S. R. Snyder and H. S. White, J. Electroanal. Chem. *394*:177 (1995).
47. A. J. Bard and F-R. F. Fan, Acc. Chem. Res. *29*:572 (1996).
48. S. P. Sharma and J. H. Thomas, J. Vac. Sci. Techol. *14*:825 (1977).
49. M. Salmeron, L. Xu, J. Hu, and Q. Dai, MRS Bull. *22*:8, 36 (1997).
50. F-R. F. Fan and A. J. Bard, Science *270*:1850 (1995).
51. R. Guckenberger, M. Heim, G. Ceve, H. F. Knapp, W. Wiegrabe, and A. Hillebrand, Science *266*:1538 (1994).
52. M. Heim, R. Eschrich, A. Hillebrand, H. F. Knapp, R. Guckenberger, and G. Cevc, J. Vac. Sci. Technol. B *14*:1498 (1996).
53. N. Patel, M. C. Davies, M. Lomas, C. J. Roberts, S. J. B. Tendler, and P. M. Williams, J. Phys. Chem. B *101*:5138 (1997).
54. H. Ibach, C. E. Bach, M. Giesen, and A. Grossmann, Surf. Sci. *375*:107 (1997).
55. W. Haiss and J-K. Sass, Langmuir *12*:4311 (1996).

56. S. Ciraci, E. Tekman, A. Baratoff, and I. P. Batra, Phys. Rev. B *46*:10411 (1992).
57. *Scanning Tunneling Microscopy III* (R. Wisendanger and H. J. Guntherodt, eds.), Springer-Verlag, New York, 1993.
58. P. Sautet, Chem. Rev. *97*:1097 (1997).
59. C. J. Chen, J. Vac. Sci. Tech. A*9*:44 (1991).
60. J. Wintterlin, J. Wiechers, H. Brune, T. Gritsch, H. Hofer, and R. J. Behm, Phys. Rev. Lett. *62*:59 (1989).
61. C. J. Chen, Phys. Rev. Lett. *65*:448 (1990).
62. T. Kuk, in *Scanning Tunneling Microscopy I* (H. J. Gunterodt and R. Wiesendanger, eds.), Springer-Verlag, New York, 1992.
63. A. Davies, J. A. Stroscio, D. T. Pierce, and R. J. Celotta, Phys. Rev. Lett. *76*:4175 (1996).
64. S. Ciraci, A. Baratoff, and I. P. Batra, Phys. Rev. B *41*:2763 (1990).
65. S. Ciraci, A. Baratoff, and I. P. Batra, Phys. Rev. B *42*:7618 (1990).
66. U. Durig, O. Zuger, and D. W. Pohl, Phys. Rev. Lett. *65*:349 (1990).
67. S. N. Magonov and M-H. Whangbo, *Surface Analysis with STM and AFM, Experimental and Theoretical Aspects of Image Analysis*, VCH, New York, 1996.
68. R. J. Behm, in *Scanning Tunneling Microscopy and Related Methods* (R. J. Behm, N. Garcia and H. Rohrer, eds.), NATO ASI Series E, Vol. 184, Kluwer Academic Publishers, Boston, 1990.
69. J. A. Stoscio, D. T. Pierce, A. Davies, and R. J. Celotta, Phys. Rev. Lett. *75*(16):2960 (1995).
70. P. Sautet, Surf. Sci. *374*:406 (1997).
71. D. M. Eigler, P. S. Weiss, E. K. Schweizer, and N. D. Lang, Phys. Rev. Lett. *66*:1189 (1991).
72. N. D. Lang, Phys. Rev. Lett. *55*:230 (1985)
73. N. D. Lang, Phys. Rev. Lett. *58*:45 (1987).
74. N. D. Lang, Phys. Rev. Lett. *56*:1164 (1986).
75. N. D. Lang, Phys. Rev. B *34*:5947 (1986).
76. U. Mazur and K. W. Hipps, J. Phys. Chem. *99*:6684 (1995).
77. R. Guckenberger, T. Hartmann, W. Wiegrabe, and W. Baumeister, in *Scanning Tunneling Microscopy II* (R. Wisendanger and H. J. Guntherodt, eds.), Springer-Verlag, New York, 1992.
78. C. Joachim and P. Sautet, in *Scanning Tunneling Microscopy and Related Methods* (R. J. Behm, N. Garcia, and H. Rohrer, eds.), NATO ASI Series, Norwell, MA, 1990.
79. S. Richter and Y. Manassen, J. Phys. Chem. *98*:2941 (1994).
80. S. Chiang, in *Scanning Tunneling Microscopy I* (H. J. Gunterodt and R. Wiesendanger, eds.), Springer-Verlag, New York, 1992.
81. D. M. Cyr, B. Venkataraman, G. W. Flynn, A. Black, and G. M. Whitesides, J. Phys. Chem. *100*:13747 (1996).

82. C. L. Clypool, F. Faglioni, W. A. Goddard, H. B. Gray, N. S. Lewis, and R. A. Marcus, J. Phys. Chem. B *101*:5978 (1997).
83. X. Lu and K. W. Hipps, J. Phys. Chem. B *101*:5391 (1992).
84. S. M. Lindsey, O. F. Sankey, Y. Li, C. Herbst, and A. Rupprecht, J. Phys. Chem. *94*:4655 (1990).
85. U. Durig, O. Zuger, B. Michel, L. Haussling, and H. Ringsdorf, Phys. Rev. B *48*:1711 (1993).
86. R. C. Jaklevic and J. Lambe, Phys. Rev. Lett. *17*:1139 (1966).
87. J. L. Tomlin, Prog. Surf. Sci. *31*:131 (1989).
88. R. Berndt, J. K. Gimzewski, and P. Johansson, Phys. Rev. Lett. *67*:3791 (1991).
89. R. Berndt, J. K. Gimzewski, and R. R. Schlittler, Ultramicroscopy *42–44*: 355 (1992).
90. R. Berndt, R. Gaisch, J. K. Gimzewski, B. Reihl, R. R Schlittler, W. D. Schneider, and M. Tschudy, Science *262*:1425 (1993).
91. I. I. Smolyaninov, M. S. Khaikin, and V. S. Edelman, Phys. Lett. A *149*: 410 (1990).
92. R. Christoph, H. Siegenthaler, H. Rohrer, and H. Wiese, Electrochem. Acta *34*:1011 (1989).
93. M. Binggeli, D. Carnal, R. Nyffenegger, and H. Siegenthaler, J. Vacuum Sci. Technol. B *9*:1985 (1991).
94. J. Halbritter, G. Repphun, S. Vinzelberg, G. Staikov, and W. J. Lorenz, Electrochem. Acta *40*:1385 (1995).
95. W. Schmickler and D. Henderson, J. Electroanal. Chem. *290*:283 (1990).
96. J. K. Sass and J. K. Gimzewski, J. Electroanal. Chem. *308*:333 (1991).
97. Y. V. Pleskov and Z. A. Rotenberg, in *Advances in Electrochemistry and Electrochemical Engineering*, Vol. 11 (H. Gerischer and C. W. Tobias, eds.), Wiley, New York, 1978.
98. K. L. Sebatian and G. Doyen, J. Chem. Phys. *99*:6677 (1993).
99. K. I. Sebastian and G. Doyen, Surf. Sci. *290*:L703 (1993)
100. D. Rostkier-Edelstein, M. Urbakh, and A. Nitzan, J. Chem. Phys. *101*:8224 (1994).
101. A. A. Kornyshev, J. Electroanal. Chem. *376*:9 (1994).
102. W. Schmickler, Surf. Sci. *335*:416 (1995).
103. O. Pecina, W. Schmickler, K. Y. Chan, and D. J. Henderson, J. Electroanal. Chem. *396*:303 (1995).
104. A. Vaught, T. W. Jing, and S. M. Lindsay, Chem. Phys. Lett. *236*:306 (1995).
105. O. M. Magnussen, J. Hagebock, J. Hotlos, and R. J. Behm, Farday Discuss. *94*:329 (1992).
106. G. J. Edens, X. Gao, and M. J. Weaver, J. Electroanal. Chem. *375*:357 (1994).
107. X. Gao and M. J. Weaver, J. Electroanalytical Chem. *367*:259 (1994).
108. K. Ataka and M. Osawa, Langmuir *14*:951 (1998).

109. P. Thiel and T. E. Madey, Surf. Sci. Rep. *7*:211 (1987).
110. W. Schmickler, in *Nanoscale Probes of the Solid/Liquid Interface* (A. A. Gewirth and H. Siegenthaler, eds.), NATO ASI Series E, Vol. 288, Kluwer Academic Publishers, Boston, 1995.
111. X. Gao, A. Hamelin, and M. J. Weaver, J. Chem. Phys. *95*:6993 (1991).
112. J. D. Porter and A. S. Zinn, J. Phys. Chem. *97*:1190 (1993).
113. M. F. Toney, J. N. Howard, J. Richer, G. L. Borges, J. G. Gordon, O. R. Melroy, D. G. Wiesler, D. Yee, and L. B. Sorensen, Nature *368*:444 (1994).
114. K. J. Schweighofer, X. Xia, and M. L. Berkowitz, Langmuir *12*:3747 (1996).
115. I. Danielewicz-Ferchmin and A. R. Ferchmin, J. Phys. Chem. *100*:17281 (1996).
116. D. L. Patrick and R. M. Lyden-Bell, Surf. Sci. *380*:224 (1997).
117. E. Weilandt, A. Menck, M. Binggeli, and O. Marti, in *Nanoscale Probes of the Solid/Liquid Interface* (A. A. Gewirth and H. Siegenthaler, eds.), NATO ASI Series E, Vol. 288, Kluwer Academic Publishers, Boston, 1995.
118. H-J. Butt, Biophys. J. *60*:1438 (1991).
119. H-J. Butt, Nanotechnology *3*:60 (1992).
120. Y. Q. Li, N. J. Tao, J. Pan, A. A. Garcia, and S. M. Lindsey, Langmuir *9*: 637 (1993).
121. T. J. Seden, C. J. Drummond, and P. Kelicheff, Langmuir *10*:358 (1994).
122. A. C. Hillier, S. Kim, and A. J. Bard, J. Phys. Chem. *100*:18808 (1996).
123. T. Arai and M. Fujihira, J. Vac. Sci. Technol. B *14*:1378 (1996).
124. D. Henderson and K-Y. Chan, J. Electroanal. Chem. *330*:395 (1992).
125. S. M. Lindsay, T. W. Jing, J. Pan, D. Lampner, A. Vaught, J. P. Lewis, and O. F. Sankey, in *Nanoscale Probes of the Solid/Liquid Interface* (A. A. Gewirth and H. Siegenthaler, eds.), NATO ASI Series E, Vol. 288, Kluwer Academic Publishers, Boston, 1995.
126. N. Batina and D. Kolb, in *The Liquid/Solid Interface at High Resolution* The Royal Society of Chemistry, London, 1992, p. 238.
127. X-P. Gao and M. J. Weaver, J. Electroanal. Chem. *367*:259 (1994).
128. N. Breuer, U. Stimming, and R. Vogel, Electrochim. Acta *40*:1401 (1995).
129. S. A. Hendricks, Y. T. Kim, and A. J. Bard, J. Electrochem. Soc. *139*:2818 (1992).
130. A. M. Bittner, J. Wintterlin, and G. Ertl, Surf. Sci. *376*:267 (1997).
131. Y-Q. Li, O. Chailapakul, and R. M. Crooks, J. Vac. Sci. Tech. B *13*:1300 (1995).
132. J. Divisek, B. Steffer, U. Stimming, and W. Schmickler, J. Electroanal. Chem. *440*:169 (1997).
133. J. F. T. Conroy, K. Caldwell, C. J. Bruckner-Lea, and J. Janata, J. Phys. Chem. *100*:18222 (1996).
134. J. F. T. Conroy, V. Hlady, C. J. Bruckner-Lea, and J. Janata, J. Phys. Chem. *100*:18229 (1996).

135. J. Buisset, H.-P. Rust, E. K. Schweizer, L. Cramer, and A. M. Bradshaw, J. Vac. Sci. Technol. B *14*:1117 (1996).
136. J. H. Schott and H. S. White, Langmuir *8*:1955 (1992).
137. W. Schmickler, J. Electroanal. Chem. *296*:283 (1990).
138. W. Schmickler and J. W. Schultze, *Modern Aspects of Electrochemistry*, (J. O'M Bockris, B. E. Conway, and R. E. White, eds.), Plenum Press, New York, 1986.
139. A. M. Kuznetsov, P. Sommer-Larsen, and J. Ulstrup, Surf. Sci. *275*:52 (1992).
140. W. Schmickler and C. Widrig, J. Electroanal. Chem. *336*:213 (1992).
141. J. E. T. Andersen, A. A. Kornyshev, A. M. Kuznetsov, L. L. Madsen, P. Moller, and J. Ulstrup, Electrochim. Acta *42*:819 (1997).
142. E. P. Friis, J. E. T. Andersen, L. L. Madsen, P. Moller, R. J. Nichols, K. G. Olesen, and J. Ulstrup, Electrochim. Acta *43*:2889 (1998).
143. W. Schmickler and N. J. Tao, Electrochem. Acta *42*:2809 (1997).
144. W. Schmickler, Surf. Sci. *295*:43 (1993).
145. J. Clavilier, R. Faure, G. Guinet, and R. Dunand, J. Electroanal. Chem. *107*: 205 (1980).
146. J. Clavilier, J. Electroanal. Chem. *107*:211 (1980).
147. A. Hamelin, *Modern Aspects of Electrochemistry*, Vol. 16 (B. E. Conway, R. E. White, and J. O'M. Bockris, eds.), Plenum Press, New York, 1985.
148. T. Will, M. Dietterle, and D. M. Kolb, in *Nanoscale Probes of the Solid/ Liquid Interface* (A. A. Gewirth and H. Siegenthaler, eds.), Kluwer Academic, Boston, 1995, p. 137.
149. C. P. Wade, H. Luo, W. L. Dunbar, M. R. Linford and C. E. D. Chidsey, MRS Symp. Proc. V 451, MRS, Pittsburgh, 1997.
150. F. T. Wagner, Structure of Electrified Interfaces (J. Lipkowski and P. N. Ross, eds.), VCH, New York, 1993, p. 309.
151. W. Schmidt, Ph.D. thesis, Dept. of Chemical Engineering, University of Illinois, Urbana-Champaign, IL, 1994.
152. W. Schmidt and R. C. Alkire, J. Electrochem. Soc. *141*:L85 (1994).
153. Q. Hong, M. F. Suarez, B. A. Coles, and R. G. Compton, J. Phys. Chem. B *101*:5557 (1997).
154. L. H. Jenkins and U. Bertocci, J. Electrochem. Soc. *112*:517 (1965).
155. R. J. Nichols, W. Beckmann, H. Meyer, N. Batina, and D. M. Kolb, J. Electroanal. Chem. *330*:381 (1992).
156. H. Dai, J. H. Hafner, A. G. Rinzler, D. T. Colbert, and R. E. Smalley, Nature *384*:147 (1996).
157. A. G. Rinzler, J. H. Hafner, P. Nikolaev, L. Lou, S. G. Kim, D. Tomanek, P. Nordlander, D. T. Colbert, and R. E. Smalley, Science *269*:1550 (1995).
158. W. A. de Heer, A. Chatelain, and D. Ugarte, Science *270*:1179 (1995).

159. W. Haiss, D. Lackey, J. K. Sass, and K. H. Besocke, J. Chem. Phys. *95*: 2193 (1991).
160. A. J. Melmed, J. Vac. Sci. Technol. B *9*:601 (1991).
161. L. A. Nagahara, T. Thundat, and S. M. Lindsay, Rev. Sci. Instrum. *60*:3128 (1989).
162. B. Zhang and E. Wang, Electrochim. Acta *39*:103 (1994).
163. C. E. Bach, R. J. Nichols, W. Beckmann, H. Meyer, A. Schulte, J. O. Besenhard, and P. D. Jannakoudakis, J. Electrochem. Soc. *140*:1281 (1993).
164. K. Itaya, T. Abe, and I. Uchida, J. Electrochem. Soc. *134*:1191 (1987).
165. R. Penner, M. J. Heben, T. L. Longin, and N. S. Lewis, Science *250*:1118 (1990).
166. S-L. Yau, C. M. Vitus, and B. C. Schardt, J. Am. Chem. Soc. *112*:3677 (1990).
167. K. F. Kelly, D. Sapkar, G. D. Hale, S. J. Oldenberg, and N. J. Halas, Science *273*:1371 (1996).
168. C. D. Frisbie, L. F. Rozsnyai, A. Noy, M. S. Wrighton, and C. M. Lieber, Science *265*:2071 (1994).
169. K. Sashikata, N. Furuya, and K. Itaya, J. Vac. Sci. Technol. B*9*:457 (1991).
170. C. E. D. Chidsey, D. N. Loiacono, T. Sleator, and S. Nakahara, Surf. Sci. *200*:45 (1988).
171. A. A. Baski and H. Fuchs, Surf. Sci. *313*:275 (1994).
172. G. L. Borges, M. G. Samant, and K. Ashley, J. Electrochem. Soc. *139*:1565 (1992).
173. *Epitaxial Growth* (J. W. Matthews, ed.), Academic Press, New York, 1975.
174. R. Piontelli, G. Poli, and G. Serravalle, *Transactions of the Symposium on Electrode Processes* (E. Yeager, ed.), The Electrochemical Society, John Wiley & Sons, 1959.
175. U. Bertocci and C. Bertocci, J. Electrochem. Soc. *118*:1287 (1971).
176. M. Hopfner, W. Obretenov, K. Juttner, W. J. Lorenz, G. Staikov, V. Bostanov, and E. Budevski, Surf. Sci. *248*:225 (1991).
177. W. Obretenov, M. Hopfner, W. J. Lorenz, E. Budevski, G. Staikov, and H. Siegenthaler, Surf. Sci. *271*:191 (1992).
178. M. Dietterle, T. Will, and D. M. Kolb, Surf. Sci. *342*:29 (1995).
179. R. Gomez and M. J. Weaver, J. Electroanal. Chem. *435*:205 (1997).
180. N. Batina, A. S. Dakkouri, and D. M. Kolb, J. Electroanal. Chem. *370*:87 (1994).
181. D. Zurawski, L. Rice, M. Hourani, and A. Wieckowski, J. Electroanal. Chem. *230*:221 (1987).
182. C. M. Vitus, S-C. Chang, B. C. Schardt, and M. J. Weaver, J. Phys. Chem. *95*:7559 (1991).
183. X. Jiang and M. J. Weaver, Surf. Sci. *275*:237 (1992).
184. K. Wang, G. S. Chottiner, and D. A. Scherson, J. Phys. Chem. *97*:10108 (1993).

185. D. Landolt, Electrochim. Acta *32*:1 (1987).
186. W. J. McTegart, *The Electrolytic and Chemical Polishing of Metals*, Pergamon Press, London, 1956.
187. R. Vidal and A. C. West, J. Electrochem. Soc. *142*:2682, 2689 (1995).
188. G. M. Brisard, E. Zenati, H. A. Gasteiger, N. N. Markovic, and P. N. Ross, Langmuir *11*:2221 (1995).
189. I. Villegas, C. B. Ehlers, and J. L. Stickney, J. Electrochem. Soc. *137*:3143 (1990).
190. R. R. Adzic, M. E. Hanson, and E. B. Yeager, J. Electrochem. Soc. *131*: 1730 (1984).
191. C. Gunther, S. Gunther, E. Kopatzki, R. Q. Hwang, J. Schroder, J. Vrijmoeth, and R. J. Behm, Ber. Bunsenges. Phys. Chem. *97*:522 (1993).
192. J. F. Wolf, B. Vicenzi, and H. Ibach, Surf. Sci. *249*:233 (1991).
193. M. Giesen-Seibert, F. Schmitz, R. Jentjens, and H. Ibach, Surf. Sci. *329*:47 (1995).
194. N. Batina, T. Will, and D. M. Kolb, *The Liquid/Solid Interface at High Resolution*, The Royal Society of Chemistry, London, 1992, p. 93.
195. B. S. Swartzentruber, Phys. Rev. Lett. *76*:459 (1996).
196. E. D. Williams and N. C. Bartelt, Science *251*:393 (1991).
197. M. Poensgen, J. F. Wolf, J. Frahn, M. Giesen, and H. Ibach, Surf. Sci. *274*: 430 (1992).
198. A. Pimpinelli, J. Villain, D. E. Wolf, J. J. Metois, J. C. Heyraud, I. Elkinani, and G. Uimin, Surf. Sci. *295*:143 (1993).
199. N. C. Bartelt, T. L. Einstein, and E. D. Williams, Surf. Sci. *312*:411 (1994).
200. M. Giesen-Seibert and H. Ibach, Surf. Sci. *316*:205 (1994).
201. M. Giesen, Surf. Sci. *370*:55 (1997).
202. A-L. Barabasi and H. E. Stanley, *Fractal Concepts in Surface Growth*, Cambridge University Press, New York, 1995.
203. W. M. Tong and R. Stanley Williams, Annu. Rev. Phys. Chem. *45*:401 (1994).
204. H. Ibach, M. Giesen, T. Flores, M. Wuttig, and G. Treglia, Surf. Sci. *364*: 453 (1996).
205. J. W. Evans and M. C. Bartelt, Langmuir *12*:217 (1996).
206. W. U. Schmidt, R. C. Alkire, and A. A. Gewirth, J. Electrochem. Soc. *143*: 3122 (1996).
207. M. Giesen, M. Dietterle, D. Stapel, H. Ibach, and D. M. Kolb, *Electrochemical Synthesis and Modification of Materials*, Mat. Res. Soc. Symp., MRS, Pittsburgh, 1997, p. 9.
208. M. Giesen, M. Dietterle, D. Stapel, H. Ibach, and D. M. Kolb, Surf. Sci. 384:168 (1997).
209. G. Aloisi, A. M. Funtikov, and R. Guidelli, Surf. Sci. *296*:291 (1993).
210. D. M. Kolb, Prog. Surf. Sci. *51*:2, 109 (1996).
211. D. M. Kolb, in *Frontiers in Electrochemistry*, Vol. 2 *Structure of Electrified*

Interfaces (J. Lipkowski and P. N. Ross, eds.), VCH Publishers, Inc., New York, 1993, p. 65.
212. R. C. Cammarata, Prog. Surf. Sci. *46*:1 (1994).
213. K. P. Bohnen and D. M. Kolb, Surf. Sci. *407*:L629 (1998).
214. O. M. Magnussen, J. Wiechers, and R. J. Behm, Surf. Sci. *289*:139 (1993).
215. W. Polewska, C. M. Vitus, B. M. Ocko, and R. R. Adzic, J. Electroanal. Chem. *364*:265 (1994).
216. D. M. Kolb, A. S. Dakkouri, and N. Batina, Nanoscale Probes of the Solid/Liquid Interface (A. A. Gewirth and H. Siegenthaler, eds.), NATO ASI Series E, Vol. 288, Kluwer Academic Publishers, Boston, 1995.
217. O. M. Magnussen, J. Hotlos, R. J. Behm, N. Batina, and D. M. Kolb, Surf. Sci. *296*:310 (1996).
218. X. Gao, G. J. Edens, A. Hamelin, and M. Weaver, Surf. Sci. *296*:333 (1993).
219. X. Gao and M. Weaver, J. Phys. Chem. *97*:8685 (1993).
220. Z. Gai, Y. He, X. Li, J. F. Jia, and W. S. Wang, Surf. Sci. *365*:96 (1996).
221. X. Gao, A. Hamelin, and M. J. Weaver, Phys. Rev. Lett. *67*:618 (1991).
222. X. Gao, A. Hamelin, and M. J. Weaver, Phys. Rev. B *46*:7096 (1992).
223. G. J. Edens, X. Gao, M. J. Weaver, N. M. Markovic, and P. N. Ross, Surf. Sci. Lett. *302*:L275 (1994).
224. F. Sette, T. Hashizume, F. Comin, A. A. MacDowell, and P. H. Citrin, Phys. Rev. Lett. *61*:1384 (1988).
225. X. Gao, A. Hamelin, and M. J. Weaver, J. Chem. Phys. *95*:6993 (1991).
226. N. J. Tao and S. M. Lindsay, J. Appl. Phys. *70*:5141 (1991).
227. N. J. Tao and S. M. Lindsey, Surf. Sci. *274*:L546 (1992).
228. X. Gao, A. Hamelin, and M. J. Weaver, Phys. Rev. B. *44*:10983 (1991).
229. X. Gao and M. J. Weaver, Surf. Sci. *313*:L775 (1994).
230. X. Gao, G. J. Edens, and M. J. Weaver, J. Phys. Chem. *98*:8074 (1994).
231. X. Gao, G. J. Edens, A. Hamelin, and M. J. Weaver, Surf. Sci. *318*:1 (1994).
232. X. Gao, A. Hamelin, and M. J. Weaver, Surf. Sci. *274*:L588 (1992).
233. A. M. Bittner, J. Wintterlin, and G. Ertl, J. Electroanal. Chem. *388*:225 (1995).
234. M. S. Zei, N. Batina, and D. M. Kolb, Surf. Sci. *306*:L519 (1994).
235. H. Baltruschat, U. Bringemeier, and R. Vogel, Faraday Discuss. *94*:317 (1992).
236. B. E. Conway, Prog. Surf. Sci. *49*:331 (1995).
237. K. Itaya, S. Sugawara, K. Sashikata, and N. Furuya, J. Vac. Sci. Technol. *A8*:515 (1990).
238. H. Honbo, S. Sugawara, and K. Itaya, Anal. Chem. *62*:2424 (1990).
239. C. M. Vitus and A. J. Davenport, J. Electrochem. Soc. *141*:1291 (1994).
240. R. J. Nichols, O. M. Magnussen, J. Hotlos, T. Twomey, R. J. Behm, and D. M. Kolb, J. Electroanal. Chem. *290*:21 (1990).

241. L. Vazquez, J. Gomez, A. M. Baro, N. Garcia, M. L. Marcos, J. Gonzalez Velasco, J. M. Vara, A. J. Arvia, J. Presa, A. Garcia, and M. Aguilar, J. Am. Chem. Soc. *109*:1730 (1987).
242. A. J. Arvia, R. C. Salvarezza, and W. E. Triaca, Electrochim. Acta *34*:1057 (1989).
243. L. Vazquez, A. Bartolome, A. M. Baro, C. Alonso, R. C. Salvarezza, and A. J. Arvia, Surf. Sci. *215*:171 (1989).
244. J. M. Dona and J. Gonzalez-Velasco, J. Phys. Chem. *97*:4714 (1993).
245. G. Andreasen, M. Nazzarro, J. Ramirez, R. C. Salvarezza, and A. J. Arvia, J. Electrochem. Soc. *143*:466 (1996).
246. H. Brune, J. Wintterlin, J. Trost, G. Ertl, J. Wiechers, and R. J. Behm, J. Chem. Phys. *99*:2128 (1993).
247. M. Baumer, D. Cappus, H. Kulenbeck, H. J. Freund, G. Wilhelmi, A. Brodde, and H. Neddermeyer, Surf. Sci. *253*:116 (1991).
248. L. Eierdal, F. Besenbacher, E. Laegsgaard, and I. Stensgaard, Surf. Sci. *312*: 31 (1994).
249. M. P. Ryan, R. C. Newman, and G. E. Thompson, J. Electrochem. Soc. *142*: L177 (1995).
250. P. Marcus, H. Talah, and V. Maurice, in Proc. of the 12th International Corrosion Congress, NACE, Houston, 1993, p. 2105.
251. V. Maurice, H. Talah, and P. Marcus, Surf. Sci. *284*:L431 (1993).
252. V. Maurice, H. Talah, and P. Marcus, Surf. Sci. *304*:98 (1994).
253. T. Suzuki, T. Yamada, and K. Itaya, J. Phys. Chem. *100*:8954 (1996).
254. S-L. Yau, F-R. F. Fan, T. P. Moffat, and A. J. Bard, J. Phys. Chem. *98*:5493 (1994).
255. C. Vitus and A. J. Davenport, in Proc. of the 12th International Corrosion Congress, NACE, Houston, 1993, p. 2091.
256. V. Maurice, W. P. Yang, and P. Marcus, J. Electrochem. Soc. *141*:3016 (1994).
257. T. P. Moffat, F-R. F. Fan, S. L. Yau, and A. J. Bard, Proc. of the H. H. Uhlig Memorial Symposium, Corrosion Monograph Series, V94-26, (F. Mansfeld, A. Asphahani, H. Bohni, and R. Latanision, eds.), The Electrochemical Society, Inc., Pennington, NJ, 1995, p. 76.
258. P. Marcus, W. Yang, and V. Maurice, Proc. of the H. H. Uhlig Memorial Symposium, Corrosion Monograph Series, V94-26, (F. Mansfeld, A. Asphahani, H. Bohni, and R. Latanision, eds.), The Electrochemical Society, Inc. Pennington, NJ, 1995, p. 102.
259. V. Maurice, W. P. Yang, and P. Marcus, J. Electrochem. Soc. *143*:1182 (1996).
260. M. P. Ryan, R. C. Newman, and G. E. Thompson, Phil. Mag. B *70*:241 (1994).
261. M. P. Ryan, R. C. Newman, and G. E. Thompson, J. Electrochem. Soc. *141*: L164 (1994).

262. P. Marcus, W. Yang, and V. Maurice, in *Surface Oxide Films* (J. A. Bardwell, ed.), PV 96-18, The Electrochemical Society, Pennington, NJ, 1997.
263. A. A. Gewirth and A. J. Bard, J. Phys. Chem. *92*:5563 (1988).
264. R. Garcia, J. J. Saenz, and N. Garcia, Phys. Rev. B *33*:4439 (1986).
265. T. Bertrams, A. Brodde, and H. Neddermeyer, J. Vac. Sci. Technol. B *12*: 2122 (1994).
266. M. C. Gallagher, M. S. Fyfield, J. P. Cowin, and S. A. Joyce, Surf. Sci. *339*: L909 (1995).
267. A. Schreyer, L. Eng, and H. Bohni, J. Vac. Sci. Technol. B *14*:1162 (1996).
268. N. Casillas, S. R. Snyder, W. H. Smyrl, and H. S. White, J. Phys. Chem. *95*:7002 (1995).
269. K. Azumi, K. Araki, and M. Seo, J. Electroanal. Chem. *427*:15 (1997).
270. Y. Zhu and D. E. Williams, J. Electrochem. Soc. *143*:L43 (1997).
271. J. W. Still and D. O. Wipf, J. Electrochem. Soc. *144*:2657 (1997).
272. O. M. Magnussen, B. M. Ocko, J. X. Wang, and R. R. Adziz, J. Phys. Chem. *100*:5500 (1996).
273. X. Wang, R. Chen, Y. Wang, T. He, and F-C. Lin, J. Phys. Chem. B *102*: 7568 (1998).
274. M. F. Toney, J. G. Gordon, G. L. Borges, O. R. Melroy, D. Yee, and L. B. Sorensen, Phys. Rev. B *49*:7793 (1994).
275. R. L. McCarley, Y-T. Kim, and A. J. Bard, J. Phys. Chem. *97*:211 (1993).
276. T. P. Moffat, in MRS Symp. Proc. Vol. 451, MRS, Pittsburgh, 1997.
277. D. Wayne Suggs and A. J. Bard, J. Phys. Chem. *99*:8349 (1995).
278. T. P. Moffat, in Nanostructured Materials in Electrochemistry, (P. Searson and J. Meyer, eds.), PV 95-8, The Electrochemical Society, Inc., Pennington, NJ, 1995, p. 225.
279. T. P. Moffat, Mat. Res. Soc. Symp. Vol. 404, MRS, Pittsburgh, 1996, p. 3.
280. M. R. Vogt, F. A. Mollers, C. M. Schilz, O. M. Magnussen, and R. J. Behm, Surf. Sci. *367*:L33 (1996).
281. B. M. Ocko, J. X. Wang, and T. Wandlowski, Phys. Rev. Lett. *79*:1511 (1997).
282. M. T. M. Koper, J. Electroanal. Chem. *450*:189 (1998).
283. T. Teshima, K. Ogaki, and K. Itaya, J. Phys. Chem. B *101*:2046 (1997).
284. K. Sashikata, Y. Matsui, K. Itaya, and M. P. Soriaga, J. Phys. Chem. *100*: 20027 (1996).
285. Y-G. Kim and M. P. Soriaga, J. Phys. Chem. B *102*:6188 (1998).
286. W. F. Temesghen, J. B. Abreu, R. J. Barriga, E. A. Lafferty, M. P. Soriaga, K. Sashikata, and K. Itaya, Surf. Sci. *385*:336 (1997).
287. S. Ando, T. Suzuki, and K. Itaya, J. Electroanal. Chem. *412*:139 (1996).
288. X. Gao and M. J. Weaver, J. Am. Chem. Soc. *114*:8544 (1992).
289. X. Gao and M. J. Weaver, Ber. Bunsenges. Phys. Chem. *97*:507 (1993).
290. S-C. Chang, S-L. Yau, B. C. Schardt, and M. J. Weaver, J. Phys. Chem. *95*: 4787 (1991).

291. X. Gao, G. J. Edens, and M. J. Weaver, J. Phys. Chem. *98*:8074 (1994).
292. B. Keita and L. Nadjo, J. Electroanal. Chem. *354*:295 (1993).
293. T. Yamada, N. Batina, and K. Itaya, Surf. Sci. *335*:204 (1995).
294. N. J. Tao and S. M. Lindsey, J. Phys. Chem. *96*:5213 (1992).
295. W. Haiss, J. K. Sass, X. Gao, and M. J. Weaver, Surf. Sci. Lett. *274*:L593 (1992).
296. N. Batina, T. Yamada, and K. Itaya, Langmuir *11*:4568 (1995).
297. X. Gao, G. J. Edens, and M. J. Weaver, J. Electroanal. Chem. *376*:21 (1994).
298. B. M. Ocko, O. M. Magnussen, J. X. Wang, and R. R. Adzic, in *Nanoscale Probes of the Solid/Liquid Interface* (A. A. Gewirth and H. Siegenthaler, eds.), Kluwer, Boston, 1995, p. 103.
299. U. Demir and C. Shannon, Langmuir *10*:2794 (1994).
300. X. Gao, Y. Zhang, and M. Weaver, J. Phys. Chem. *96*:4156 (1992).
301. T. Sawaguchi, T. Yamada, Y. Okinaka, and K. Itaya, J. Phys. Chem. *99*: 14149 (1995).
302. M. Ge, B. K. Niece, C. G. Wall, W. G. Klemperer, and A. A. Gewirth, MRS Symp. Proc. V451, Pittsburgh, 1997.
303. X. Gao, G. J. Edens, F-C. Liu, A. Hamelin, and M. J. Weaver, J. Phys. Chem. *98*:8086 (1994).
304. L. B. Goetting, B. M. Huang, T. E. Lister, and J. L. Stickney, Electrochim. Acta *40*:143 (1995).
305. X. Gao and M. J. Weaver, Surf. Sci. *313*:L775 (1994).
306. J. H. Schott and H. S. White, J. Phys. Chem. *98*:291 (1994).
307. J. H. Schott and H. S. White, J. Phys. Chem. *98*:297 (1994).
308. M. Kawasaki and H. Ishii, Langmuir *11*:832 (1995).
309. M. L. Foresti, G. Aloisi, M. Innocenti, H. Kobayashi, and R. Guidelli, Surf. Sci. *335*:241 (1995).
310. N. Shinotsuka, K. Sashikata, and K. Itaya, Surf. Sci. *335*:75 (1995).
311. G. D. Aloisi, M. Cavallini, M. Innocenti, M. L. Foresti, G. Pezzatini, and R. Guidelli, J. Phys. Chem. B *101*:4774 (1997).
312. A. M. Funtikov, U. Linke, U. Stimming, and R. Vogel, Surf. Sci. *324*:L343 (1995).
313. A. M. Funtikov, U. Stimming, and R. Vogel, J. Electroanal. Chem. *428*:147 (1997).
314. R. Vogel, I. Kamphausen, and H. Baltruschat, Ber. Bunsenges. Phys. Chem. *96*:252 (1992).
315. J. Inukai, Y. Osawa, M. Wakisaka, K. Sashikata, Y. G. Kim, and K. Itaya, J. Phys. Chem. B *102*:3498 (1998).
316. A. M. Bittner, J. Wintterlin, B. Beran, and G. Ertl, Surf. Sci. *335*:291 (1995).
317. S. Tanaka, S-L. Yau, and K. Itaya, J. Electroanal. Chem. *396*:125 (1995).
318. C. Stuhlmann, I. Villegas, and M. Weaver, Chem. Phys. Lett. *219*:319 (1994).

319. I. Villegas, X. Gao, and M. J. Weaver, Electrochim. Acta *40*:1267 (1995).
320. I. Villagas and M. J. Weaver, J. Chem. Phys. *101*:1648 (1994).
321. K. Yoshimi, M-B. Song, and M. Ito, Surf. Sci. *368*:389 (1996).
322. I. Villegas and M. J. Weaver, J. Electroanal. Chem. *373*:245 (1994).
323. S. Zou, I. Villegas, C. Stuhlmann, and M. J. Weaver, Electrochim. Acta *43*: 2811 (1998).
324. R. Vogel and H. Baltruschat, Surf. Sci. Lett. *259*:L739 (1991).
325. R. Vogel and H. Baltruschat, Ultramicroscopy *42–44*:562 (1992).
326. J. Clavilier, J. M. Orts, and J. M. Feliu, J. Phys. IV *4*:C1–303 (1994).
327. J. M. Orts, R. Gomez, J. M. Feliu, A. Aldaz, and J. Clavilier, Langmuir *13*: 3016 (1997).
328. V. Maurice and P. Marcus, Surf. Sci. Lett. *262*:L59 (1992).
329. V. Maurice and P. Marcus, J. Phys. IV *4*:C1–309 (1994).
330. G. Beitel, O. M. Magnussen, and R. J. Behm, Surf. Sci. *336*:19 (1995).
331. L-J. Wan, S-L. Yau, G. M. Swain, and K. Itaya, J. Electroanal. Chem. *381*: 105 (1995).
332. L-J. Wan, S-L. Yau, and K. Itaya, J. Phys. Chem. *99*:9507 (1995).
333. S. L. Yau, X. Gao, S. C. Chang, B. C. Schardt, and M. J. Weaver, J. Am. Chem. Soc. *113*:6049 (1991).
334. X. Gao, S-C. Chang, X. Jiang, A. Hamelin, and M. J. Weaver, J. Vac. Sci. Technol. *A10*:2972 (1992).
335. D. Wayne Suggs and A. J. Bard, J. Am. Chem. Soc. *116*:10725 (1994).
336. M. Kruft, B. Wohlmann, C. Stuhlmann, and K. Wandelt, Surf. Sci. *377–379*:601 (1997).
337. M. Wilms, P. Broekmann, M. Kruft, Z. Park, C. Stuhlmann, and K. Wandelt, Surf. Sci. *402–404*:83 (1998).
338. L-J. Wan and K. Itaya, Meeting Abstract #1015, Joint 192nd ECS Meeting and 48th Annual ISE Meeting, The Electrochemical Society, Pennington, NJ, 1997.
339. T. Suzuki, T. Yamada, and K. Itaya, J. Phys. Chem. *100*:8954 (1996).
340. P. N. Ross, in *Structure of Electrified Interfaces* (J. Lipkowski and P. N. Ross, eds.), VCH, New York, 1993, p. 35.
341. *Electrochemical Phase Formation and Growth* (E. Budevski, G. Staikov, and W. J. Lorenz, eds.), VCH Publishers, New York, 1996.
342. U. Muller, D. Carnal, H. Siegenthaler, E. Schmidt, W. J. Lorenz, W. Obretenov, U. Schmidt, G. Staikov, and E. Budevski, Phys. Rev. B *46*:12899 (1992).
343. U. Muller, D. Carnal, H. Siegenthaler, E. Schmidt, W. J. Lorenz, W. Obretenov, U. Schmidt, G. Staikov, and E. Budevski, Phys. Rev. B *49*:7795 (1994).
344. M. G. Samant, M. F. Toney, G. L. Borges, L. Blum, and O. R. Melroy, J. Phys. Chem. *92*:220 (1988);

345. O. R. Melroy, M. F. Toney, G. L. Borges, M. G. Samant, J. B. Knight, P. N. Ross, and L. Blum, Phys. Rev. B *38*:10962 (1988).
346. T. Hachiya, H. Honbo, and K. Itaya, J. Electroanal. Chem. *315*:275 (1991).
347. N. Batina, T. Will, and D. M. Kolb, Faraday Discuss. *94*:93 (1994).
348. Z. Shi, S. Wu, and J. Lipkowski, Electrochim. Acta *40*:9 (1995).
349. M. S. Zei, G. Qiao, G. Lehmpfuhl, and D. Kolb, Ber. Bunsenges. Phys. Chem. *91*:349 (1987).
350. L. Blum, D. A. Huckaby, and M. Legault, Electrochim. Acta *41*:2207 (1996).
351. M. F. Toney, J. N. Howards, J. Richer, G. L. Borges, J. G. Gordon, O. R. Melroy, D. Yee, and L. B. Sorensen, Phys. Rev. Lett. *75*:4472 (1995).
352. H. Matsumoto, J. Inukai, and M. Ito, J. Electroanal. Chem. *379*:223 (1994).
353. I. Oda, Y. Shingaya, H. Matsumoto, and M. Ito, J. Electroanal. Chem. *409*: 95 (1996).
354. O. M. Magnussen, F. Moller, M. R. Vogt, and R. J. Behm, in *In Situ Local Probe Techniques for Studies of Electrochemical Interfaces* (W. J. Lorenz and W. Plieth, eds.), VCH-Wiley, New York, 1998, p. 159.
355. J. Hotlos, O. M. Magnussen, and R. J. Behm, Surf. Sci. *335*:129 (1995).
356. M. H. Holzle, V. Zwing, and D. M. Kolb, Electrochim. Acta *40*:1237 (1995).
357. M. P. Green, K. J. Hanson, R. Carr, and I. Lindau, J. Electrochem. Soc. *137*: 3493 (1990).
358. E. Ammann, Masters thesis, University of Bern, 1995.
359. U. Schmidt, S. Vinzelberg, and G. Staikov, Surf. Sci. *348*:261 (1996).
360. N. J. Tao, J. Pan, Y. Li, P. I. Oden, J. A. Rose, and S. M. Lindsey, Surf. Sci. Lett. *271*:L338 (1992).
361. M. P. Green and K. J. Hanson, Surf. Sci. Lett. *259*:L743 (1991).
362. D. Carnal, P. I. Oden, U. Muller, E. Schmidt, and H. Siegenthaler, Electrochem. Acta *40*:1223 (1995).
363. J. Tersoff, Phys. Rev. Lett. *74*:434 (1995).
364. D. W. Suggs and J. L. Stickney, Surf. Sci. *290*:375 (1993).
365. T. E. Lister and J. L. Stickney, J. Phys. Chem. *100*:19568 (1996).
366. A. Gichuhi, B. E. Boone, U. Demir, and C. Shannon, J. Phys. Chem. B *102*: 6499 (1998).
367. B. E. Hayden and I. S. Nandhakumar, J. Phys. Chem. B *102*:4897 (1998).
368. M. L. Foresti, G. Pezzatini, M. Cavallini, G. Aloisi, M. Innocenti, and R. Guidelli, J. Phys. Chem. B *102*:7413 (1998).
369. B. M. Huang, T. E. Lister, and J. L. Stickney, Surf. Sci. *392*:27 (1997).
370. R. R. Adzic, Adv. Electrochem. Electrochem. Eng. *13*:159 (1984).
371. T. Abe, G. M. Swain, K. Sashikata, and K. Itaya, J. Electroanal. Chem. *382*: 73 (1995).
372. S. G. Corcoran, G. S. Chakarova, and K. Sieradzki, Phys. Rev. Lett. *71*:1585 (1993).

373. S. G. Corcoran, G. S. Chakarova, and K. Sieradzki, J. Electroanal. Chem. *377*:85 (1994).
374. K. Ogaki and K. Itaya, Electrochim. Acta *40*:1249 (1995).
375. G. Staikov, K. Juttner, W. J. Lorenz, and E. Budevski, Electrochim. Acta *39*:1019 (1994).
376. G. Staikov and W. J. Lorenz, in *Nanoscale Probes of the Solid/Liquid Interface* (A. A. Gewirth and H. Siegenthaler, eds.), Kluwer, Boston, 1995, p. 215.
377. S. Garcia, D. Salinas, C. Mayer, E. Schmidt, G. Staikov, and W. J. Lorenz, Electrochem. Acta *43*:3007 (1998).
378. T. Hachiya and K. Itaya, Ultramicroscopy *42–44*:445 (1992).
379. K. Itaya, Nanotechnology *3*:185 (1992).
380. S. Sugita, T. Abe, and K. Itaya, J. Phys. Chem. *97*:8780 (1993).
381. C. Chen, K. D. Kepler, A. A. Gewirth, B. M. Ocko, and J. Wang, J. Phys. Chem. *97*:7290 (1993).
382. J. C. Bondos, A. A. Gewirth, and R. G. Nuzzo, J. Phys. Chem. *100*:8617 (1996).
383. O. M. Magnussen, J. Hotlos, G. Beitel, D. M. Kolb, and R. J. Behm, J. Vac. Sci. Technol. *B9*:969 (1991).
384. H. Matsumoto, I. Oda, J. Inukai, and M. Ito, J. Electroanal. Chem. *356*:275 (1993).
385. W. Haiss, D. Lackey, J. K. Sass, H. Meyer, and R. J. Nichols, Chem. Phys. Lett. *200*:343 (1992).
386. J. Inukai, S. Sugita, and K. Itaya, J. Electroanal. Chem. *403*:159 (1996).
387. J. L. Bubendorff, L. Cagnon, V. Costa-Kieling, J. P. Bucher, and P. Allongue, Surf. Sci. *384*:L836 (1997).
388. K. Uosaki, S. Ye, Y. Oda, T. Haba, and K. Hamada, Langmuir *13*:594 (1997).
389. W. Polewska, J. X. Wang, B. M. Ocko, and R. R. Adzic, J. Electroanal. Chem. *376*:41 (1994).
390. S. G. Garcia, D. Salinas, C. Mayer, J. R. Vilche, H. J. Pauling, S. Vinzelberg, G. Staikov, and W. J. Lorenz, Surf. Sci. *316*:143 (1994).
391. F. A. Moller, O. M. Magnussen, and R. J. Behm, Phys. Rev. B *51*:2484 (1995).
392. U. Schmidt, S. Vinzelburg, and G. Staikov, Surf. Sci. *348*:261 (1996).
393. W. J. Lorenz and G. Staikov, Surf. Sci. *335*:32 (1995).
394. F. Moller, O. M. Magnussen, and R. J. Behm, Electrochim. Acta *40*:1259 (1995).
395. W. Obretenov, U. Schmidt, W. J. Lorenz, G. Staikov, E. Budevski, D. Carnal, U. Muller, H. Siegenthaler, and E. Schmidt, J. Electrochem. Soc. *140*: 692 (1993).
396. W. Obretenov, U. Schmidt, W. J. Lorenz, G. Staikov, E. Budevski, D. Car-

nal, U. Muller, H. Siegenthaler, and E. Schmidt, Faraday Discuss. *94*:107 (1992).

397. G. Staikov and W. J. Lorenz, in *Electrochemical Technology, Innovation and New Developments* (N. Masuko, T. Osaka, and Y. Ito, eds.), Gordon and Breach Publ., Amsterdam, 1996, p. 241.
398. S. Vinzelberg, G. Staikov, and W. J. Lorenz, in Proc. of the Second Symposium on Electrochemically Deposited Thin Films, Vol. 93-26, The Electrochemical Society, Pennington, NJ, 1995, p. 1.
399. U. Schmidt, S. Vinzelberg, G. Repphuhn, G. Staikov, W. J. Lorenz, and J. Halbritter, Proc. of the Second Symposium on Electrochemically Deposited Thin Films, Vol. 93-26, The Electrochemical Society, Pennington, NJ, 1995, p. 107.
400. D. Carnal, U. Muller, and H. Siegenthaler, J. Phys. IV *4*:C1–297 (1994).
401. J. Sackmann, A. Bunk, R. T. Potzschke, G. Staikov, and W. J. Lorenz, Electrochim. Acta *43*:2863 (1998).
402. R. Christoph, H. Siegenthaler, H. Rohrer, and H. Weise, Electrochem. Acta *34*:1011 (1989).
403. N. Kimizuka and K. Itaya, Faraday Discuss. *94*:117 (1992).
404. K. Sashikata, N. Furuya, and K. Itaya, J. Electroanal. Chem. *316*:361 (1991).
405. Y. Shingaya, H. Matsumoto, H. Ogasawara, and M. Ito, Surf. Sci. *335*:23 (1995).
406. R. R. Adzic, J. Wang, C. M. Vitus, and B. M. Ocko, Surf. Sci. Lett. *293*: L876 (1993).
407. Y. Shingaya, H. Matsumoto, H. Ogasawara, and M. Ito, Surf. Sci. *335*:23 (1995).
408. T. P. Moffat, J. Phys. Chem. B, 102, 10020 (1998).
409. M. Nakamura, A. Aramata, A. Yamagishi, and M. Taniguchi, J. Electoanal. Chem. *446*:227 (1998).
410. A. M. Bittner, J. Electroanal. Chem. *431*:51 (1997).
411. J. G. Amar and F. Family, in *Evolution of Epitaxial Structure and Morphology* (A. Zhangwill, D. Jesson, D. Chambliss, and R. Clarke, eds.), MRS, Pittsburgh, 1995.
412. W. Wulfhekel, N. N. Lipkin, J. Kliewer, G. Rosenfeld, L. C. Jorritsma, B. Poelsema, and G. Comsa, Surf. Sci. *348*:227 (1996).
413. R. Q. Hwang and M. C. Bartelt, Chem. Rev. *97*:1063 (1997).
414. Z. Zhang and M. G. Lagally, Science *276*:377 (1997).
415. H. Iwasaki and T. Yoshinobu, Phys. Rev. B *48*:8282 (1993).
416. M. J. Armstrong and R. H. Muller, J. Electrochem. Soc. *138*:2303 (1991).
417. R. J. Nichols, C. E. Bach, and H. Meyer, Ber. Bunsenenges. Phys. Chem. *97*:1012 (1993).
418. R. J. Nichols, E. Bunge, H. Meyer, and H. Baumgartel, Surf. Sci. *335*:110 (1995).

419. N. Batina, D. M. Kolb, and R. J. Nichols, Langmuir *8*:2572 (1992).
420. M. H. Holzle, C. W. Apsel, T. Will, and D. M. Kolb, J. Electrochem. Soc. *142*:3741 (1995).
421. R. J. Nichols, W. Beckmann, H. Meyer, N. Batina, and D. M. Kolb, J. Electroanal. Chem. *330*:381 (1992).
422. R. J. Nichols, in *Nanoscale Probes of the Solid/Liquid Interface* (A. A. Gewirth and H. Siegenthaler, eds.), NATO ASI Series E, Vol. 288, Kluwer, Boston, 1995, p. 163.
423. M. R. Vogt, W. Polewska, O. M. Magnussen, and R. J. Behm, J. Electrochem. Soc. *144*:L113 (1997).
424. E. Bunge, R. J. Nichols, B. Roelfs, H. Meyer, and H. Baumgartel, Langmuir *12*:3060 (1996).
425. K-N. Tu, J. W. Mayer, and L. C. Feldman, *Electronic Thin Film Science*, Macmillan Publishing Co., New York, 1992.
426. J. Y. Tsao, *Materials Fundamentals of Molecular Beam Epitaxy*, Academic Press, New York, 1993.
427. R. L. Schwoebel, J. Appl. Phys. *40*:614 (1969).
428. R. L. Schwoebel and E. J. Shipsey, J. Appl. Phys. *37*:3682 (1966).
429. E. Altman and R. J. Colton, Surf. Sci. Lett. *304*:L400 (1994).
430. R. Sonnenfeld and B. C. Schardt, Appl. Phys. Lett. *49*:1172 (1986).
431. R. S. Robinson, J. Vac. Sci. Technol. A *8*:511 (1990).
432. L. Vazquez, A. Hernandez Creus, P. Carro, P. Ocon, P. Herrasti, C. Palacio, J. M. Vara, R. C. Salvarezza, and A. J. Arvia, J. Phys. Chem. *96*:10454 (1992).
433. R. T. Potzschke, C. A. Gervasi, S. Vinzelberg, G. Staikov, and W. J. Lorenz, Electrochim. Acta *40*:1469 (1995).
434. J. L. Zubimendi, L. Vazquez, P. Ocon, J. M. Vara, W. E. Triaca, R. C. Salvarezza, and A. J. Arvia, J. Phys. Chem. *97*:5095 (1993).
435. N. Breuer, U. Stimming, and R. Vogel, in *Nanoscale Probes of the Solid/Liquid Interface* (A. A. Gewirth and H. Siegenthaler, eds.), NATO ASI Series E, Vol. 288, Kluwer, Boston, 1995, p. 121.
436. H. Martin, P. Carro, A. Hernandez Creus, S. Gonzalez, R. C. Salvarezza, and A. J. Arvia, Langmuir *13*:100 (1997).
437. Z. Chen, J. Li, and E. Wang, J. Electroanal. Chem. *373*:83 (1994).
438. D. M. Schaefer, A. Ramachandra, R. P. Andres, and R. Reifenberger, Z. Phys. D *26*:249 (1993).
439. U. Hartmann, in *Scanning Tunneling Microscopy III* (R. Wisendanger and H. J. Guntherodt, eds.), Springer-Verlag, New York, 1993, p. 293.
440. J. V. Zoval, R. Steiger, P. R. Biernacki, and R. M. Penner, J. Phys. Chem. *100*:837 (1996).
441. D. Lu, Y. Okawa, K. Suzuki, and K. Tanaka, Surf. Sci. *325*:L397 (1995).
442. M. Baldauf and D. M. Kolb, Electrochim. Acta *38*:2145 (1993).
443. H. Naohara, S. Ye, and K. Uosaki, J. Phys. Chem. B *102*:4366 (1998).

444. M. H. Holzle, C. W. Apsel, T. Will, and D. M. Kolb, J. Electrochem. Soc. *142*:3741 (1995).
445. B. Ke, J. J. Heekstra, B. C. Sason Jr., and D. Trivich, J. Electrochem. Soc. *106*:382 (1959).
446. Li. Sun and R. M. Crooks, J. Electrochem. Soc. *138*:L23 (1991).
447. O. Cavalleri, S. E. Gilbert, and K. Kern, Surf. Sci. *377–379*:931 (1997).
448. S. E. Gilbert, O. Cavalleri, and K. Kern, J. Phys. Chem. *100*:12123 (1996).
449. F. P. Zamborini, J. K. Campbell, and R. M. Crooks, Langmuir *14*:640 (1998).
450. F. P. Zamborini and R. M. Crooks, Langmuir *14*:3279 (1998).
451. F. A. Moller, O. M. Magnussen, and R. J. Behm, Phys. Rev. Lett. *77*:5249 (1996).
452. F. A. Moller, O. M. Magnussen, and R. J. Behm, Phys. Rev. Lett. *77*:3165 (1996).
453. D. D. Chambliss, R. J. Wilson, and S. Chiang, Phys. Rev. Lett. *66*:1721 (1991).
454. O. M. Magnussen, F. A. Moller, A. Lachenwitzer, and R. J. Behm, Mat. Res. Soc. Symp., Materials Research Society, Pittsburgh, 1997.
455. A. Lachenwitzer, M. R. Vogt, O. M. Magnussen, and R. J. Behm, Surf. Sci. *382*:107 (1997).
456. C. Gunther, J. Vrijmoeth, R. Q. Hwang, and R. J. Behm, Phys. Rev. Lett. *74*:754 (1995).
457. B. Muller, B. Fischer, L. Nedelmann, A. Fricke, and K. Kern, Phys. Rev. Lett. *76*:2358 (1996).
458. H. Brune, H. Roder, C. Boragno, and K. Kern, Phys. Rev. B *49*:2997 (1994).
459. J. A. Meyer, P. Schmid, and R. J. Behm, Phys. Rev. Lett. *74*:3864 (1995).
460. C. B. Carter and R. Q. Hwang, Phys. Rev. B *51*:4730 (1995).
461. N. Frank, G. Springholz, and G. Bauer, Phys. Rev. Lett. *73*:2236 (1994).
462. D. M. Kolb, in Mat. Res. Soc. Symp. Proc., V 451, MRS, Pittsburgh, 1997.
463. M. Dietterle, T. Will, and D. M. Kolb, Surf. Sci. *396*:189 (1998).
464. W. Haiss, R. J. Nichols, and J-K. Sass, Surf. Sci. *388*:141 (1997).
465. S. Chiang, Chem. Rev. *97*:1083 (1997).
466. G. E. Poirier, Chem. Rev. *97*:1117 (1997).
467. S-L. Yau, Y-G. Kim, and K. Itaya, J. Am. Chem. Soc. *118*:7795 (1996).
468. L-J. Wan and K. Itaya, Langmuir *13*:7173 (1997).
469. P. Sautet and M. L. Bocquet, Phys. Rev. B *53*:4910 (1996).
470. G. Treboux and M. Aono, J. Phys. Chem. B *101*:4620 (1997).
471. S-L. Yau, Y-G. Kim, and K. Itaya, J. Phys. Chem. B *101*:3547 (1997).
472. N. J. Tao, J. A. DeRose, and S. M. Lindsey, J. Phys. Chem. *97*:910 (1993).
473. M. H. Holzle, T. Wandlowski, and D. M. Kolb, Surf. Sci. *335*:281 (1995).
474. T. Wandlowski, D. Lampner, and S. M. Lindsay, J. Electroanal. Chem. *404*: 215 (1996).

475. T. Dretschkow, A. S. Dakkouri, and T. Wandlowski, Langmuir *13*:2843 (1997).
476. T. Dretschkow and T. Wandlowski, Electrochim. Acta *43*:2991 (1998).
477. F. Cunha and N. J. Tao, Phys. Rev. Lett. *75*:2376 (1995).
478. G. Andreasen, M. E. Vela, R. C. Salvarezza, and A. J. Arvia, Langmuir *13*: 6814 (1997).
479. A. S. Dakkouri, D. M. Kolb, R. Edelstein-Shima, and D. Mandler, Langmuir *12*:2849 (1996).
480. Y. He, Y. Wang, G. Zhu, and E. Wang, J. Electroanal. Chem. *440*:65 (1997).
481. E. Bunge, R. J. Nichols, B. Roelfs, H. Meyer, and H. Baumgartel, Langmuir *13*:85 (1997).
482. M. Cavallini, G. Aloisi, M. Bracali, and R. Guidelli, J. Electroanal. Chem. *444*:75 (1998).
483. Y. Zhang, X. Gao, and M. J. Weaver, J. Phys. Chem. *96*:510 (1992).
484. M. Kunitake, N. Batina, and K. Itaya, Langmuir *11*:2337 (1995).
485. N. Batina, M. Kunitake, and K. Itaya, J. Electroanal. Chem. *405*:245 (1996).
486. K. Ogaki, N. Batina, M. Kunitake, and K. Itaya, J. Phys. Chem. *100*:7185 (1996).
487. K. Sashikata, T. Sugata, M. Sugimasa, and K. Itaya, Langmuir *14*:2896 (1998).
488. N. J. Tao, G. Cardenas, F. Cunha, and Z. Shi, Langmuir *11*:4445 (1995).
489. N. J. Tao and Z. Shi, J. Phys. Chem. *98*:1464 (1994).
490. N. J. Tao and Z. Shi, Surf. Sci. *321*:L149 (1994).
491. Y-T. Kim, R. L. McCarley, and A. J. Bard, J. Phys. Chem. *96*:7416 (1992).
492. R. Heinz and J. P. Rabe, Langmuir *11*:506 (1995).
493. G. E. Poirier, Science *272*:1145 (1996).
494. G. E. Poirier, Langmuir *13*:2019 (1997).
495. I. Touzov and C. B. Gorman, J. Phys. Chem. B *101*:5263 (1997).
496. M. Rohwerder, K. de Weldige, E. Vago, H. Viefhaus, and M. Stratmann, Thin Solid Films *264*:240 (1995).
497. R. Yamada and K. Uosaki, Langmuir *14*:855 (1998).
498. D. Hobara, K. Miyake, S. Imabayashi, K. Niki, and T. Kakiuchi, Langmuir *14*:3590 (1998).
499. Li-Jun Wan, Y. Hara, H. Noda, and M. Osawa, J. Phys. Chem. B *102*:5943 (1998).
500. D. Oyamatsu, M. Nishizawa, S. Kuwabata, and H. Yoneyama, Langmuir *14*: 3298 (1998).
501. H. Kind, A. M. Bitter, O. Cavalleri, K. Kern, and T. Greber, J. Phys. Chem. B *102*:7582 (1998).
502. F. P. Zamborini and R. M. Crooks, Langmuir *13*:122 (1997).
503. R. L. McCarley, Y-T. Kim, and A. J. Bard, J. Phys. Chem. *97*:212 (1993).

504. M. R. Vogt, A. Lachenwitzer, O. M. Magnussen, and R. J. Behm, Surf. Sci. *399*:49 (1998).
505. M. P. Soriaga, J. A. Schimpf, A. Carrasquillo, J. B. Abreu, W. Temesghen, R. J. Barriga, J-J. Jeng, K. Sashikata, and K. Itaya, Surf. Sci. *335*:273 (1995).
506. M. Hopfner, W. Obretenov, K. Juttner, W. J. Lorenz, G. Staikov, V. Bostanov, and E. Budevski, Surf. Sci. *248*:225 (1991).
507. M. R. Vogt, R. J. Nichols, O. M. Magnussen, and R. J. Behm, J. Phys. Chem. B *102*:5859 (1998).
508. J. Scherer, M. R. Vogt, O. M. Magnussen, and R. J. Behm, Langmuir *13*: 7045 (1997).
509. I. C. Oppenheim, D. J. Trevor, C. E. D. Chidsey, P. L. Trevor, and K. Sieradzki, Science *254*:687 (1991).
510. K. Wagner, S. R. Brankovic, N. Dimitrov, and K. Sieradzki, J. Electrochem. Soc. *144*:3545 (1997).
511. T. P. Moffat, Fu-Ren Fan, and A. J. Bard, J. Electrochem. Soc. *138*:3225 (1991).
512. S. J. Chen, F. Sanz, D. F. Olgetree, V. M. Hallmark, T. M. Devine, and M. Salmeron, Surf. Sci. *292*:289 (1993).
513. F. Besenbacher, F. Laegsgaard, L. Plieth Nielsen, L. Ruan, and I. Stensgaard, J. Vac. Sci. Technol. B *12*:1758 (1994).
514. C. Nagel, O. Haller, E. Platzammer, M. Schmid, and P. Varga, Surf. Sci. *321*:237 (1994).
515. C. Nagel, E. Platzammer, O. Haller, M. Schmid, and P. Varga, Surf. Sci. *331*:831 (1995).
516. D. Eigler and E. K. Schweizer, Nature *344*:524 (1990).
517. M. Crommie, C. P. Lutz, D. M. Eigler, Science *262*:218 (1993).
518. T. A. Jung, R. R. Schlittenand, J. K. Gimzewski, H. Tang, and C. Joachim, Science *271*:181 (1996).
519. M. T. Cuberes, R. R. Schlitter, and J. K. Gimzewski, Surf. Sci. *371*:L231 (1997).
520. P. Avouris, Acc. Chem. Res. *28*:95 (1995).
521. U. Staufer, *Scanning Tunneling Microscopy II* (R. Wisendanger and H. J. Guntherodt, eds.), Springer-Verlag, Berlin, 1992, p. 273.
522. R. M. Nyffenegger and R. M. Penner, Chem. Rev. *97*:1195 (1997).
523. R. J. Hamers, J. Phys. Chem. *100*:13103 (1996).
524. B. J. McIntyre, M. Salmeron, and G. A. Somorjai, Science *265*:1415 (1994).
525. D. M. Kolb, R. Ullmann, and T. Will, Science *275*:1097 (1997).
526. R. Ullmann, T. Will, and D. M. Kolb, Ber. Bunsenges. Phys. Chem. *99*:1414 (1995).
527. D. M. Kolb, R. Ullmann, and J. C. Ziegler, Electrochim. Acta *43*:2751 (1998).

528. G. E. Engelmann, J. C. Ziegler, and D. M. Kolb, Surf. Sci. *401*:L420 (1998).
529. L. Kuipers, M. S. Hoogeman, J. W. M. Frenken, Surf. Sci. *340*:231 (1995).
530. U. Landman and W. D. Luedtke, in *Scanning Tunneling Microscopy III* (R. Wisendanger and H. J. Guntherodt, eds.), Springer-Verlag, New York, 1993, p. 207.
531. C. Z. Li and N. J. Tao, Appl. Phys. Lett. *72*:894 (1998).
532. W. Li, J. A. Virtanen, and R. M. Penner, Appl. Phys. Lett. *60*:1181 (1992).
533. W. Li, G. S. Hsiao, D. Harris, R. M. Nyffenegger, J. A. Virtanen, and R. M. Penner, J. Phys. Chem. *100*:20103 (1996).
534. R. M. Nyffenegger and R. M. Penner, J. Phys. Chem. *100*:17041 (1996).
535. G. Ertl, Electrochim. Acta *43*:2743 (1998).
536. T. R. Albrecht, M. M. Dovek, M. D. Kirk, C. A. Lang, C. E. Quate, and D. P. E. Smith, Appl. Phys. Lett. *55*:1727 (1989).
537. R. M. Penner, M. J. Heben, N. S. Lewis, and C. F. Quate, Appl. Phys. Lett. *58*:1389 (1991).
538. J. H. Thomas and P. Sharma, J. Vac. Sci. Technol. *13*:549 (1976).
539. S. L. Yau, T. P. Moffat, F. R. Fan, and A. J. Bard, Chem. Phys. Lett. *198*: 383 (1992).
540. C. Lebreton and Z. Z. Wang, J. Vac. Sci. Technol. B *14*:1356 (1996).
541. C. Lebreton and Z. Z. Wang, Surf. Sci. *382*:193 (1997).
542. C. B. Ross, L. Sun, and R. M. Crooks, Langmuir *9*:632 (1993).
543. J. K. Schoer, F. P. Zamborini, and R. M. Crooks, Langmuir *100*:11086 (1996).
544. Y. T. Kim and A. J. Bard, Langmuir *8*:1096 (1992).
545. H. Sugimura, T. Uchida, N. Kitamura, and H. Masuhara, Appl. Phys. Lett. *63*:1288 (1993).
546. H. J. Song, M. J. Rack, K. Abugharbieh, S. Y. Lee, V. Khan, D. K. Ferry, and D. R. Allee, J. Vac. Sci. Technol. B *12*:3720 (1994).
547. K. Matsumoto, M. Ishii, and K. Segawa, J. Vac. Sci. Technol. B *14*:1331 (1996).
548. E. S. Snow and P. M. Campbell, Science *270*:1639 (1995).
549. E. F. Cerrina and C. Marrian, MRS Bull. *21*:12, 56 (1996).

AUTHOR INDEX

Italicized pages locate the full title of the work cited.

SUBJECT INDEX